Industrial and Manufacturing Designs

Scrivener Publishing
100 Cummings Center, Suite 541J
Beverly, MA 01915-6106

Publishers at Scrivener
Martin Scrivener (martin@scrivenerpublishing.com)
Phillip Carmical (pcarmical@scrivenerpublishing.com)

Industrial and Manufacturing Designs

Quantitative and Qualitative Analysis

Edited by

Atul Kumar Sahu

Department of Industrial and Production Engineering, SOSE&T, Guru Ghasidas Central University, Bilaspur, India

Rakesh D. Raut

Department of Operations and Supply Chain Management, Indian Institute of Management, Mumbai, India

Rohit Raja

Department of Information Technology, SOSE&T, Guru Ghasidas Central University, Bilaspur, India

Anoop Kumar Sahu

Department of Mechanical Engineering, SOSE&T, Guru Ghasidas Central University, Bilaspur, India

and

Nitin Kumar Sahu

Department of Industrial and Production Engineering, SOSE&T, Guru Ghasidas Central University, Bilaspur, India

Scrivener Publishing

WILEY

This edition first published 2024 by John Wiley & Sons, Inc., 111 River Street, Hoboken, NJ 07030, USA and Scrivener Publishing LLC, 100 Cummings Center, Suite 541J, Beverly, MA 01915, USA

For more information about Scrivener publications please visit www.scrivenerpublishing.com.

Wiley Global Headquarters
111 River Street, Hoboken, NJ 07030, USA

For details of our global editorial offices, customer services, and more information about Wiley products visit us at www.wiley.com.

Library of Congress Cataloging-in-Publication Data

ISBN 978-1-394-21174-6

Cover image: Pixabay.Com
Cover design by Russell Richardson

Set in size of 11pt and Minion Pro by Manila Typesetting Company, Makati, Philippines

Printed in the USA

10 9 8 7 6 5 4 3 2 1

Contents

Preface

The main aim of this book is to disseminate creative thinking to leading academics, scientists, researchers, and research scholars to enhance their experiences and work on qualitative and quantitative analysis for reinforcing engineering designs related to industrial and manufacturing boundaries. The book provides a platform for researchers, practitioners, and educators to learn about and understand many dimensions, recent innovations, trends, and concerns, as well as the practical challenges encountered and solutions adopted in the fields of industrial and manufacturing engineering.

The present book:

- Explains of qualitative and quantitative research for exploiting system characteristics and attaining system efficiency, as well as what decision-making tools are available for attaining sustainability in industrial fields;
- Discusses the utility of qualitative and quantitative models and analytical frameworks for enduring sustainability and the exploration of new ideas, critical theories, and methods related to qualitative and quantitative research;
- Disseminates industrial and manufacturing knowledge and allied boundaries, and demonstrates of case studies and sustainable ways to retain excellence in industrial and manufacturing domains based on qualitative and quantitative analysis;
- Presents critical aspects related to lean manufacturing and lean management tools, and demonstrates pure research and practical solutions to manufacturing problems;
- Offers theoretical content and demonstrations of manufacturing applications, as well as critical thinking and methodological support for optimizing resources and consumption;
- Reviews the development of supply chain network designs based on qualitative and quantitative aspects.

By focusing on the development of the theoretical foundations, applications of diverse tools, and support techniques for evaluating the design and operation of systems in manufacturing industries and businesses, this book meets the demands of scenarios, stakeholders, and markets under the sphere of manufacturing management. It discusses the utility of integration of critical components and systems engineering for effective systems design. The book presents research directions and can be used as a tool to understand in-depth possible ways related to the management of industrial and manufacturing activities, processes, actions, benchmarks, and so on. The new ideas, theories, and methods related to qualitative and quantitative research that are presented herein will improve manufacturing knowledge linked to the manufacturing process.

This book is intended for students in the areas of industrial engineering, manufacturing engineering, supply chain management, qualitative analysis, quantitative analysis, and to practitioners working under allied boundaries of industrial and manufacturing practices. Additionally, individuals with interdisciplinary knowledge of statistical quality control, production planning, and control, performance measurement, mathematical modeling, decision-making approaches, and framework development with the generation of performance index will find value in this book. Application developers, business professionals, and researchers who seek information on industrial and manufacturing dimensions can utilize this book, as well.

We are deeply grateful to everyone who helped with this book and greatly appreciate the dedicated support and valuable assistance rendered by Martin Scrivener and the Scrivener Publishing team during its publication.

Dr. Atul Kumar Sahu
Dr. Rakesh D. Raut
Dr. Rohit Raja
Dr. Anoop Kumar Sahu
Dr. Nitin Kumar Sahu

1

Demonstrating the Role of Qualitative and Quantitative Information in Industrial and Manufacturing Designs

Nitin Kumar Sahu[1]*, Sanju Kumar Nishad[1], Atul Kumar Sahu[1] and Anoop Kumar Sahu[2]

[1]*Department of Industrial and Production Engineering, Guru Ghasidas Central Vishwavidyalaya, Bilaspur, India*
[2]*Department of Mechanical Engineering, Guru Ghasidas Central Vishwavidyalaya, Bilaspur, India*

Abstract

In the present chapter, the role of qualitative and quantitative (Q&Q) information for evaluating industrial and manufacturing designs is presented, where a group of application cases to reveal the importance of Q&Q analysis is demonstrated. Here, application cases related with the selection of solar panels, evaluation of automatic/robotic welding system, selection of smart alloys, identification of logistic service provider, evaluation of machine tool, and election of industrial robot are presented to represent the utility and importance of aforesaid information in evaluation. Various cases under aforesaid aspects are presented to report the importance of Q&Q information. The chapter will help readers in understanding the worth and values of Q&Q information in analysis. The chapter describes the developed multi-criteria decision-making (MCDM) methods that can be used for reinforcing industrial and manufacturing practices based on the utilization of Q&Q information. The present chapter will assist in creating a learning atmosphere and developing capabilities in effectively evaluating decisions that involve multiple criteria and factors. Additionally, the chapters will assist learners in prioritizing and ranking different alternatives based on their alignment with specific criteria and objectives. The understanding of Q&Q information to quantify and incorporate subjective preferences into decision-making and the ability to analyze complex

**Corresponding author*: nitin83sahu@gmail.com

Atul Kumar Sahu, Rakesh D. Raut, Rohit Raja, Anoop Kumar Sahu and Nitin Kumar Sahu (eds.) Industrial and Manufacturing Designs: Quantitative and Qualitative Analysis, (1–44) © 2024 Scrivener Publishing LLC

data related to multiple criteria for making better decisions with a large amount of information can be attained through the present chapter. The chapter will help in understanding how Q&Q criteria values impact the final decision outcomes and can help in enhancing your ability to make robust choices. Familiarization with different MCDM methods to choose the most suitable approach for different decision contexts can be attained through this chapter. The chapter will help in handling decision complexity related toward analyzing and addressing complex decision challenges, especially in scenarios where there are conflicting objectives or numerous alternatives. Moreover, the chapter will assist in modeling a structured framework for strategic planning to enable learners to align your decisions with long-term objectives based on Q&Q information.

Keywords: Qualitative and quantitative, decision-making, industrial and manufacturing designs, multi-criteria decision-making (MCDM), evaluation

1.1 Introduction

In the present scenario, manufacturers around the world are moving toward manufacturing their product in sustainable manner. Moreover, sustainable practices are highly demandable to be inculcated in the manufacturing activities for streamlining manufacturing operations. In addition, various regulatory norms, regulation policies, acts, etc., are imposed on manufacturing organizations to produce their products via proper utilization of resources and minimal waste generation for respecting environmental structure. Manufacturing is the wide expression and is typically used to illustrate machining of materials to attain desired workpiece. It comprises various activities like metal forming, customized operations, cutting, grinding, unconventional machining, foundry practices, and sheet metal working. Nowadays, modern manufacturing facilities are incorporating computer software platforms in manufacturing information systems to produce desired shapes with ease and effectiveness. The same are incorporating to facilitate the manufacturing managers and judgment makers to appraise and assess the conditions of the production process and assist in understanding "how the production system can be optimized and enhanced". Today, there is a need for manufacturing information systems with real-time data and feedback support with miscellaneous information related with the process, machines, inputs, personnel, and many more. Similarly, it will help in monitoring the tendency of faults, operational accuracy, precision, fruitful utility of production time, worker time, right repairs, and quality concerns, as well as in increasing safety and in decreasing business threats throughout the operation, which demands the exploitation of Q&Q factors.

The available data with the manufacturing systems are needed to evaluate for observing effectiveness, organizing inventory, and goods; taking action against consumer demands; developing financial execution; and delivering essential and judicious information to banks, associates, sponsors, and key stakeholders. Thus, it is needed to capture more data from the manufacturing system characteristics that can be supported by qualitative and quantitative (Q&Q) measures. The more data will be collected, the more accuracy will be attained by the system. It is found that categorization of the data can be done by considering two aspects, i.e., primary data or secondary data based on the source of collection. Primary data are directly gathered from the main source, i.e., experiments, monitoring, inspection, surveys, and questionnaires. Secondary data are the data that were gathered by other indirect sources, i.e., government organization records, research foundation reports, previous investigations, manuscripts, journals, newspapers, reporters, publications, data records, and catalogs. Moreover, one can say that data are the spirit and essence to monitor and evaluate the system, and it only provides information when it is in the structured form. The Q&Q data play a significant role in evaluating and understanding the behavior of any systems. The authentic, accurate, and trustworthy data help in revealing and enhancing the performance of your intrusion and support the decision-making (DM) process.

It is ascertained that the utilization of the combination of Q&Q data can improve the evaluation and DM process by assuring that the restrictions of one type of data are influenced by the potency of another data. Qualitative data collection incorporates intangible factors like knowledge, judgment, inspiration, behaviors, and explanations of activity, occasion, or a particular significant situation. Hence, in further expressions, one can say that a qualitative approach uses public chronicle as well as their knowledge and outlooks to evaluate transformation; is more open, informal, and unstructured; and supplies more freedom toward the data collection. The frequent qualitative data gathering means and equipments are unrestricted surveys, unrestricted interviews, society interviews/gathering, centred assembly discussions, case learning, examination/surveillance, techno science, visual procedures, literature, document review, spoken history, etc. The quantitative methodology is more prearranged, uncomplicated, clear-cut, and always prescribed and employed to derive responses for multiple locations assessment that engage huge group of respondents. Typical quantitative data collection approaches comprise experiments, collecting appropriate facts from management information systems, administering surveys with closed-ended queries, as well as monitoring and recording clear described events. It is easier to collect and analyze quantitative data, and there are fewer chances of biasness in the interpretation of results. Results of the quantitative data are mathematical, goal-oriented, decisive, and, to the point, easier to conclude and helpful for building comparative analysis. The frequent quantitative data gathering approaches are structured closed-ended interview, investigations and opinion polls, investigational research, correlation study, comparative analysis, statistical data examination, laboratory testing, etc.

The process of data collection for Q&Q methods has its own advantages. Moreover, modern industrial and manufacturing practices are complex as well as intricate and cannot be disentangled through one method alone. Therefore, integration of Q&Q methods is too important to enable researchers to gain a more holistic understanding of the intervention. Mixed methods introduce precision in the monitoring and assessment. Accordingly, it is advocated to develop Q&Q methods toward focusing on the deficiencies and restrictions of each method in order to provide more logical, consistent, and constructive conclusions that augment the overall confidence in the soundness of the assessment outcomes. Accordingly, in the present chapter, the role of Q&Q information for evaluating industrial and manufacturing designs is presented, where a group of application cases to reveal the importance of Q&Q is presented. Various cases are presented to report the importance of Q&Q information. The chapter will help readers in understanding the worth and values of Q&Q information in analysis.

1.2 Literature Review

Today, it is needed to explore new opportunities in industrial and manufacturing designs for crucially satisfying the dynamic needs of world, which requires mental and physical abilities to make vital DM. The same is needed to explore different solutions for handling the manufacturing situation and to receive elevated outcomes. A dignified decision in manufacturing is needed to purposely select a best answer from a set of substitutes to attain organizational or managerial objectives. Many researchers are continuously analyzing and evaluating their engineering domain in order to increase their business benefit. A successful DM plays a decisive responsibility in constructing the long period associations and augments profitability. Moreover, DM for reinforcing industrial and manufacturing designs needs engineering principles, technological tools, and information for processing and requires a specialized form of management skills, which is sometimes called as engineering management to determine the right choice and integrated DM structure. Various researchers have applied diverse multi-criteria decision-making (MCDM) techniques for tackling various manufacturing applications, where Hernández *et al.* [1] have applied analytic hierarchy process (AHP) and analytic network process (ANP) techniques under qualitative study to determine the influence of reverse logistics (RL) practices in automotive corporate performance of Brazilian automotive sector. Huang *et al.* [2] have proposed the concept of key performance indicator (KPI) under qualitative information to evaluate RL practices based on the collection of studies and expert interviews and used ANP to obtain relative weights of KPI to form the final performance evaluation model. Sahu *et al.* [3] have explored the TOPSIS (Technique for Ordering Preference by Similarity to Ideal Solution) technique in conjunction with the trapezoidal fuzzy number set to evaluate computer numerical control (CNC) machine tool under qualitative information. Here, the authors have determine the preference orders considering the decreasing value of the "collective index" and identified that the elevated value of "collective index" under qualitative premises is crucial and replicates the elevated extent of performance by the machine tool, which, in turn, assisted in enabling advanced manufacturing environment toward the organization and helped the organization toward attaining sustainability and competitiveness worldwide. Sheu [4] utilized quantitative information to develop a multi-objective optimization approach using a linear multi-objective optimization model to optimize the operations of nuclear power generation and the accompanying RL process. A cluster approach based on ratio analysis, reference point analysis, and full multiplication form under type-2 fuzzy sets to handle the right selection of industrial robot (IR) for manufacturing based on multiple qualitative measures is found built for DM under Q&Q information [5]. Here, the authors have utilized qualitative and quantitative approach considering 59-dimensional measures to select IRs for manufacturing firms operations. Additionally, many approaches like Simple Additive Weighting (SAW) and TOPSIS methodology under quantitative information for examining turning operations in CNC lathe machine to evaluate the impact of manufacturing process parameters, i.e., cutting speed, feed, and depth of cut, are found [6]. Sahu *et al.* [7] presented a replacement model combined with straight-line depreciation method to determine the economic life of the productive machines and equipments under quantitative medium. The proposed methodology enables the organization to impart efficiency to operators, machines, equipment, parts, etc. Their approach determined the replacement period of machines considering quantitative information under uneconomic circumstances, breakdowns, sudden failures, and gradual deterioration with the passage of time. Thus, one can say that the collection of Q&Q information n is crucial.

Moreover, the MCDM techniques have also been applied with Q&Q, which is generally a process of selecting the right choice from available alternatives or ranking of alternatives under Q&Q conflicting criteria. Under MCDM, a performance appraisement platform under chain of mixed green-lean-agile (G-L-A) logistic activities is found, and fuzzy performance index model is developed with qualitative medium to estimate the overall performance of organization [8]. Here, the authors have fruitfully used centroid method with fuzzy number set considering Q&Q information for understanding ill and strong G-L-A metrics. Guo *et al.* [9] fabricated a novel integrated computational approach under manufacturing to operate a mild steel plate workpiece for optimizing weld bead geometry in MIG (metal inert gas) welding. Here, the authors have investigated optimum setting of MIG welding process parameters to attain high tensile strength with quality characteristics, i.e., bead width, reinforcement, penetration, and dilution. It is found that a zone of various prominent methods, i.e., Simple Additive Weighting (SAW) method [10], compromise programming [11], AHP method [12], Technique for Ordering Preference by Similarity to Ideal Solution (TOPSIS) method [13], Preference Ranking Organization Method for Enrichment Evaluations (PROMETHEE) method [14], DEMATEL [15], Elimination and Choice Expressing Reality (ELECTRE) method [16], Complex Proportional Assessment (COPRAS) method [17], VIKOR (Visekriterijumska optimizacijai Kompromisno Resenje in Serbian, means Multicriteria Optimization and Compromise Solution) method [18], Additive Ratio Assessment (ARAS) method [19], Multi-Objective Optimization on the basis of Ratio Analysis (MOORA) method [20], and ANP [21], is developed by the researchers for attaining quality facts and outputs by the processes, whose implication depends on the attainment of Q&Q information. It is found that various academic researchers, decision-makers, and experts have applied these methods in supply chain management (SCM), logistics management, supplier selection, machine tool selection, and on various other fields. Additionally, researchers has also stated concentrating on the application of these methods on the production and manufacturing arena, which will facilitates in the DM process. The same is demonstrated in this chapter, where application cases related with the selection of solar panels (SPs), evaluation of automatic/robotic welding system (WS), selection of smart alloys, identification of logistic service provider, evaluation of machine tool, and election of IR are presented to represent the utility of Q&Q information in evaluation. Various cases under aforesaid aspects are presented to report the importance of Q&Q information.

It is admired that the multi-level MCDM algorithmic technique for appraising the economic values of the material handling system under qualitative and quantitative information is applied to build robust selection decision [22]. Here, the authors have evaluated the performance of logistic activities in the realm of industrial management. He *et al.* [23] developed the decision support framework (DSF) under qualitative premises to evaluate the case of polar robot evaluation, where the authors measured the overall economic worth of advanced manufacturing machines under quantitative information and reported the effectiveness of Q&Q information in making effective decision. The implication of four MCDM techniques under Q&Q is reported for making effective decision-making (DM) framework. It is found significant to utilize qualitative information to explore DM capabilities and to establish a link between digital transformation and firms and supply chains' capabilities [24]. Here, the authors have suggested to utilize Q&Q information under STRATH (strategies and threats)–based modeling to reinforce capabilities of supply chains (SCs). The authors have used qualitative measure to define SWOT (strength, weakness, opportunity, and threats) architectures for elevating the fertility in SCM.

Lee *et al.* [25] developed dynamic location and allocation models under Q&Q medium to handle the issues related with RL network configurations and various associated factors, where an extended two stage stochastic programming is used to develop RL network design. Here, elevated outcomes in terms of sustainable indicators are demonstrated to underline the utility of the developed stochastic model. A DSF based on qualitative factors is found significant to select robust supplier under the boundaries of Lean-Agile-Resilient-Green (LARG) practices for developing capabilities of an Indian automotive manufacturing company [26]. A gray set–based scorecard model is found admired to measure the performances of fruit supply bazaars [27]. Here, the advanced hierarchical structural model is constructed on the basis of the chain of Q&Q macro-micro parameters among fruit retailers. Accordingly, one can say that Q&Q information can discharge fruitful results in exploiting system characteristics.

Today, the manufacturing capabilities of large industries require the introduction and adoption of highly automated and computer-controlled strategies to satisfy the ever-changing market requirement [28]. The same is needed to economically operate manufacturing systems efficiently and to receive elevated pay back and to demonstrate the success of these manufacturing systems and machines, which depends on selection of a range of decisive parameters in the form of Q&Q factors [29]. The right judgment depends on the right selection of Q&Q information. Moreover, it is also needed to develop efficient techniques to analyze these Q&Q information, where many researchers have started focusing and taking initiatives to develop tools, critical ways, and techniques under industrial and manufacturing fields in diverse engineering areas such as welding, SCM, agile manufacturing, production planning, and quality control. The same is required in industrial and manufacturing domains for attaining sustainable industrial manufacturing practices. Sahu *et al.* [30] have developed one DM tool based on quantitative experimental information for machining nickel-based super alloy for optimizing output responses, i.e., power consumption, machining time, and material removal rate. Here, the authors have integrated two DM techniques with Taguchi technique and found imperative to develop tools, critical ways, and techniques under manufacturing sphere to receive eminence outputs. A decision support system (DSS) under qualitative factors to extract customer's preferences is found fruitful to prevent replacement of mobiles and e-waste generation for supporting green issues [31]. Here, qualitative factors are analyzed by exploring generalized interval-valued trapezoidal fuzzy numbers with a degree of similarity approach to attain sustainability. Arya *et al.* [32] have examined manufacturing characteristics, i.e., tensile strength, bead width, bead height, penetration, and Heat Affected Zone (HAZ) in MIG welding to ascertain the synergy among inputs such as wire diameter, welding current, arc voltage, welding speed, and gas flow rate by using Signal to Noise (S/N) ratio and ANOVA analysis to find significant input parameter, where the role of quantitative information is appraised by the authors. Additionally, Mukhraiya *et al.* [33] have examined the torsional rigidity of weld in MIG welding in respect to ascertain the synergy among inputs such as wire feed rate, welding voltage, and welding current by using S/N ratio and ANOVA analysis to find significant input parameter. Moreover, Sivasakthivel *et al.* [34] have examined tensile strength of weld in MIG welding in respect to ascertain the synergy among inputs such as speed, current, and voltage by using S/N ratio and ANOVA analysis to find significant input parameter. Thus, one can say that quantitative information is desired to obtain the precise understanding of optimum process parameters. The study conducted by Perumal *et al.* [35] utilized quantitative information to examine the effect of pulse on duration, wire tension, and wire feed on metal removal rate and surface roughness (SR) for machining. Here, grey relational approach (GRA) is used there to analyze the collected quantitative information with the intension to receive significant facts and to review the process parameters for sustainable outcomes. Nas *et al.* [36] utilized quantitative information to machine AISI D2 cold work tool steel and thrust to utilize GRA to categorize the influence of machining parameters on output parameters, where it is found that machining parameters, i.e., amperage and pulse on time, significantly influence the SR and hole diameter in electric discharge machining (EDM). Considering quantitative information, Abed *et al.* [37] proposed a sequential discharge model for EDM machining, and Guo *et al.* [38] also utilized quantitative information to machine Ti-6Al-4V work material considering compressed air, kerosene, and water-based emulsion as a dielectric medium to investigate the material removal process in machining, where results are experimentally validated and a need is found to develop appropriate mathematical modeling to minimize the extent of errors from machining parameters and to receive elevated characteristics in output materials. Accordingly, it is investigated that the extraction of quantitative information is needed to receive the right understanding of optimum process parameters.

Manufacturing is the keen driver for the economy and for its sustainability in the present scenario; manufacturing is every day facing new challenges for the never-ending upcoming demands. Moreover, due to various constraints, i.e., design complexity, material hardness, cost, production efficiency, government laws, and environmental responsibility, it is also not possible to manufacture the components in cost effective manner by traditional manufacturing methods, and, hence, manufacturing organizations are evolving, changing, and shifting from conventional manufacturing methods to modern manufacturing practices and trying to becoming cleaner, leaner, and greener for gaining competitive advantage, but that demands DSF, which relies on the Q&Q information. Today, the whole manufacturing world is looking for the possibilities for becoming progressive manufacturing companies with the perspective of future growth. Now, industrial and manufacturing practices are shifting to take advantage of the better operational efficiencies, better utilization of resources, minimal wastage, waste reduction, effective and environmental production, etc., for the overall societal development. Accordingly, it is imperative to extract quality information from the ocean of Q&Q characteristics to receive progression and success.

1.3 Decision-Making (DM) and Framework

The DM is always crucial and involves a procedure that is systematically used for evaluating options, taking factors into account, and chooses the best from available alternatives. DM process collects and gathers the information for defined objectives and performs brainstorming and assessment for the available options against the each selected criteria. Afterward, selection of the best option among the available options is carried out with respect to the goals and criteria. Thereafter, the selected alternatives are implemented into the action and continuous monitoring and tracking of the implemented outcomes has to be assessed in order to recognize the impact of the chosen alternative on organizational performance. A schematic representation of the DM procedure is depicted in Figure 1.1. It is important to note that the intricacy of the DM procedure can vary on the basis of the nature of the decision, the amount of available information, the risk involved, and the number of people or stakeholders affected by the decision. Additionally, DM can be influenced by cognitive biases, emotions, and external pressures. Developing a structured approach to DM helps to minimize these influences and make more informed choices.

It is highlighted that the MCDM framework is typically composed of several key elements, which are explored by the decision-maker in situations having multiple criteria and alternatives. The complexity of MCDM frameworks can vary, and they can encompass diverse mathematical models and techniques tailored to the precise context and demands of the decision. The selection of a MCDM framework is contingent upon factors like the quantity of alternatives, the number of criteria, data availability, and the inclinations of the decision-makers, where Q&Q information plays a

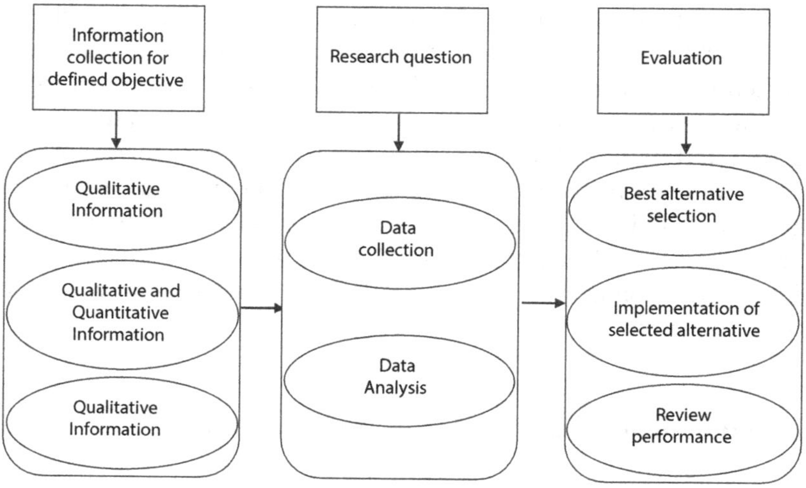

Figure 1.1 Decision-making procedure.

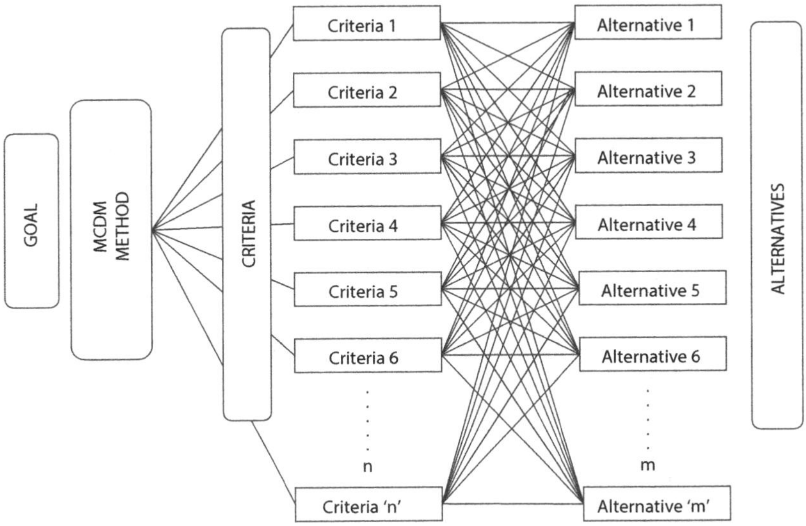

Figure 1.2 General framework for the MCDM model.

very crucial role in analyzing solutions from MCDM framework. A general framework for the MCDM model is depicted in Figure 1.2.

1.4 Directory of Cases

In the present chapter, various cases are evaluated under Q&Q information aspects to report the importance of Q&Q information and to create right understanding. The decision is basically a choice that the mind makes up for the best solution on the basis of various alternatives. DM derives from the Latin word "decidere" that means "to decide, determine", and it can be considered as the cognitive practices that result in the assortment of the most excellent scenario among the numerous alternate scenarios. Application cases related with the role of Q&Q information under selection of SPs, automatic/robotic WS, smart alloys, logistic service provider, and machine tool, IR are presented in Sections 1.4.1 to 1.4.6 to represent the utility and importance of Q&Q information in evaluation.

1.4.1 Role of Qualitative or Quantitative Criteria Toward Solar Panel Evaluation

Solar energy is one of the most competent, clean, trustworthy, and environmentally responsive power sources. It has begun gaining importance in the present scenario and is outstanding between renewable power sources. The evaluation of SP can help to decrease greenhouse gas emissions. The evaluation is needed to deliver low emissions and environmental sustainability in contrast to the conventional fossil fuel–based energy utilization systems. Nowadays, it is obligatory to select a proper solar system based on multiple aspects, criteria, and parameters to harvest utmost benefits for sustainable development. Numerous Q&Q criteria under the sphere of fiscal, societal, ecological, and scientific realm are needed to be examined for amalgamation and accurate selection of the suitable alternative of SP for binding maximum benefits.

The sustainable energy appointments can be achieved by the evaluation of Q&Q criteria related to solar sources. The same may lead in attaining less emission of contaminants, CO_2, and greenhouse gases into the environment. SP systems are enormously acknowledged as it offers contribution like cutback in CO_2 emission in cities and rural regions, offers social adequacy, supports increasing energy demand, and offers utmost efficiency with least losses. Today, extensive diversity of SP exists in the market, whose selection is thus hard and useful for the practitioners. Solar energy

Table 1.1 Solar panel evaluation criteria for modeling.

S. no.	Criteria	S. no.	Criteria
1.	Maintenance and handling	9.	Capital cost
2.	Technological drag	10.	Climate
3.	Payback period	11.	Reliability
4.	Average temperature	12.	Power factor and capacity factor
5.	Operation costs	13.	Transmission problem
6.	Service support	14.	Working automation
7.	Installation cost	15.	Convenience of use
8.	Availability of land		

alternative evaluations can support sustainable satisfaction of the power necessities due to its elevated accessibility, environmental friendliness, and huge availability under cheap segments. The exploitation of technical factors for evaluation is needed for right DM. Fifteen Q&Q factors, i.e., maintenance and handling, technological drag, payback period, average temperature, operation costs, service support, installation cost, availability of land, capital cost, climate, reliability, power factor and capacity factor, transmission problem, working automation, and convenience of use, can be utilized to evaluate SP. A model related with the 15 Q&Q factors can be significant to evaluate wide varieties of SP. The same can deliver a high performance for the selected SP machinery assets. The 15 criteria as presented in Table 1.1 can be exploited to evaluate SP.

1.4.1.1 Summary and Discussions Toward Evaluating Solar Panels

It is found that multi-criteria evaluation approaches are important to build DSF for evaluation. Multi-criteria evaluation is now becoming a prime choice in assessing the system behavior [39–41]. Amin *et al.* [42] have evaluated four dissimilar kinds of SP, i.e., as mono-crystalline silicon, multi-crystalline silicon, amorphous silicon, and copper–indium–diselenide and carried out a practical field study to investigate and compare their performance beneath Malaysia's weather. Kengpol *et al.* [43] have applied integrated fuzzy AHP and TOPSIS to formulate DSS for solar power plant site selection in Thailand. The authors have utilized geographic information system (GIS) as the qualitative criteria to decide the optimal site location for installation of solar power plant. Beltran [44] has applied AHP and ANP techniques to establish a decision model for a Spanish solar power investment company related to determine to invest in a particular solar-thermal power plant project or not. Cavallaro [45] has utilized fuzzy multi-criteria method approach to evaluate dissimilar heat transfer fluids to examine the possibility of employing molten salt in the concentrated solar power systems. Khaparde and Mukerjee [46] have presented reasonable clarifications for the deployment of renewable energy technologies as adequate infrastructure development affirms them as critical importance to exploit power from them. Moreover, diverse options are presented by them to highlight the possible transformations by renewable energy for the upcoming power system. Tavana *et al.* [47] introduced a three-stage fuzzy evaluation framework based on ANFIS (Adaptive Neuro-Fuzzy Inference System), AHP, and FIS for selecting convenient locations for solar farm installation. Here, the authors highlighted the installation of SP as a straightforward problem for finding the requisite investment considering Q&Q information, i.e., dissimilarities in climatic aspects, the type and availability of land, transportation infrastructures, and the quality of power lines. Uyan [48] applied GIS and AHP to determine the suitable site selection for the solar farms based on qualitative criteria, i.e., elevated transmission capacity lines, local weathering factors, farming facilities, quality of topography, and ecological preservation concerns for large investments in Turkey. Lozano *et al.* [49] applied GIS and MCDM methods to acquire the appraisal of the most advantageous position of photovoltaic solar power plants in Cartagena area of Murcia, in Spain. The authors have applied AHP under Q&Q criteria to calculate the weights of the criteria and TOPSIS for the assessment of the alternatives of solar farms locations according to their degree of adequacy. Salah *et al.* [50] have presented an assessment framework based on real time with regard to multi-energy save criteria and the photovoltaic panel utmost obtainable energy and equipment states. Their decision model is capable to identify whether the connection of domestic appliances was made to the electrical network or a photovoltaic panel in the north of Tunisia.

Hence, it can be assumed that the mathematical modeling techniques and MCDM evaluation having Q&Q information have been found capable to assess the overall selection of SP among the conflicting criteria. MCDM-based decision model facilitates the decision-makers to perform precise and faster computation via inculcating both subjective and objective information. These methods provide efficient and effective DSS that are simpler and user-friendly. One can say that the ocean of Q&Q information plays a crucial role in the decision related to evaluations and selections of SPs for progression, success, and sustainability.

1.4.2 Role of Qualitative or Quantitative Criteria Toward Optimization of Automatic/Robotic Welding Systems

Welding engineering plays a comprehensive role in manufacturing products for different engineering applications. In welding, coalescence of materials is done by heating them to appropriate temperatures, with or without the application of pressure, and with or without the use of filler material. Here, the projected metal is firstly cleaned from dirt, grease, or other contaminants on the surface. Moreover, the edge preparation/joint preparation depends on the type of WS and the plate thickness. Proper heat generation at the surface being welded is keen important as too high-heat generation can spoil the welded metals being joined, whereas too low-heat supply results in improper melting and fusing of weld metal.

Moreover, the application of pressure also highly impacts the weld strength and weld quality. Too high weld pressure can ruin the weld metal, whereas too low pressure will not permit them to fuse together properly. Welding engineering is a comprehensive field that necessitates expertise in an extensive range of engineering applications, i.e., nuclear, petrochemical, automotive, health, shipbuilding, aerospace, power generation, and heavy equipment sectors. Moreover, proper selection of WS and machine plays a key role in overall product quality, weldability, production rate, efficiency, etc. However, improper utilization of welding power source and equipment results in the welding defects and causes dissatisfaction during the service period in the engineering application. Thus, the need for the proper election of welding equipments and machine becomes essential for having good welded joint, which relies on the Q&Q measures and characteristics.

Today's automation world is focusing and needs sophisticated WS having various features to enable the welders to have efficient working atmosphere. But, on the other hand, selection of these sophisticated WSs having various features also becomes crucial as the novel tools are continuously developed. The same has attracted attention of many researchers to develop DSF for selecting welding machines, which demands Q&Q welding parameters. In this section, the WS selection has been appraised, and recent developments toward applying MCDM methods are reviewed for selecting welding machines in the manufacturing and other industries. Arc WSs are employing power sources of the constant current–type or constant voltage–type characteristics. Moreover, modern information technologies, i.e., IoT, big data, AI, cloud computing, and intelligent manufacturing, have transformed WS into intelligent welding systems (IWSs). Welding is a hazardous job when adequate provisions are not taken to guard the welder from inherent accidents. Moreover, technological breakthroughs and higher employment of sensor driven command have facilitated manual welding processes to be replaced with automatic welding. Also, overlooking the protection and security procedures may cause welders to encounter such hazards like electric shock, overexposure to fumes and gases, arc rays, fire, and explosion. Exposure to the evaporate gases in welding process has high effects on welders' health. Hence, it is quite essential and recommended to use adequate provisions such as mounting of proper ventilation arrangement, workplace atmosphere supervision, implementing suitable respiratory protection, and instructing the welders and personnel for their protection. Recently, for making DM effective under welding, both quantitative and qualitative data are utilizing to build DSF and to assess sustainable manufacturing practices. It is found that the form of information that is supplied for evaluation plays an essential role in DM. The information may be in the form of scientifically derived hard data, subjective interpretations, uncertain information, reasoning based information, etc., or their combination.

This section aims to present "how one can utilize MCDM applications in sustainable welding and manufacturing". The 23 Q&Q factors, i.e., weld current, capital cost, welding speed, weld quality, accuracy, exposure to operator unrest/comfort, efficiency, power factor and capacity factor, maintenance cost, payback period, operating cost, maintainability/regular maintenance, noise and vibration, duty cycle, functionality, man–machine interface ability, overall accuracy, life expectancy, depreciation rate of machine, simplicity, health and safety, stability and reliability, and precision can be utilized to evaluate WSs. The evaluation of suitable alternative in the welding environment is keen essential because of the availability of an extensive variety of welding alternatives and similarity among them, where the aforesaid Q&Q factors can assist in proper selection. Therefore, there

Table 1.2 Welding system evaluation criteria for modeling.

S. no.	Criteria	S. no.	Criteria
1	Weld current	13	Capital cost
2	Welding speed	14	Weld quality
3	Accuracy	15	Exposure to operator unrest/ comfort
4	Efficiency	16	Power factor and capacity factor
5	Maintenance cost	17	Payback period
6	Operating cost	18	Maintainability/regular maintenance
7	Noise and vibration	19	Duty cycle
8	Functionality	20	Man–machine interface ability
9	Overall accuracy	21	Life expectancy
10	Depreciation rate of machine	22	Simplicity
11	Health and safety	23	Precision
12	Stability and reliability		

is a need for straightforward, uncomplicated, organized, and reasonable process or mathematical procedure to direct decision-makers in considering selection characteristic and their interrelations. The 23 criteria as presented in Table 1.2 can be exploited for evaluation of WS.

1.4.2.1 Summary and Discussions Toward Evaluating Welding System

It is highlighted that various MCDM methods are, at present, existing to assist organizations in deciding the best decisive course of actions. It is found that Kumar *et al.* [51] have designed and developed a programmable system-on-chip–based embedded weld monitoring system with suitable software package to measure all the dynamic variations of welding voltage and current. Banga *et al.* [52] have analyzed the robot movement to augment the interaction between four robots using offline robot simulation software "DELMIA-V5". Kah *et al.* [53] have discussed the role of sensors in robotic arc WSs with adaptive and intelligent control system features that can track the joint, in-process weld quality and variation in joint position, and geometry. The authors have described various aspects and programming of robotic WSs in order to attain better competence to control welding parameters, robotic motion, and improved fault detection and correction. Wang *et al.* [54] have examined elemental means and practices that are necessary to make IWS. Hong *et al.* [55] have reported the safe practices to be followed by the welders and practitioners while welding. The authors studied the impact of welding over personal safety and instructed about the safety and security conditions that must have to be pursued in welding processes. Mehrifar *et al.* [56] have evaluated and assessed the health risk of exposure of welders toward welding gases in seven types of welding in the steel industry. Dehghan *et al.* [57] have investigated the impact of concentrated lead fumes generated during welding on reproductive hormones of exposed welders. The authors have found that occupational exposure to metal fumes containing lead among welders may alter the level of sexual hormones and potentially harm the reproductive system. Ravisankar *et al.* [58] have employed MIG, TIG (tungsten inert gas), and PAW (plasma arc welding) for making butt joints in AA 7075 aluminum alloy and utilized AHP to recognize the chief WS. Singh and Rao [59] have proposed a hybrid DM-GTAMA and AHP for answering industrial environment problem as a choice of WS to attain a decisive solution toward an unstructured assessment problem having manifold characteristics. Chakraborty and Zavadskas [60] have explored the weighted aggregated sum product assessment (WASPAS) method over eight manufacturing DM problems and found that WASPAS has the potential of correctly ranking the substitutes. Omar and Soltan [61] have developed integrated support system to differentiate welding processes for industrial purposes by using FUZZY-AHP and FUZZY-TOPSIS in portable software. Bagheri *et al.* [62] studied the impact of CO_2 laser welding, friction stir welding, and friction stir vibration welding on tailor welded blank (TWB) to identify the effects of these different welding methods on mechanical and formability properties of developed TWBs. Jamwal *et al.* [63] highlighted "how the modern engineering organizations can gain competitive advantage from MCDM methods in their DM".

There is a perceptible augmentation found in the employment of theMCDM techniques for machine selection problems that serve as decision-support tool for the welding or design engineer. Evolving technological landscape has been shifting the production scenario into modern organizations, where it is needed to take prompt, swift, and accurate decisions concerning with the effectual employment of machineries resources. Additionally, to sustain in the present comprehensive worldwide competitive environment, sustainable production practices are needed to provide a better solutions to synchronize manufacturing activities and biodiversity preservation, which can be easily done by identifying a zone of allied Q&Q information for related sources. The more the Q&Q measures information will be collected, the more the accuracy will be attained by the system, which, in turn, will assist in providing satisfactory solution for the specified problem on the basis of strategic and technical knowledge.

1.4.3 Role of Qualitative or Quantitative Criteria Toward Selection of Smart Alloys and Materials

Engineering designs are often constrained with lots of objectives and selection of proper engineering materials for a particular engineering field in accordance with the desired objective is too crucial, very tedious, and intellectual task. The selection of material may be such that it fulfills the desired objectives such as high performance, minimum cost, technically sound, good machinability, and being environmentally friendly. Selection of the most suitable material for miscellaneous and crucial engineering applications is often regarded as a MCDM problem, which governs with the sets of Q&Q criteria and sub-criteria. Most suitable material selection for miscellaneous engineering applications has strong relationships with the MCDM methods. Selection of proper materials and constructing decision matrix consisting of a restricted set of feasible alternatives is the most complicated task in designing and developing novel products. Various super alloys are widely developed and gaining importance in current era of industry 4.0 (I-4.0), but what should be the medium or way of selection of an appropriate supper alloy is needed for convincing accuracy and desired attainments. Here, Q&Q assessment can impart fruitful solutions. The modern technological and competitive requirement has imposed the business to additionally concentrate on assortment of materials that accomplish the

Table 1.3 Smart alloy evaluation criteria for modeling.

S. no.	Criteria	S. no.	Criteria
1	Hardness	18	Fire resistance
2	Young's modulus	19	Compactness
3	Elastic recovery	20	Specific stiffness
4	Friction coefficient	21	Corrosion resistance
5	Critical load for adhesion to substrate	22	Material cost
6	Wear resistance	23	Toughness index
7	Yield strength	24	Thermal expansion coefficient
8	Thermal conductivity	25	Specific heat
9	Fabrication cost	26	Repairability
10	Ability for weight reduction	27	Risk
11	Work hardening	28	CO_2 emission
12	Shock resistance	29	Melting point
13	Fatigue resistance	30	Surface texture
14	Machinability	31	Environmental load
15	Human health	32	Geographical location
16	Accessibility	33	Political stability and foreign policy
17	Availability	34	Durability

design criterion while providing the uppermost level of performance at the most reasonable cost. Accordingly, it is presented that the below mentioned 34 Q&Q factors, i.e., hardness, Young's modulus, elastic recovery, friction coefficient, critical load for adhesion to substrate, wear resistance, yield strength, thermal conductivity, fabrication cost, ability for weight reduction, work hardening, shock resistance, fatigue resistance, machinability, human health, fire resistance, compactness, specific stiffness, corrosion resistance, material cost, toughness index, thermal expansion coefficient, specific heat, repairability, risk, CO_2 emission, melting point, surface texture, environmental load, accessibility, availability, geographical location, political stability and foreign policy, and durability under the origins of material selection can be utilized to evaluate smart alloys (Table 1.3).

1.4.3.1 Summary and Discussions Toward Selection of Smart Alloys and Materials

Numerous scholars have highlighted the link between the MCDM techniques and the challenges associated with material selection. Material selection constitutes a multifaceted DM process, where the choice of the most appropriate materials from various available options is made by considering multiple criteria that can be qualitative or quantitative or qualitative and quantitative. Zakeri *et al.* [64] proposed the simple ranking process for addressing intricate material selection challenges and highlighted that the selection of an effective material choice plays a pivotal role in engineering and will aid in satisfying all the functional requirements of the object. Bhaskar and Khan [65] have demonstrated the proficiency of five different hybrid MCDM methods, i.e., AHP-VIKOR, AHP-TOPSIS, AHP-MOORA, AHP-ELECTRE, and AHP-PROMTHEE, to recognize the most excellent polymer based biomaterial used in dentistry. Farid *et al.* [66] have proposed single-valued neutrosophic and utilized Einstein interactive weighted averaging and geometric operators for material selection of cryogenic storage containers in engineering design. Aksakal *et al.* [67] have considered various criteria in the MCDM domain in order to evaluate insulation materials, i.e., polyurethane (1), polyurethane (2), recycled textile (1), recycled textile (2), wood fiber (1), wood fiber (2), stone wool (1), stone wool (2), cellulose (1), cellulose (2), vermiculite (1), vermiculite (2), recycled glass, fiber (commercial), recycled glass fiber, XPS (1), XPS (2), straw bale, cork, perlite, kenaf, hemp, flax, recycled PET (commercial), cotton (recycled), mineralized wood fiber, EPS, sheep wool, coir, jute, polyisocyanurate, glass wool, and reeds for healthier environment. Bhadra and

Dhar [68] have utilized fuzzy integrated AHP- and TOPSIS-based appraisement considering Evaluated Distance from Average Solution (EDAS) and COPRAS techniques in order to evaluate natural fibers, which is to be used in the interior of aerospace cabins. Agrawal [69] has utilized four MCDM techniques, i.e., SAW, MOORA, TOPSIS, and VIKOR, to analyze and compare sustainable material selection for three AM technologies, i.e., FDM, SLS, and SLA. Chatterjee and Chakraborty [70] have applied DEMATEL technique for a gear material selection process. Figueiredo [71] has applied FAHP and integrated life cycle sustainability assessment approach under MCDM and built information modeling to select appropriate materials for constructions. Prasad [72] has utilized fuzzy AHP and TOPSIS for evaluation of coating material for the magnesium AZ31B alloy substrate.

Accordingly, one can say that material selection process has miscellaneous information range and one has to utilize sophisticated tools, methods, etc., that help in converting and transforming this diverse information into better decisions. Hence, one can say that DM is indeed required in this segmental field, where one should apply their skills and knowledge to achieve the informative and significant results in said field. Additionally, in many other fields, i.e., basic science, engineering, social science, medical field, and managerial science, DM is also required to receive significant ingredient and to flag success and progression. It is highlighted that many challenges are frequently emerged in practical scenarios during collection of information, which can be easily settled by the utility of Q&Q factors and measures. The same can assist individuals in ease DM and to select favorable choice from a range of options that exhibit variations across numerous criteria or aspects. It is ascertained that the MCDM techniques are highly favored among the available tools for analysing Q&Q information. It is informed that MCDM have the capacity and constitutes a category of mathematical techniques that proves value in tackling intricate DM challenges related to material selection, serving as efficient instruments for the proficient management of material selection quandaries that encompass diverse material attributes and a range of Q&Q criteria.

1.4.4 Role of Qualitative or Quantitative Criteria Toward Logistic Service Provider Evaluation

Over the last few decades, the organizations have only made their concerns and focus toward the forward logistics activities of SC in order to maximize their revenues, incomes, and not realizing their liability toward their manufactured components after components of life are over [73].

Modern developing world and technological advancements have made the life of the products as short-lived durations, which have reasonably resulted in rapid rising amount of dispose. Each significant task starting from procurement to distribution results in the waste generation. The inefficient disposal of this waste results in the environmental loads and leads organization toward unethical environmental respect with unconscious business practices. Now, government is more focused for ecology and accordingly; organizations are focusing their intension toward continuously increasing interest in environmentalism and recycling of the waste resulting during manufacturing. The increasing global awareness related to environmental issues has led the establishment of numerous guidelines, regulations, policies, and legislations worldwide. Organizations are taking initiatives and focusing to achieve green issues in reply, so as to survive

Table 1.4 Logistic service provider evaluation criteria for modeling.

S. no.	Criteria	S. no.	Criteria
1	International scope	13	Quality of interrelation with vendor
2	Facility and technology	14	Security and insurance
3	Sensitivity to consumer requirements	15	Environmental consideration
4	Reliability	16	Strategic alliances
5	Security	17	Cost
6	Professionalism of salesperson	18	Topographical positions and span of services
7	Logistics information system	19	Logistics equipment
8	General reputation	20	Cancellation service
9	On-time consignments and deliveries	21	Service quality
10	Availability of trained/ skilled personnel	22	Value added/customized services
11	Vendor's service contract	23	Learning
12	Adaptability	24	Compliance

and become sustainable. Currently, the waste reductions and its recycling toward environmental protection have prompted manufacturers to intensify their focus toward recognizing RL as an indicator of competitiveness, where Q&Q information is reasonably required for collecting appropriate information and for building DSF.

RL aids companies in boosting their competitive edge and establishing a positive business image by efficiently gathering, recycling, refurbishing, and restoring used components for reprocessing and remanufacturing. It is highlighted that the collection of widely dispersed return products and components is too sophisticated and requires building of critical RL system under inventory management, logistics, transportation, warehousing, product returns service, product packaging, logistics information system, etc., each of which is surrounded by a cluster of driving measures that are Q&Q in nature. These measures aim to address waste management and material recuperation via recycling and recovery of parts and products through remanufacturing, renovating, reprocessing, and reusing. In this section, the 24 Q&Q factors, i.e., international scope, facility and technology, sensitivity to consumer requirements, reliability, security, professionalism of salesperson, logistics information system, general reputation, on-time consignments and deliveries, availability of trained and skilled personnel, vendor's service contract, adaptability, quality of interrelation with vendors, security and insurance, environmental consideration, strategic alliances, cost, topographical positions and span of services, logistics equipment, cancellation service, service quality, value added/customized services, learning, and compliance, are presented to be utilized for logistic service provider evaluation (Table 1.4).

1.4.4.1 Summary and Discussions Toward Evaluation of Logistic Service Provider

It is found that Dev *et al.* [74] have examined two facets under RL, namely, the real-time exchange of information with the RL system and the proliferation of environmentally friendly products in the market. The same are examined with the intension to affect the financial and green performances for making RL system sustainable. Here, the authors have implemented the principles of I-4.0 and prepared a model of circular economy to carry out the simulation of the RL model, remanufacturing model, and Taguchi model for attaining sustainable RL system. Cao *et al.* [75] have examined the situation of remanufacturing and its utility in China under three perspectives, i.e., the government guidelines, enterprise, and community consciousness. The authors have found that the manufacturing degree and yearly production are improving due to investment and technological assistance received from the government and with the implementation of the latest guidelines toward remanufacturing enterprises. Govindan *et al.* [76] have applied the ANP process to evaluate 3PRLP (third-party logistic provider) in an automobile company. Zarbakhshnia *et al.* [77] have utilized fuzzy SWARA, fuzzy COPRAS, and COPRAS-G (grey numbers) techniques to evaluate 3PRLP based on multiple Q&Q criteria. Sharma *et al.* [78] have applied fuzzy TOPSIS technique in order to evaluate the impact of RL practices and its influence on retailers' performance. Govindan *et al.* [79] have applied Interpretive Structural Modeling (ISM) and fuzzy TOPSIS technique in multi-criteria model to evaluate the selection process of best 3PRLP in the context of manufacturing industry in India. Tavana *et al.* [80] have developed a hybrid model, which is merged with ANP and IFG-SIR toward selection of the best 3PRLP in order to help an industrial production group. Jayant *et al.* [81] have applied hybrid approach by applying AHP and TOPSIS in order to facilitate top management of cell phone manufacturing company for the evaluation and assessment of various 3PRLP. Agrawal *et al.* [82] have explored the various disposition criteria to evaluate decision alternatives based on the consideration of revert components for repair or reclaim for sale as new, revamped, or renovated products in a mobile manufacturing firm. The disposition index is proposed to provide effectual DM process for making disposition decision in RL. Prakash *et al.* [83] have demonstrated the application of fuzzy AHP and VIKOR technique for the evaluation and assessment of the RL partner in an Indian electronics organization. Bouzon *et al.* [84] have utilized a grey-based DEMATEL technique for assessing barriers for RL implementation under manifold perspective in the Brazilian context. Accordingly, one can say that RL DM can help in finding the ways for the generation of economical as well as ecological assets that arrived in attainment of a specified manufacturing goal. The same demands the need of a sophisticated tools, techniques, and data collection sources that bring an ultimate choice or output in the form of an action or a judgment of preference. In the past few decades, it has been found that RL has drowned the interest of industries and that the assessment of RL is too challenging as it has been surrounded with different domains of criteria either in the form of qualitative or quantitative or combination of both. Hence, it is found to develop specific MCDM tools under industrial pace that can satisfy the capability to assess and appraise alternatives using a solitary metric as these measures may have dissimilar units of measurement. The proper evaluation of RL system is becoming an imperative area and also indeed necessary as the market has become more comprehensive and it enables the organizations to cut their product costs and advance their customer service quality.

1.4.5 Role of Qualitative or Quantitative Criteria Toward Machine Tool Evaluation

In the past decades, manufacturing organizations are producing product by conventional manufacturing methods and fulfilling the market demands. It has been realized that conventional manufacturing practices are preventing them from efficient working and also from profitable goals. Today, it is the demand of growing world to produce new and sophisticated product that can meet the requirements of various engineering applications and

Table 1.5 Machine tool evaluation criteria for modeling.

S. no.	Criteria	S. no.	Criteria
1	Imbalance in mechanical parts	13	Bent shafts
2	Bearing failures/alignment fault	14	Gearbox failure
3	Mechanical looseness	15	Empty space
4	Misalignment	16	Critical speeds
5	Resonance and natural frequencies	17	Electrical motor faults
6	Part delivery position accuracy	18	Control technologies
7	Drive system performance aspect	19	Force-position (compliance) control
8	Noise delivery	20	Favorable appearance and proper structure
9	Tool change time	21	Convertibility (design for Functionality Changes
10	Max spindle torque	22	Cost
11	Failure rate	23	Space requirement (dimensions and weight)
12	Work support		

daily needs. Choosing the right machinery in the manufacturing sector has now been a critical concern because making an incorrect choice can detrimentally impact a company's productivity, efficiency, preciseness, adaptability, easiness, and its ability to respond swiftly in manufacturing operations. Here, in this section, 23 Q&Q factors, i.e., imbalance in mechanical parts, bearing failures/alignment fault, mechanical looseness, misalignment, resonance and natural frequencies, part delivery position accuracy, drive system performance aspect, noise delivery, tool change time, max spindle torque, failure rate, work support, bent shafts, gearbox failure, empty space, critical speeds, electrical motor faults, control technologies, force-position (compliance) control, favorable appearance and proper structure, convertibility (design for functionality changes, cost, and space requirement (dimensions and weight), are presented to be utilized for the selection and evaluation of machine tool (Table 1.5).

1.4.5.1 Summary and Discussions Toward Evaluation of Machine Tool

Presently, the MCDM approaches are prevailing to make crucial decisions [85]. MCDM approaches can be employed for making decisions about machine tool selection predominantly based on the erection of Q&Q factors with emphasis on the biased viewpoints. Since, the objective assessment reflects the real-world effectiveness of machine tools. Hence, it is imperative to take into account both subjective and objective viewpoints, when opting for the appropriate machine tool. With the growing array of available machine options in the market and the ongoing advancements in technology, the challenge of selecting the most suitable machine has garnered significant attention from researchers. Li *et al.* [86] have proposed a novel fused MCDM model based on fuzzy DEMATEL–based entropy weighting and VIKOR techniques for the evaluation problem of machine tool. It is found that the manufacturers can achieve cost-effective machining operations over the course of a year through replacement analysis based on qualitative features [7]. This approach can also assist manufacturers in avoiding obsolescence resulting from technological advancements or changes that can unexpectedly arise. Hagag *et al.* [87] have identified various MCDM tools that have been utilized in numerous machine tool selection problems of different areas of engineering. Cetinkaya *et al.* [88] have applied fused fuzzy AHP and PROMETHEE technique for 3D printer machine evaluation decisions. Çakır [89] has proposed an amalgamated technique consist of fuzzy simple multi-attribute rating technique and fuzzy weighted axiomatic design to select the best optimal continuous fluid bed tea dryer machine on the basis of overall performance for a Turkey-based private-owned tea plant. Hutahaean *et al.* [90] have proposed a DSF for equipment assortment based on AHP for the evaluation of biomass separation equipment. Accordingly, one can say that the process of choosing machine is both crucial and challenging for organizations, as it exerts a significant influence on the overall performance of their manufacturing systems [91–95]. An appropriate machine assortment among feasible options has been a decisive concern for any manufacturing firms as it may result to more problems than anticipated [96–98]. An incorrect machine tool selection may reduce the productivity and increase the overall costing of the manufacturing systems [99–104]. Before proceeding for the selection of machine systems, it is hugely necessary that the engineers should first scrutinize the measures and sub-measures for recording Q&Q information according to their process requirements [105, 106]. Moreover, DM problems related to machine tool have multiple conflicting criteria, and choosing the right machine tool based on the measures and sub-measures may vary according to the type of the manufacturing facility, market necessities, and consumer perspective. The nature of criteria for the selection of an appropriate machine tool may be qualitative or quantitative or both, and, thus, one can regard machine tool selection as a MCDM problem [107–111]. Consequently, it is indeed necessary to concentrate on Q&Q measures for retaining best machine tool selection decision. Proper machine selection with manufacturing goals, in response to organization requirements and competitive positioning, extremely impacts the manufacturing costs and influences products excellence, productivity market share, and environmental aspects. Customer satisfaction should be the most important paramount criteria that should be considered as it directly influences product demands based on product performances, prices, and technological equipments.

1.4.6 Role of Qualitative or Quantitative Criteria Toward Industrial Robot Selection

Effective IR selection is critical for optimizing production efficiencies and acquires a competitive edge for blooming in the global industrial scenario. A general-purpose, versatile, and reprogrammable mechanism designed with particular anthropometric characteristics is described as an IR. The DM process for choosing IR is complex due to the requirement of diverse characteristics, manufacturing needs, precise thresholds, and accurate projections, and, thus, IR-DM requires a critical assessment of factors toward impacting functional success. The huge number of Q&Q factors impacts the usability and suitability of the chosen robots for various tasks among the wide array of available alternatives. Optimizing manufacturing efficiency by the use of IR requires the right selection of IR, which is too complex for the industrial organizations as the vast number of manufacturing characteristics and provisions are being continuously indulged and offered by the robot manufacturers. Right robot selection considering optimum capacity and functionality is a crucial decision as it improves product quality and increases productivity for most manufacturing companies and, thus, demands exploitation of Q&Q measures.

In the current situation, the primary reason for the augmented employment of robots in industrial works is attributed by the technological advancement done in last decades as well as ease availability of different capabilities and specifications of IR in market, which has led the adaptability and utility of IR for performing recurring, complicated, or risky tasks with accuracy. Consequently, the trend for using IRs in numerous industrial functions, including machining, assembling, materials processing, forming, painting, and fabricating, is done by robots as it drastically improved quality and productivity, when robots are utilized correctly. Over the past decades, there has been a substantial rise in both the wide range of applications and the available number of IR, resulting in enhanced product qualities, quicker manufacturing rates, deliveries, elevated product flexibilities, and enhanced effectiveness. Due to the increased complexities and requirement of inclusion of highly developed features and facilities in IR for customized operations, it is required to vigilantly assess and evaluate the essential variants of IR from the available range of IR, which demands critical criteria, multiple characteristics, and allied Q&Q information for evaluation. Factors such as ability toward product design, operation system suitability, and financial feasibility must be taken into account, prior to selecting an appropriate robot that can be opted. Inadequate DM in robot selection will have an adverse impact on a company's ability in terms of production efficiency and quality. Here, in this section, 19 Q&Q factors, i.e., vendors' service quality, power supply/energy consumption, scheduling utilization, risk, productivity, overall flexibility, quality aspect, overall performance, high-level structured object-oriented programming language, spline interpolation, teach-in play-back programming technology, program steps, online program correction, technical features, geometrical dexterity, programming flexibility, path correction facilities, implementation easiness, and user support, can be utilized for the evaluation of IR (Table 1.6).

Table 1.6 Industrial robot evaluation criteria for modeling.

S. no.	Criteria	S. no.	Criteria
1	Vendors' service quality	11	Teach-in play-back programming technology
2	Power supply/energy consumption	12	Program steps
3	Scheduling utilization	13	Online program correction
4	Risk	14	Technical features
5	Productivity	15	Geometrical dexterity
6	Overall flexibility	16	Programming flexibility
7	Quality aspect	17	Path correction facilities
8	Overall performance	18	Implementation easiness
9	High-level structured object-oriented programming language	19	User support
10	Spline interpolation		

1.4.6.1 Summary and Discussions Toward Selection of Industrial Robot

To satisfy the requirements of the fast-growing market, to induce increased accuracies, and to intensify competition, it is required to identify momentous criteria to be utilized in the selection of robot among available alternatives. Rao *et al.* [112] introduced an integrated DM method based on fuzzy logic, which is combined with subjective and objective factors to address the challenge of selecting a suitable robot from an extensive group of options based on various capabilities and specifications for specific applications and production environments. Chatterjee *et al.* [113] found to perfectly choose a tailored robot to a specific manufacturing application to satisfy the dynamic manufacturing landscape. The selection of a tailored robot has thereby found intricate by the authors due to rising intricacy, availability of sophisticated features, and diverse manufacturers' technologies. The authors address the challenge of selection of a tailored robot by employing VIKOR and ELECTRE techniques and demonstrated the effectiveness of said techniques in selection decision through real-world examples and validated it with prior research in robot selection. Abdullah *et al.* [114] have employed a hybrid AHP-fuzzy TOPSIS approach to rank the impact of I-4.0 technologies on manufacturing strategy outputs, revealing cost as the crucial factor and additive manufacturing (AM) as the top technology for enhancing global competitiveness. Zhou *et al.* [115] have utilized fuzzy extended VIKOR approach that integrated fuzzy with AHP for mobile robot selection for hospital pharmacies in the context of developing the Internet of Health Things and health care robotic automation. Ghorabaee [116] has utilized interval type-2 fuzzy sets and the VIKOR method for selecting IR in complex and uncertain environment. Fu *et al.* [117] have utilized four experts under objective weight determination method and found the utility of stochastic multi-criteria acceptability analysis crucial for robot evaluation problem. Liu *et al.* [118] have integrated Quality Function Deployment (QFD) theory and QUALIFLEX under interval-valued Pythagorean uncertain linguistic to address the uncertainty of decision-makers' opinions in robot selection for auto manufacturing company. Nasrollahi *et al.* [119] have addressed the complex challenge of selecting the optimal IR for specific manufacturing needs through the application of the fuzzy best-worst method and PROMETHEE. Bairagi [120] has introduced a novel methodology called TARO, which employs a sophisticated entropy weighting technique to ensure proper and accurate DM under the MCDM environment for real-life robot selection. Shanmugasundar *et al.* [121] have employed Method Based on the Removal Effects of Criteria (MEREC) approach for weight calculation of quantitative factors and applied five MCDM techniques under them to choose the optimal spray-painting robot among 12 substitutes based on the seven decisive factors and found that HY1010A-143 is the preferred option for the task.

Various manufacturing firms across different industries are increasingly using IR. The challenge lies in selecting appropriate robot to be deployed for specific application and manufacturing environment. The process of selection is complicated due to the growing complexity in terms of technology, size, materials, input programming, geometrical flexibility, and advanced features introduced by diverse manufacturers. Thus, decision-makers must identify and choose the most appropriate robot considering multiple criteria to achieve desired output.

1.5 Critical Aspects

Manufacturing is defined as an interrelated, organized activities, processing, and operations that are indulged in creating a solitary element or manifold elements that are needed to be brought together to create a final component for utility by the society. In manufacturing, raw materials are converted into products to society with the support of technology, machinery, chemical practices, human resources, and equipments into components or final components within manufacturing industries. MCDM has proven the identification of optimal alternatives based on preferences for selected problems that encompasses numerous competing criteria and constraints. As compared to previous scenarios, now MCDM problems are more subjective and objective and require concrete choices of higher-level managerial decisions with clear objectives. Accordingly, it is suggested to employ MCDM techniques under Q&Q information to select appropriate choice. MCDM is found qualified to evaluate various versions of Q&Q criteria due to the ease evaluation capacities and due to the development of large number of mathematical techniques in this domain that prove value in tackling intricate DM challenges related to material selection, logistic selection, quality management, production planning, facility valuation, and so on, and, thus, MCDM serves as an efficient instruments for the proficient management of aforesaid quandaries that encompass diverse material attributes and a range of Q&Q criteria. The more the Q&Q measures information will be collected, the more the accuracy will be attained by the system, which, in turn, will assist in providing satisfactory solution for the specified problem on the basis of strategic and technical knowledge. Definitely, MCDM tools are comprehensive evaluation tools and provides systematic and objective approach that assists in identifying critical criteria, comparing their performance, and enhancing transparency in DM. Moreover, MCDM effectiveness relied more on human preferences, which is based on their expertise, experience, etc. Optimal selection of criteria in MCDM is challenging as wrong selection may result in wrong judgment. Moreover, MCDM encompasses Q&Q weighting methods and diverse ranking methods for conflicting criteria and sub-criteria in order to attain multiple goals associated with these criteria and sub-criteria effectually. It is found in this chapter that, nowadays, MCDM problems are more subjective, are semi-structured, and requires top-level administrative assessment in comparison to the previous MCDM problems, where the objectives are defined and the criteria for assessment are quantifiable. Hence, the success-ability and the efficiency of the MCDM rely on the decision-makers' judgment and the ability to systematically analyze and evaluate multiple criteria to facilitate informed DM.

1.6 Implication and Discussions

The present chapter advocated its existence to assist in suggesting a wide range of Q&Q characteristics, which can be used to evaluate the system performance in the case of industrial and manufacturing designs. Manufacturing is the process of converting a set of input into output component with economic value. Manufacturing process starts with the initial product designing phase and depends on the proper identification of the Q&Q factors for attaining good outputs. Afterward, proper material is selected on the basis of Q&Q information to create the product. Thereafter, several intermediate process selection Q&Q parameters are evaluated and then end with the testing and quality checking. The outcome of the production may be in any form: either it may be tangible or intangible good (production by service), but, in every phase, it is needed to exploit appropriate factors and allied Q&Q information. The progression of choosing the best course of action from a set of accessible choices in DM can only be effective, if appropriate efforts are implicated in understanding Q&Q characteristics. Within a manufacturing facility, the significant Q&Q information is needed toward guaranteeing the smooth and efficient operation of the production process. Planned DM is assumed to be successful only if Q&Q characteristics are exploited fruitfully. The same will particularly assist in defining a rapidly paced manufacturing setting and aid in reaching the tight deadlines and uphold elevated levels of quality. The role of DM in manufacturing is paramount and multifaceted, encompassing various aspects that directly impact the efficiency, quality, and overall success of production processes. It helps in various domains of production process such as process optimization, quality control, resource allocation, capacity planning, risk management, continuous improvement, environmental impact, human resource management, and new product introductions, but the DM will only be successful only if Q&Q characteristics are exploited effectively.

1.7 Conclusions

DM holds a pivotal role within the realm of production and its impacts the efficiency, quality, and ultimate outcomes of production processes. In the present chapter, various cases are presented from manufacturing perspective to explore the DM utilization in the current manufacturing scenario. In essence, one can say that DM in manufacturing is a planned, intended, and operational necessity and needs Q&Q information to attain the efficiency, effectiveness, and competitiveness in manufacturing operations while adapting to changing market dynamics and technological advancements. It is affirmed by the present chapter that the MCDM methods are crucial for evaluating Q&Q information and can involve certainty and human interventions, i.e., psychological and behavioral factors are important in MCDM and a crucial care is needed to be taken for making the effective decisions. The chapter undertakes the ability to make informed decisions considering diverse factors and variables based on Q&Q information, which can significantly contributes to the success of manufacturing operations. By optimizing Q&Q factors under processes, resource allocation, and technology adoption, DM empowers manufacturing to meet market demands, to navigate complexities, and to achieve sustainable growth. Today, to evolve sustainable tools for industrial and manufacturing decisions, Q&Q information is necessary and plays an importance role in effective DM for innovation, competitiveness, and operational excellence. Hence, in essence, the present chapter can be utilized to understand Q&Q information to influence every facet of production, manufacturing, and business, which can assist in influencing minor daily choices to major strategic moves and to shape the trajectory of outcomes, shaping success, growth, and well-being worldwide.

References

1. Hernández, C.T., Marins, F.A.S., da Rocha, P.M., Using AHP and ANP to evaluate the relation between reverse logistics and corporate performance in Brazilian automotive industry. *Proceedings of the 10th International Symposium on the Analytic Hierarchy/Network Process Multi-criteria Decision Making*, Pennsylvania, USA, July 29 - August 1, pp. 1–14, 2009.
2. Huang, R.H., Yang, C.L., Wuang, M.S., Tsui, C.S., Constructing a performance evaluation model for reverse logistics-Cases of recycled tire traders. *The 5th IEEE International Conference on Management of Innovation and Technology (ICMIT)*, pp. 606–611, 2010, 10.1109/ICMIT.2010.5492749.
3. Sahu, A.K., Sahu, A.K., Sahu, N.K., Benchmarking of advanced manufacturing machines based on fuzzy-TOPSIS method, in: *Theoretical and Practical Advancements for Fuzzy System Integration*, pp. 309–350, 2017.
4. Sheu, J., Green supply chain management, reverse logistics and nuclear power generation. *Transp. Res. Part E: Logist. Transp. Rev.*, *44*, 1, 19–46, 2008.

5. Sahu, N.K., Sahu, A.K., Sahu, A.K., Cluster approach integrating weighted geometric aggregation operator to appraise industrial robot: Knowledge based decision support system. *Kybernetes*, *47*, 3, 487–524, 2018.
6. Sahu, A.K., Sahu, N.K., Sahu, A.K., Rajput, M.S., Narang, H.K., T-SAW methodology for parametric evaluation of surface integrity aspects in AlMg3 (AA5754) alloy: comparison with T-TOPSIS methodology. *Measurement*, *132*, 309–323, 2019.
7. Sahu, A.K., Narang, H.K., Sahu, A.K., Sahu, N.K., Machine economic life estimation based on depreciation-replacement model. *Cogent Eng.*, *3*, 1, 1249225, 2016.
8. Sahu, A.K., Sahu, N.K., Sahu, A.K., Performance estimation of firms by GLA supply chain under imperfect data, in: *Theoretical and Practical Advancements for Fuzzy System Integration*, pp. 245–277, 2017.
9. Guo, X., Sahu, A., Sahu, N.K., Sahu, A.K., A novel integrated computational TRIFMRG approach with grey relational analysis toward parametric evaluation of weld bead geometry of ms-grade: IS 2062. *Grey Syst.: Theory Appl.*, *12*, 1, 117–141, 2022.
10. Afshari, A., Mojahed, M., Yusuff, R.M., Simple Additive Weighting approach to Personnel Selection problem. *Int. J. Innov. Manage. Technol.*, *1*, 5, 511–515, 2010.
11. Manoliadis, O.G., Pantouvakis, J.P., Tsolas, L.E., Compromise programming model in site selection for construction temporary facilities. *Oper. Res.*, *7*, 381–400, 2007.
12. Roig, T.N., Puig, A.B., Vera, B.J., Verdu, M.F., The retail site location decision process using GIS and the analytical hierarchy process. *Appl. Geogr.*, *40*, 191–198, 2013.
13. Yang, T., Xu, S., Xiong, N.N., A Novel Machine Selection Method Combining Group Eigen value Method with TOPSIS Method. *Int. J. Future Gener. Commun. Netw.*, *9*, 6, 201–210, 2016.
14. Maity, S.R. and Chakraborty, S., Tool steel material selection using PROMETHEE II method. *Int. J. Adv. Manuf. Technol.*, *78*, 1537–1547, 2015.
15. Li, Y. and Mathiyazhagan, K., Application of DEMATEL approach to identify the influential indicators towards sustainable supply chain adoption in the auto components manufacturing sector. *J. Clean. Prod.*, *172*, 2931–2941, 2018.
16. Jacek, Z. and Szymon, W., The selection of the logistics center location based on MCDM/A methodology. *Transp. Res. - Proc.*, *3*, 555–564, 2014.
17. Hashemkhani, Z.S. and Bahrami, M., Investment prioritizing in high tech industries based on SWARA-COPRAS approach. *Technol. Econ. Dev. Econ.*, *20*, 534–553, 2014.

18. Tavana, M., Mavi, R.K., Arteaga, F.J.S., Doust, E.R., An extended VIKOR method using stochastic data and subjective judgments. *Comput. Ind. Eng.*, *97*, 240–247, 2016.
19. Zavadskas, E.K. and Turskis, Z., A new additive ratio assessment (ARAS) method in multicriteria decision-making. *Technol. Econ. Dev. Econ.*, *16*, 2, 159–172, 2010.
20. Chakraborty, S., Applications of the MOORA method for decision making in manufacturing environment. *Int. J. Adv. Manuf. Technol.*, *54*, 9, 1155–1166, 2011.
21. Wua, K.J., Tseng, M.L., Chiu, A.S.F., Using the Analytical Network Process in Porter's Five Forces Analysis – Case Study in Philippines. *Proc. Soc. Behav. Sci.*, *57*, 1–9, 2012.
22. Sahu, A.K., Sahu, A.K., Sahu, N.K., Appraisements of material handling system in context of fiscal and environment extent: a comparative grey statistical analysis. *Int. J. Logist. Manage.*, *28*, 1, 2–28, 2017.
23. He, Z., Ma, X., Luo, J., Sahu, A.K., Sahu, A.K., Sahu, N.K., Exploitation of the advanced manufacturing machine tool evaluation model under objective-grey information: a knowledge-based cluster with the grey relational analysis approach. *Grey Syst.: Theory Appl.*, *11*, 3, 394–417, 2020.
24. Sahu, A.K., Sahu, N.K., Sahu, A.K., Laminating STRATH block chain technology-SWOT architectures to endure business strategy between digital transformation, firms and supply chains capabilities for sustainability. *J. Clean. Prod.*, *383*, 135531, 2023.
25. Lee, D. and Dong, M., Dynamic network design for reverse logistics operations under uncertainty. *Transp. Res. Part E: Logist. Transp. Rev.*, *45*, 3, 61–71, 2009.
26. Sahu, A.K., Sharma, M., Raut, R.D., Sahu, A.K., Sahu, N.K., Antony, J., Tortorella, G.L., Decision-making framework for supplier selection using an integrated MCDM approach in a lean-agile-resilient-green environment: evidence from Indian automotive sector. *TQM J.*, *35*, 4, 964–1006, 2023.
27. Sahu, A.K., Sahu, N.K., Sahu, A.K., Narang, H.K., Rajput, M.S., Grey-based scorecard model for opting fruit supply bazaar locality under advanced chain of macro-micro parameter. *Br. Food J.*, *120*, 1, 59–79, 2018.
28. Vinodh, S. and Joy, D., Structural equation modeling of sustainable manufacturing practices. *Clean Technol. Environ. Policy*, *14*, 79–84, 2012.
29. Moldavskaab, A. and Welob, T., The concept of sustainable manufacturing and its definitions: A content-analysis based literature review. *J. Clean. Prod.*, *166*, 10, 744–755, 2017.
30. Sahu, N.K., Singh, M.K., Mutono-Mwanza, B.G., Sahu, A.K., Investigation of machinability characteristics of EDMed inconel 825 alloy under

multidimensional parametric modeling by using holistic grey-PCA statistical models. *Adv. Mater. Sci. Eng.*, *2022*, 1–19, 2022.
31. Sahu, A.K., Sahu, N.K., Sahu, A.K., Rajput, M.S., Narang, H.K., An Investigation Tool for Mounting Sustainable Practice: Modeling Using GIVTFNs in an Indian Context. *Int. J. Decis. Support Syst. Technol.*, *12*, 2, 25–49, 2020.
32. Arya, D.M., Chaturvedi, V., Vimal, J., Application of signal to noise ratio methodology for optimization of MIG welding process parameters. *Int. J. Eng. Res. Appl.*, *3*, 4, 1395–1399, 2013.
33. Mukhraiya, V., Yadav, R.K., Jathar, S., Parametric optimization of MIG welding process with the help of Taguchi method. *Int. J. Eng. Res. Technol.*, *3*, 1, 1407–1410, 2014.
34. Sivasakthivel, K., Janarthanan, K., Rajkumar, R., Optimization of Welding Parameter in MIG Welding by Taguchi Method. *Int. J. Adv. Res. Mech. Eng. Technol.*, *1*, 1, 36–38, 2015.
35. Perumal, A., Kailasanathan, C., Stalin, B., Kumar, S.S., Rajkumar, P.R., Gangadharan, T., Venkatesan, G., Nagaprasad, N., Dhinakaran, V., Krishnaraj, R., Multiresponse Optimization of wire electrical discharge machining parameters for Ti-6Al-2Sn-4Zr-2Mo (α-β) alloy using Taguchi-grey relational approach. *Adv. Mater. Sci. Eng.*, 2022, 1–13, 2022.
36. Nas, E., Özbek, O., Bayraktar, F., Kara, F., Experimental and Statistical Investigation of Machinability of AISI D2 Steel Using Electroerosion Machining Method in Different Machining Parameters. *Adv. Mater. Sci. Eng.*, 2021, 1–17, 2021.
37. Abed, F.N., Ramesh, V., Jwaid, M.F., Agarwal, N., Koundal, D., Ibrahim, A.M., Enhancement modelling based on electrical discharge machining successive discharges. *Adv. Mater. Sci. Eng.*, 2022, 1–7, 2022.
38. Guo, C., Sun, S., Di, S., Wei, D., Study of Working Medium Performance by Acoustic Emission in EDM Machining of Ti6Al4V. *Adv. Mater. Sci. Eng.*, 2020, 1–10, 2020.
39. Jangde, K., Singh, A., Arjunan, T.V., Efficient solar drying techniques: A review. *Environ. Sci. Pollut. Res.*, 29, 34, 50970–50983, 2022, https://doi.org/10.1007/s11356-021-15792-4.
40. Jangde, P.K. and Arjunan, T.V., Thermal performance assessment of the coaxial pipe evacuated tube solar air heater in Bilaspur climatic conditions. *Proc. Inst. Mech. Eng., Part E: J. Process Mech. Eng.*, 2023, 1–11, 2023, doi: 10.1177/09544089231154528.
41. Jangde, P.K., Selvaraj, V., Arjunan, T.V., Kanagavel, P., Solar Thermal Energy for Industrial Process Heating: Potentials and Challenges, in: *Solar Thermal Conversion Technologies for Industrial Process Heating*, pp. 77–98, CRC Press, Boca Raton, 2023.

42. Amin, N., Lung, C.W., Sopian, K., A practical field study of various solar cells on their performance in Malaysia. *Renew. Energy J.*, *34*, 8, 1939–1946, 2009.
43. Kengpol, A., Rontlaong, P., Tuominen, M., A Decision Support System for Selection of Solar Power Plant Locations by Applying Fuzzy AHP and TOPSIS: An Empirical Study. *J. Software Eng. Appl.*, *6*, 470–481, 2013.
44. Beltran, P., An AHP (Analytic Hierarchy Process)/ANP (Analytic Network Process)-based multi-criteria decision approach for the selection of solar-thermal power plant investment projects. *Energy*, *17*, 645–658, 2013.
45. Cavallaro, F., Fuzzy TOPSIS approach for assessing thermal-energy storage in concentrated solar power (CSP) systems. *Appl. Energy*, *87*, 496–503, 2010.
46. Khaparde, S.A. and Mukerjee, A., Infrastructure for Sustainable Renewable Energy in India: A Case Study of Solar PV Installation. *IEEE Power and Energy Society General Meeting - Conversion and Delivery of Electrical Energy in the 21st Century*, *68*, 289–301, Pittsburgh, PA, USA, 2008.
47. Tavana, M., Arteaga, F.J.S., Mohammadi, S., Alimohammadi, M., A fuzzy multicriteria spatial decision support system for solar farm location planning. *Energy Strategy Rev.*, *18*, 93–105, 2017.
48. Uyan, M., GIS-based solar farms site selection using analytic hierarchy process (AHP) in Karapinar region, Konya/Turkey. *Renew. Sustain. Energy Rev.*, *28*, 11–17, 2013.
49. Lozano, J.M.S., Teruel-Solano, J., Soto-Elvira, P.L., García-Cascales, M.S., Geographical Information Systems (GIS) and Multi-Criteria Decision Making (MCDM) methods for the evaluation of solar farms locations: case study in south-eastern Spain. *Renew. Sustain. Energy Rev.*, *24*, 544–556, 2013.
50. Salah, C.B., Chaabene, M., Ammar, M.B., Multi-criteria fuzzy algorithm for energy management of a domestic photovoltaic panel. *Renew. Energy*, *33*, 5, 993–1001, 2008.
51. Kumar, V., Albert, S.K., Chanderasekhar, N., Development of programmable system on chip-based weld monitoring system for quality analysis of arc welding process. *Int. J. Comput. Integr. Manuf.*, *33*, 9, 925–935, 2020.
52. Banga, H.K., Kalra, P., Kumar, R., Singh, S., Pruncu, C., II, Optimization of the cycle time of robotics resistance spot welding for automotive applications. *J. Adv. Manuf. Process.*, 3, 1–11, 2021. https://doi.org/10.1002/amp2.10084.
53. Kah, P., Shrestha, M., Hiltunen, E., Martikainen, J., Robotic arc welding sensors and programming in industrial applications. *Int. J. Mech. Mater. Eng.*, *10*, 1, 1–13, 2015, https://doi.org/10.1186/s40712-015-0042-y.
54. Wang, S.J., Hu, L., Sun, T., Freiheit, T., Intelligent welding system technologies: State-of-the-art review and perspectives. *J. Manuf. Syst.*, *56*, 373–391, 2020.
55. Hong, T.S. and Ghobakhloo, M., Safety and security conditions in welding processes, in: *Comprehensive Materials Processing*, vol. 6, Bayraktar, E. (Ed.), pp. 213–225, 2014, 10.1016/B978-0-08-096532-1.00608-7.

56. Mehrifar, Y., Zeverdegani, S.K., Faraji, M., Rismanchian, M., Risk assessment of welders exposure to the released contaminated gases in different types of welding processes in a steel industry. *Health Scope*, *7*, 4, 1176, 1–6, 2018, 10.5812/jhealthscope.58267.
57. Dehghan, S.F., Mehrifar, Y., Ardalan, A., The relationship between exposure to lead-containing welding fumes and the levels of reproductive hormones. *Ann. Glob. Health*, *85*, 1, 1–6, 2019. 10.5334/aogh.2617.
58. Ravisankar, V., Balasubramanian, V., Muralidharan, C., Selection of welding process to fabricate butt joints of high strength aluminium alloys using analytic hierarchic process. *Mater. Des.*, *27*, 5, 373–380, 2006.
59. Singh, D. and Rao, R.V., A hybrid multiple attribute decision making method for solving problems of industrial environment. *Int. J. Ind. Eng. Comput.*, *2*, 3, 631–644, 2011.
60. Chakraborty, S. and Zavadskas, E.K., Applications of WASPAS method in manufacturing decision making. *Informatica*, *25*, 1, 1–20, 2014.
61. Omar, M. and Soltan, H., A framework for welding process selection. *SN Appl. Sci.*, *2*, 3, 346, 2020.
62. Bagheri, B., Abbasi, M., Hamzeloo, R., Comparison of different welding methods on mechanical properties and formability behaviors of tailor welded blanks (TWB) made from AA6061 alloys. *Proc. Inst. Mech. Eng., Part C: J. Mech. Eng. Sci.*, *235*, 12, 2225–2237, 2021.
63. Jamwal, A., Agrawal, R., Sharma, M., Kumar, V., Review on multi-criteria decision analysis in sustainable manufacturing decision making. *Int. J. Sustain. Eng.*, *14*, 3, 202–225, 2021.
64. Zakeri, S., Chatterjee, P., Konstantas, D. *et al.*, A decision analysis model for material selection using simple ranking process. *Sci. Rep.*, *13*, 1, 1–34, 2023, doi: 10.1038/s41598-023-35405-z.
65. Bhaskar, A.S. and Khan, A., Comparative analysis of hybrid MCDM methods in material selection for dental applications. *Expert Syst. Appl.*, *209*, article ID 118268, 2022, https://doi.org/10.1016/j.eswa.2022.118268.
66. Farid, H.M.A. and Riaz, M., Single-valued neutrosophic Einstein interactive aggregation operators with applications for material selection in engineering design: Case study of cryogenic storage tank. *Complex Intell. Syst.*, *8*, 2131–2149, 2022, https://doi.org/10.1007/s40747-021-00626-0.
67. Aksakal, B., Ulutaş, A., Balo, F., Karabasevic, D., A new hybrid MCDM model for insulation material evaluation for healthier environment. *Buildings*, *12*, 5, 655, 2022.
68. Bhadra, D. and Dhar, N.R., Selection of the natural fiber for sustainable applications in aerospace cabin interior using fuzzy MCDM model. *Materialia*, *21*, article ID 101270, 2022, https://doi.org/10.1016/j.mtla.2021.101270.

69. Agrawal, R., Sustainable material selection for additive manufacturing technologies: A critical analysis of rank reversal approach. *J. Clean. Prod.*, *296*, 126500, 2021, https://doi.org/10.1016/j.jclepro.2021.126500.
70. Chatterjee, S. and Chakraborty, S., Material selection of a mechanical component based on criteria relationship evaluation and MCDM approach. *Mater. Today*, *44*, 1621–1626, 2021.
71. Figueiredo, K., Pierott, R., Hammad, A.W., Haddad, A., Sustainable material choice for construction projects: A life cycle sustainability assessment framework based on BIM and Fuzzy-AHP. *Build. Environ.*, 196, article ID 107805, 2021, https://doi.org/10.1016/j.buildenv.2021.107805.
72. Prasad, R.V., Rajesh, R., Thirumalaikumarasamy, D., Selection of coating material for magnesium alloy using Fuzzy AHP-TOPSIS. *Sādhanā*, *45*, 1, 1–20, 2020.
73. Sahu, A.K., Katyayan, A., Khandey, U., Jangde, P., Sahu, A.K., Sahu, N.K., Adaptation of blockchain technology in SCM for steering managerial strategies: Investigative study under Indian context. *J. Int. Logist. Trade*, *21*, 1, 18–40, 2023, https://doi.org/10.1108/JILT-06-2022-0020.
74. Dev, N.K., Shankar, R., Qaiser, F.H., Industry 4.0 and circular economy: Operational excellence for sustainable reverse supply chain performance. *Resour. Conserv. Recycl.*, *153*, article ID 104583, 2020, https://doi.org/10.1016/j.resconrec.2019.104583.
75. Cao, J., Chen, X., Zhang, X., Gao, Y., Zhang, X., Kumar, S., Overview of remanufacturing industry in China: Government policies, enterprise, and public awareness. *J. Clean. Prod.*, *242*, article ID 118450, 2020, https://doi.org/10.1016/j.jclepro.2019.118450.
76. Govindan, K., Sarkis, J., Palaniappan, M., An analytic network process-based multicriteria decision making model for a reverse supply chain. *Int. J. Adv. Manuf. Technol.*, *68*, 1-4, 863–880, 2013.
77. Zarbakhshnia, N., Soleimani, H., Ghaderi, H., Sustainable third-party reverse logistics provider evaluation and selection using fuzzy SWARA and developed fuzzy COPRAS in the presence of risk criteria. *Appl. Soft Comput.*, *65*, 307–319, 2018.
78. Sharma, N.K., Kumar, V., Verma, P., Luthra, S., Sustainable reverse logistics practices and performance evaluation with fuzzy TOPSIS: A study on Indian retailers. *Cleaner Logist. Supply Chain*, *1*, article ID 100007, 2021, https://doi.org/10.1016/j.clscn.2021.100007.
79. Govindan, K., Pokharel, S., Kumar, P., A hybrid approach using ISM and fuzzy TOPSIS for the selection of reverse logistics provider. *Resour. Conserv. Recycl.*, *54*, 1, 28–36, 2009.

80. Tavana, M., Zareinejad, M., Santos-Arteaga, F.J., An intuitionistic fuzzy-grey superiority and inferiority ranking method for third-party reverse logistics provider selection. *Int. J. Syst. Sci.*, *5*, 175–194, 2018.
81. Jayant, A., Gupta, P., Garg, S.K., Khan, M., TOPSIS-AHP based approach for selection of reverse logistics service provider: A case study of mobile phone industry. *Proc. Eng.*, *97*, 2147–2156, 2014.
82. Agrawal, S., Singh, R.K., Murtaza, Q., Disposition decisions in reverse logistics: Graph theory and matrix approach. *J. Clean. Prod.*, *137*, 93–104, 2016.
83. Prakash, C. and Barua, M.K., A combined MCDM approach for evaluation and selection of third-party reverse logistics partner for Indian electronics industry. *Sustain. Prod. Consumption*, *7*, 66–78, 2016.
84. Bouzon, M., Govindan, K., Rodriguez, C.M., Evaluating barriers for reverse logistics implementation under a multiple stakeholders' perspective analysis using grey decision making approach. *Resour. Conserv. Recycl.*, *128*, 315–335, 2018.
85. Sahu, A.K., Sahu, N.K., Sahu, A.K., Appraise the economic values of logistic handling system under mixed information, in: *Theoretical and Practical Advancements for Fuzzy System Integration*, pp. 278–308, 2017.
86. Li, H., Wang, W., Fan, L., Li, Q., Chen, X., A novel hybrid MCDM model for machine tool selection using fuzzy DEMATEL, entropy weighting and later defuzzification VIKOR. *Appl. Soft Comput. J.*, *91*, 1–14, 2020.
87. Hagag, A.M., Yousef, L.S., Abdelmaguid, T.F., Multi-Criteria Decision-Making for Machine Selection in Manufacturing and Construction: Recent Trends. *Mathematics*, *11*, 1–21, 2023.
88. Cetinkaya, C., Kabak, M., Ozceylan, E., 3D Printer Selection by Using Fuzzy Analytic Hierarchy Process and PROMETHEE. *Bilişim Teknolojileri Derg.*, *10*, 3, 371–380, 2017.
89. Çakır, S., An Integrated Approach to Machine Selection Problem Using Fuzzy SMART-Fuzzy Weighted Axiomatic Design. *J. Intell. Manuf.*, *29*, 1433–1445, 2018.
90. Hutahaean, J., Cilliers, J., Brito-Parada, P.R., A Multicriteria Decision Framework for the Selection of Biomass Separation Equipment. *Chem. Eng. Technol.*, *41*, 11, 2346–2357, 2018.
91. Ozdagoglu, A., Yakut, E., Bahar, S., Machine Selection in a Dairy Product Company with Entropy and SAW Methods Integration. *Dokuz Eylul Univ. J. Econ. Adm. Sci.*, *32*, 343–361, 2017.
92. Wang, Y., Zhong, R.Y., Xu, X., A Decision Support System for Additive Manufacturing Process Selection Using a Hybrid Multiple Criteria Decision-Making Method. *Rapid Prototyping J.*, *24*, 9, 1544–1553, 2018.

93. Uz Zaman, U.K., Rivette, M., Siadat, A., Mousavi, S.M., Integrated Product-Process Design: Material and Manufacturing Process Selection for Additive Manufacturing Using Multi-Criteria Decision Making. *Rob. Comput. Integr. Manuf.*, *51*, 169–180, 2018.
94. Stirbanovic, Z., Stanujkic, D., Miljanovic, I., Milanovic, D., Application of MCDM Methods for Flotation Machine Selection. *Minerals Eng.*, *137*, 140–146, 2019.
95. Cioca, L., II, Breaz, R.E., Racz, S.G., Reducing the Risks during the Purchase of Five-Axis CNC Machining Centers Using AHP Method and Fuzzy Systems. *Sustainability*, *11*, 2, 1–23, 2019, https://doi.org/10.3390/su11020315.
96. Justino Netto, J.M., Ragoni, I.G., Frezzatto Santos, L.E., Silveira, Z.C., Selecting Low-Cost 3D Printers Using the AHP Method: A Case Study. *SN Appl. Sci.*, *1*, 3, 1–12, 2019, https://doi.org/10.1007/s42452-019-0352-4.
97. Mathew, M. and Thomas, J., Interval Valued Multi Criteria Decision Making Methods for the Selection of Flexible Manufacturing System. *Int. J. Data Netw. Sci.*, *3*, 4, 349–358, 2019.
98. Ozdagoglu, A., Yilmaz, K., Kele¸s, M.K., Evaluation of Machine Selection Criteria with MACBETH Method in a Ginnery Factory. *Süleyman Demirel Üniv. Sosyal Bilim. Enst. Derg.*, *37*, 26–37, 2020.
99. Ayag, Z. and Ozdemir, R.G., An intelligent approach to machine tool selection through fuzzy analytic network process. *J. Intell. Manuf.*, *22*, 2, 163–177, 2011.
100. Onut, S., Kara, S.S., Efendigil, T., A hybrid fuzzy MCDM approach to machine tool selection. *J. Intell. Manuf.*, *19*, 4, 443–453, 2008.
101. Ayag, Z., A hybrid approach to machine-tool selection through AHP and simulation. *Int. J. Prod. Res.*, *45*, 9, 2029–2050, 2007.
102. Samvedi, A., Jain, V., Chan, F.T., An integrated approach for machine tool selection using fuzzy analytical hierarchy process and grey relational analysis. *Int. J. Prod. Res.*, *50*, 12, 3211–3221, 2012.
103. Ayag, Z. and Ozdemir, R.G., Evaluating machine tool alternatives through modified TOPSIS and alpha-cut based fuzzy ANP. *Int. J. Prod. Econ.*, *140*, 2, 630–636, 2012.
104. Nguyen, H.T., Dawal, S.Z.M., Nukman, Y., Aoyama, H., A hybrid approach for fuzzy multi-attribute decision making in machine tool selection with consideration of the interactions of attributes. *Expert Syst. Appl.*, *41*, 6, 3078–3090, 2014.
105. Aghdaie, M.H., Zolfani, S.H., Zavadskas, E.K., Decision making in machine tool selection: An integrated approach with SWARA and COPRAS-G methods. *Eng. Econ.*, *24*, 1, 5–17, 2013.

106. Perçin, S. and Min, H., Optimal machine tools selection using quality function deployment and fuzzy multiple objective decision making approach. *J. Intell. Fuzzy Syst.*, *24*, 1, 163–174, 2013.
107. Dagdeviren, M., Decision making in equipment selection: An integrated approach with AHP and PROMETHEE. *J. Intell. Manuf.*, *19*, 4, 397–406, 2008.
108. Yurdakul, M.M. and Iç, Y.T., Analysis of the benefit generated by using fuzzy numbers in a TOPSIS model developed for machine tool selection problems. *J. Mater. Process. Technol.*, *209*, 1, 310–317, 2009.
109. Ozgen, A., Tuzkaya, G., Tuzkaya, U.R., A multi-criteria decision making approach for machine tool selection problem in a fuzzy environment. *Int. J. Comput. Syst.*, *4*, 4, 431–445, 2011.
110. Taha, Z. and Rostam, S., A hybrid fuzzy AHP-PROMETHEE decision support system for machine tool selection in flexible manufacturing cell. *J. Intell. Manuf.*, *23*, 6, 2137–2149, 2012.
111. Ilangkumaran, M., Sasirekha, V., Anojkumar, L., Raja, M.B., Machine tool selection using AHP and VIKOR methodologies under fuzzy environment. *Int. J. Model. Oper. Manage.*, *2*, 4, 409–436, 2012.
112. Rao, R.V., Patel, B.K., Parnichkun, M., Industrial robot selection using a novel decision making method considering objective and subjective preferences. *Rob. Auton. Syst.*, *59*, 5, 367–375, 2011.
113. Chatterjee, P., Athawale, V.M., Chakraborty, S., Selection of industrial robots using compromise ranking and outranking methods. *Rob. Comput. Integr. Manuf.*, *26*, 4, 483–489, 2010.
114. Abdullah, F.M., Al-Ahmari, A.M., Anwar, S., A hybrid fuzzy multi-criteria decision-making model for evaluating the influence of Industry 4.0 technologies on manufacturing strategies. *Machines*, *11*, 10, 310, 2023.
115. Zhou, F., Wang, X., Goh, M., Fuzzy extended VIKOR-based mobile robot selection model for hospital pharmacy. *Int. J. Adv. Rob. Syst.*, *15*, 5, 1–11, 2018, https://doi.org/10.1177/1729881418787315.
116. Ghorabaee, M.K., Developing an MCDM method for robot selection with interval type-2 fuzzy sets. *Rob. Comput. - Integr. Manuf.*, *37*, 221–232, 2016.
117. Fu, Y., Li, M., Luo, H., Huang, G.Q., Industrial robot selection using stochastic multicriteria acceptability analysis for group decision making. *Rob. Auton. Syst.*, 122, article ID 103304, 2019, https://doi.org/10.1016/j.robot.2019.103304.
118. Liu, H.-C., Quan, M.-Y., Shi, H., Guo, C., An integrated MCDM method for robot selection under interval-valued Pythagorean uncertain linguistic environment. *Int. J. Intell. Syst.*, *34*, 2, 188–214, 2019.
119. Nasrollahi, A., Ramezani, J., Sadraei, M., A FBWM-PROMETHEE approach for industrial robot selection. *Heliyon*, *6*, 5, 1–6, 2020.

120. Bairagi, B., Technique of Accurate Ranking Order (TARO): A novel multi criteria analysis approach in performance evaluation of industrial robots for material handling. *Decis. Sci. Lett.*, *11*, 5, 563–589, 2022.
121. Shanmugasundar, G., Sapkota, G., Čep, R., Kalita, K., Application of MEREC in Multi-Criteria Selection of Optimal Spray-Painting Robot. *Processes*, *10*, 9, 1172, 2022.

2

Sustainable Supply Chain Management Practices in Developing Economies: A Qualitative Mapping Approach

Bupe G. Mwanza* **and Eneless Manda**

Graduate School of Business, University of Zambia, Lusaka, Zambia

Abstract

Sustainable supply chain management (SSCM) focuses on integrating economically, socially, and environmentally sound practices throughout the value chain. The integration of these sound practices contributes to value addition throughout the product's life. Using a qualitative approach, the objectives of the chapter were to ascertain the SSCM practices integration in the service and manufacturing companies, to identify the challenges encountered during the integration process, and to establish the solutions to the challenges faced by the companies. A multiple–case study approach was used in which key managers in the supply chain–related functions of a developing economy were interviewed. A sample size of 18 managers across different industries was interviewed. Using qualitative mapping, the structured interviews elicited common themes of SSCM practices. Furthermore, the results of the study demonstrated the types of challenges faced by the companies and the solutions for overcoming the challenges. The relevance of social, economic, and environmental aspects to technological investments, sustainability, strategic partnerships, strategic procurement, demand forecasting, continuous improvement programs, and organizational cultures was demonstrated in the chapter. The findings of this chapter cannot be generalized. Accordingly, this chapter was organized with the intention to show the SSCM practices on a case by case. The chapter recommends that future studies should focus on applying a quantitative approach in order to complement the qualitative perspectives to explore crucial boundaries of SSCM. The chapter focuses on a case-by-case approach of companies in a developing economy, where issues of sustainability in the supply

**Corresponding author*: bupe.mwanza@gmail.com

Atul Kumar Sahu, Rakesh D. Raut, Rohit Raja, Anoop Kumar Sahu and Nitin Kumar Sahu (eds.) Industrial and Manufacturing Designs: Quantitative and Qualitative Analysis, (45–74) © 2024 Scrivener Publishing LLC

chains are still at infancy. Lastly, the results presented in the chapter provide a referential basis for informing practitioners, researchers, and policymakers to make strategic decisions.

Keywords: Developing economies, qualitative, mapping, supply chains, value addition, management, sustainability

2.1 Introduction

Globally, companies need to adopt SSCM practices in the management of their goods and services in order to remain standing over a longer period of time. Today, global supply chains (SCs) continue to face an increasing number of SC risks in the form of failure by supplier to deliver, failure by supplier to meet quality standards, commodity and raw material prices, logistic delays, logistic damages, information loss, business interruption (theft, vandalism, and conflicts), fluctuating demand, fluctuation in fuel/oil/energy/transportation prices, machine breakdowns, and lagging technology [1]. As a result of these SC-related risks, there is need for companies to have a more sustainable way of managing their products and services [2]. A lack of commitment by companies to incorporate SSCM principles into their production processes has forced environmentalists, consumers, governments, and other stakeholders to put pressure on these organizations to integrate sustainability into their value chains [3]. Sustainability has multiple facets, including social and economic as well as environmental sustainability. The triple bottom line (TBL) is the collective term for these three dimensions [4]. Companies should thrive to implement the three aspects of sustainability in their operations. Businesses in society should focus more on the long-term effects of their operations on the environment, the economy, and people [5].

Developing economies' SCs face many challenges, and understanding SSCM practices and challenges from a service and manufacturing perspective presents an opportunity to many companies to understand and develop sustainability in their operations. In this current era, competition is among the SCs and not companies, and, therefore, it is necessary to assess these practices and challenges from both the service and manufacturing industries. Numerous studies in both developed and developing economies have been done on SSCM. For example, Wolf [6] developed a model for integrating SSCM into the German manufacturing industry using a qualitative approach. The study identified factors that impede and drive the integration of SSCM. The study only focused on the manufacturing industry from a developed economy perspective. A study by Rebs *et al.* [7] compared the influence of stakeholders in SSCM integration using qualitative and quantitative studies. Their study only focused on one factor and used published studies for identifying the influence of that factor. In Zambia, few studies have been conducted on SSCM integration, and these studies have not looked at the service and manufacturing concurrently. For instance, Nambela [8] assessed the impact of adopting SSCM practices on the performance of Zambia Air Force (a service company), whereas Mukuka [9] evaluated the effectiveness of SSCM practices in a brewery company (a manufacturing company). Lastly, Manchisi [10] and Chileshe and Phiri [11] focused on SSCM practices in small and medium enterprises (SMEs). This study focuses on the service and manufacturing industries using a qualitative approach to establish the practices and challenges of integrating SSCM. None of the studies explained above have taken an approach that can fill the existing gap of ascertaining the SSCM practices integration in the service and manufacturing companies from a Zambian perspective.

A qualitative mapping approach is accordingly applied in this chapter because it presents an opportunity to explore the SSCM practices and challenges as experienced by the different companies. It is an approach that enables the policymakers and researchers to access immeasurable insights in the root causes of issues and establish multiplex-informed intervention strategies [12]. It has been applied in many studies to understand SC-related issues [1, 6, 13, 14]. In this chapter, the qualitative approach is applied in the service and manufacturing industries in a developing economy. This approach is taken to understand the SSCM practices integration in the service and manufacturing industries as compared to single industry perspective. Different and connected practices and challenges are mapped using a qualitative approach to present a holistic picture. In order to conduct and map the findings, the chapter considers the following objectives:

1. to establish the SSCM practices integrated in the service and manufacturing companies,
2. to identify the challenges encountered during the integration process, and
3. to develop the solutions to the challenges faced by the companies.

The chapter is structured as follows: a literature review is provided in Section 2.2. This is followed by the methodology in Section 2.3. The results and discussion are presented in Section 2.4, whereas the conclusion and recommendations are presented in Section 2.5.

2.2 Literature Review

2.2.1 Sustainable Supply Chain Management

As organizations run their daily activities, they should take keen interest in looking at how they run their internal operations as well as how their cooperating partners do their business. Many players in the SC are forming more and more strategic partnerships with their SC partners in order to attain sustainability [15]. These alliances mean sharing information, materials, capital, and risks and building relationships based on trust. Therefore, organizations should look at creating strategic partnerships with parties in the SC and work together to ensure improvement in performance in demand forecasting, procurement, production, distribution, and after-sales services.

Sustainable SCs look at the impact of production processes of goods and services on the environment and ensure that strategies are employed to guarantee reduction of carbon emissions, which are harmful to the environment, ensuring preservation of energy and natural resources and engaging in proper ways of managing wastes. Economically, organizations need to remain profitable and stay up-to-date with technological advancements. To bring about social sustainability, it is important to energize the workforce and foster a safe and healthy environment [16]. Careful use of environmental, human, and capital resources is a key toward achieving a sustainable SC. Ahmad *et al.* [17] affirm that, to achieve sustainability, firms should have their concerns extended beyond their short-term profits but should go a step further and really care about the environment and stakeholders.

Supply chain management (SCM) can be explained as the coordination of all the actions taken by SC parties to deliver a product or service that meets the needs of a customer and simultaneously reduce organizational costs [18]. Customer satisfaction is the end goal; therefore, firms must continue to work in order to bring increased customer retention. In order to have more sustainable operations, organizations along the chain should coordinate their activities and work collaboratively [19]. Some of the practices integrated include recycling, refurbishing, waste management, and other environmentally, socially, and economically viable practices. These are integrated into the entire SC life cycle as part of sustainable supply chain management (SSCM) [20]. The integration of these sustainable practices should run from design and development, material selection and purchasing, manufacturing, packaging, transportation, warehousing, distribution, consumption, to return and disposal. Integrating these practices across the entire value chain is important as it brings about long-term growth. Tsai *et al.* [21] defined SSCM as managing the flow of materials, information, and capital among firms within the SC. Effective management of the SC for sustainability is what stakeholders and customers expect in today's economy. Soosay *et al.* [22] posit that SSCM means keeping an eye on the chain's finances, information, and resources while trying to maximize both the chain's profitability and the welfare of the community. This is required to be done to gain competitive advantage and to stay out of trouble with regulators. Therefore, managing information flow, collaboration, coordination, and connectivity across the SC network is a key toward achieving a higher level of organizational and SC sustainability performance [23].

The aim of SC sustainability is to generate and preserve financial, environmental, and social benefits continuously and iteratively for all parties that are involved in the process of delivering goods and services to the market. Dey *et al.* [24] posit that sustainable practices should be incorporated into today's SCs not just for financial reasons but for ethical considerations too. Zimon and Domingues [25] recognize that numerous benefits come with the implementation of sustainable solutions, which include risk reduction, enhancing the overall SC performance and preventing production issues [26, 27]. Mujkić *et al.* [28] add that strategic and operational advantages of SSCM integration include increased profits, waste reduction (through recycling and remanufacturing), and improved brand and image awareness.

Sustainable supply chain management is an important aspect as it gives industrial goods a helpful green image. The value chain of a product must prioritize environmental sustainability because it lowers risks for businesses. These risks include the possibility of increased expenses as a result of fines for non-compliance, as well as the possibility of decreased sales as a result of a decline in the company's reputation and a shift in customer preferences, which therefore has an impact on the firm's competitiveness [29]. Firms must view environmental impact of their activities as an integral part of decision-making as it impacts their overall operating cost. It is imperative that they should integrate environmental initiatives and progress toward environmentally conscious purchasing, resource efficiency, environmentally conscious product design, waste management, eco-efficiency in operations, and reverse logistics [30]. Integration of environmental concerns into the SC requires that all supply participants work together toward the cause [31]. The study done by Vila *et al.* [32] concluded that the future must have factories that are committed to a green product life cycle.

Social sustainability within and outside the organization looks at people's well-being, working conditions, quality of life, equality, diversity, and connectedness [33]. These important aspects must be incorporated into the SC of firms. Fair treatment of the people involved in production or provision of a service brings improved productivity as a result of increased morals and commitments. The social aspect of SSCM also focuses on the external environment in the name of corporate social responsibility (CSR), which generally concerns about incorporating the local community into SC operations [34]. Companies may not be obliged to engage in CSR, but if these responsibilities are ignored, then the reputation of firms could crumble, as stakeholders would have a notion that the firm is only in business for money with little to no concern for the society, in which the firm does its business. Along the SC, social sustainability also involves making sure that products are sourced, produced, and delivered to customers in an ethical manner. Finally, when it comes to the economic dimension, Marchi *et al.* [35] posit that sustainability is usually measured by a firm's ability to make sales and profit from operational efficiency. Organizations concentrate on economic sustainability, where the goal is more about "how organizations stay in business". Economic sustainability's focus is internal and looks at the firm's financial performance or how the company is managing intangible assets. Economic sustainability also influences the wider economy. Paying attention to these aspects is crucially required, if the company plans on staying in business for a longer period of time.

Therefore, management should make decisions that are aimed at ensuring that both individual companies and their SC perform well. This can be achieved through a strategic integration. Nambela [8] showed an increasing concern toward the integration of SSCM and his study revealed that, across many aspects, environmental aspects of SSCM are relatively established in comparison to other aspects of SSCM, i.e., buying from diverse suppliers, supporting human rights and ensuring safe practices in the SC. For best managing of the SC, the TBL approach should be considered, which reflect on the financial worth of the business along with the level of social and environmental responsibility. The study presented by Chowdhury *et al.* [36] related to the attainment of SC sustainability emphasizes on the incorporation of economic, social, and environmental aspects in an integrated fashion. Therefore, it is found important to have a robust integration of the three dimensions, i.e., economic, social, and environmental dimensions for sustainability. The consideration of the said three dimensions has been done by various authors for attaining oblige SSCM boundaries, which can be seen in Table 2.1.

Table 2.1 Dimensions of sustainability (triple bottom line).

ID	Author/year	Environmental	Economical	Social
1	[37]	√	√	√
2	[38]	√	√	√
3	[39]	√	x	x
4	[40]	√	√	x
5	[41]	x	x	√
6	[42]	√	√	x
7	[43]	x	x	√
8	[44]	x	x	√

2.2.2 Sustainable Supply Chain Management Practices

Optimization and continuous planning in all the activities from product design, manufacture, usability to end-of-life of a product is needed to establish SSCM in a product life cycle [32]. SSCM is the maximization of SC's profitability while, at the same time, minimizing environmental impacts and maximizing social well-being [45]. In order to have a successful business and be able to respond well to customer needs, efficient and effective SCM practices are essential.

According to Sroufe and Melnyk [46], a few examples of sustainable SC practices are waste management, transportation of finished goods and services to customers, materiality, product or process design, life cycle assessment, manufacturing procedures, end-of-life management of products, and closed-loop systems. The main management practices connected with SSCM includes risk management, sustainable supplier management, and sustainable operations [47]. Alzubi and Akkerman [48] outlined the ways to increase their economic performance and competitiveness of the companies by emphasizing on waste water treatment, renewable energy use, biodegradable and recyclable material design, and waste prevention. A study by Mulwa [49] emphasized that stakeholder engagement is key and that governments must ensure that the goods that they buy are morally right, promote fair trade, and ensure that SCs do not violate labor laws. It is found important to improve worker welfare and provide opportunities for employment and training, especially for marginalized groups like immigrants and those with disabilities. Additionally, social inclusion,

Table 2.2 Sustainable supply chain practices.

SSCM practices	Author and year
Stakeholder engagement	[46]
Product life cycle analysis	[8, 51]
Corporate social responsibility	[52, 53]
Sustainable supply network management	[54, 55]
Reverse logistics	[56, 57]
Environmental purchasing	[58]
Recycling and reuse of material	[56, 57, 59]
Green packaging	[57, 60]
Redesign of products with biodegradable materials	[61]
Implementing the ISO14001 system	[62]
Green product design and innovation	[63–65]
Strategic supplier collaboration	[66]
Energy consumption and emission reduction	[61, 67–69]
Green purchasing and raw material procurement	[19, 48]
Supplier sustainability assessment	[70]
Waste management	[71]
Emission reduction	[52]
Collaborative inventory management	[72]
Supply chain integration system	[73]
Green warehousing	[74]
Green shipping and distribution	[19]
Strategic supply chain collaboration	[19, 72, 75]
Collaborative supply chain planning	[72]
Attention to employee welfare	[49, 50]

diversity, and equality are also found significant to which the companies should adhered to. The engagement and identification of a diverse base of suppliers is also required for sustainability. The social responsibility can be attained by creating and sustaining robust, healthy communities as well as improving the welfare of the locals by creating jobs for them locally, which are crucial. In today's developing economies, maintaining a safe workplace has become important toward protecting employees' health and safety and promoting their welfare. Kang *et al.* [50] add that good health and safety conditions are primarily crucial to firms for pursuing sustainability, whereas other principles like adhering to remuneration requirements, paying attention to issues of forced and child labor within the chain, and compliance of human rights are secondarily crucial to firms for pursuing sustainability. A number of authors have highlighted various sustainable SC practices to ensure value adding in service and manufacturing companies. The said SC practices can be identified in Table 2.2.

2.2.3 Challenges of Integrating SSCM

Organizations face various challenges while incorporating sustainable SC practices in their SCs. Some of the obstacles, which are preventing the adoption of sustainable SC practice to create value, include supplier hindrances, managerial barriers, operational hindrances, cost hindrances, awareness hindrances, and political hindrances. Soda *et al.* [76] found that there is a general lack of awareness about sustainability in developing countries among suppliers and consumers and that a weak demand from consumers is responsible for sustainable products. When consumers make purchases, they care more about their basic needs rather than concentrating on the quality of the purchases [77]. Organizations possess lack of knowledge pertaining to the kind of sustainable actions that should be undertaken for SSCM. Ambiguity is identified as a problem, and firms lack clarity in terms of the specific part to play including, dimensions, measures, and criterions which each firm needs to use to have a sustainable SC and engage other stakeholders [78]. Lack of awareness and environmental initiatives hinders the top management from adopting environmental practices.

Jia *et al.* [79] highlighted the importance of green aspects to the companies, who are working in developing countries, where the initial cost of going green may be too high for them and may not yield the benefits immediately. Thus, the company's intention to attain ready profit may act as a challenge for SSCM. Some organizations lack infrastructure for recycling and lack initiatives to reuse certain products and services that too acts as a challenge for SSCM. Low-carbon technologies are also not easy to find; additionally, availability of funds and their effects on return on investment and overall price for products and services [80] also act as a challenge for SSCM. According to a study on Indian manufacturers, executives find it difficult to persuade their companies to adopt sustainable practices because of the high costs associated with doing so and their concern is that doing so might not result in immediate profitability because they are unable to accurately plan ahead more than 5 years [78].

Another challenging problem is the lack of adaptability, as businesses are hesitant to modify their practices and policies to incorporate sustainability. Li *et al.* [81] affirm that major issues with integrating sustainability in SC include shifting mindsets and cultures, as well as managing trade-offs and uncertainties. Furthermore, integrating sustainability in SC is faced with a number of other challenges such as unstable economic conditions, lack of support from upper management, challenges in modifying organizational policies and practices, lack of sufficient regulations and lax enforcement, and misaligned objectives [65]. A study done by Jia *et al.* [79] highlighted

Table 2.3 Challenges of integrating SSCM.

Challenges	**Author**
Lack of awareness	[76, 78]
Change of mindset, culture, and policies and practices	[82–86]
Lack of top management support	[86, 87]
Low customer demand for sustainable products	[88–90]
Weak regulatory checks and controls	[91, 92]
High overall cost	[93, 94]
Lack of political support in developing countries	[95]
High levels of corruption	[96, 97]
Lack of motivation toward employees (incentives)	[98]
Lack of qualified staff and training programs	[98]
Lack of infrastructure for recycling	[80]
Lack of clear direction from the organizational leaders	[99]
Lack of reward for adopting sustainability from governments	[99]

that high levels of corruption have been reported in a number of nations, and suppliers started buying certification labels from independent auditors without following the necessary procedures. This shows a high level of lack of responsibility, which is a challenge for SSCM. The sustainability goal can only be achieved through concerted efforts from all parties of the SC for it to be effect. Literature evidence reveals a number of challenges that organizations face during the SSCM integration process as shown in Table 2.3.

2.2.4 Strategies for Enhancing SSCM Integration

Despite the obstacles that exist in integrating sustainable practices in SC to enhance value addition across a products life cycle, there are several strategies that firms should embrace in order to get the most from their efforts. In order to attain sustainability, companies must restructure their existing SC to integrate sustainability objectives throughout their entire operations, ranging from procurement to distribution [100]. Businesses should be willing to take the necessary actions to safeguard the interests of stakeholders and the environment in addition to their primary concern of making profit, for them to effectively implement sustainable practices. In order to jointly optimize these sustainability aspects, the companies need to use new approaches, such as waste optimization, carbon footprint reduction, green purchasing, and green product design [101].

For companies to achieve sustainability, it is imperative to raise awareness about sustainable practices. There has to be some form of metrics and frameworks to quantify sustainability, so as to properly monitor the level of company sustainability. Local community awareness programs can be used to teach the public about sustainability and its effects. According to a study by Ibrahim and Ahmed [102], it is highlighted that awareness programs should focus on sustainable issues such as planting trees, recycling, and promoting a culture that values these practices. Strategically, employee education is important because it keeps employees up-to-date with latest technological advancements that are necessary for improving SSCM and allied procedures. Top management support and commitment is very crucial, and the company's top management's commitment, willingness, and support are the core primary elements for enduring sustainability initiative. Support and commitment from top management include supplying tools, training, supervising implementation across the board, and reviewing and adjusting the policy in light of realized outcomes. To boost public confidence, other potential options, which included conducting sustainability audits, measurements, creating a policy, and providing community training [94], should also be considered as a strategic means for enduring SSCM.

In a developed industry, engaging with stakeholders is required in the form of strategic SC partners to collaborate and to achieve long-term goals [103]. Building the capacity to work with other departments is also strategically necessary, which requires collaborative connections in all departments, i.e., purchasing, production, and engineering. In order to fully reap the benefits of sustainability in SC, there has to be concentrated efforts from all stakeholders, and there has to be increased international collaboration on issues to do with the environment [93]. Engaging in partnerships and joint projects with SC partners presents a chance to address the growing demands of stakeholders while also potentially enhancing SC efficiency [104].

A code of ethics must also be strategically implemented in order to improve the integration of SSCM practices in SC. Codes of conduct for businesses present a priceless chance for ethical businesses and to establish a favorable public image for themselves. This can result in a more favorable political and regulatory landscape as well as a rise in public confidence and trust among key stakeholders. Environmental regulation is also a strategic tool that helps SC participants to realize the consequences of their operations. Hussain *et al.* [105] conducted a study and discovered that a company's competitiveness and economic performance can be influenced by its choice of sustainable suppliers in terms of their social responsibility. For instance, choosing suppliers in relation to their sustainability performance and auditing them in line with the TBL performance can bring improved product quality and increased market share [106].

It is found that environmental dimension, waste optimization, and carbon footprint reduction have to be incorporated strategically to enhance performance and competitiveness [101]. Massaroni *et al.* [107] add that waste optimization can be realized by employing energy efficiency practices, water conservation, elimination of waste, internal recycling of waste, and so on. Sustainable SCM should incorporate sustainable purchasing strategies aimed at reducing waste, hazardous materials, and pollution. These strategies include recycling, scrapping, sorting, and using biodegradable packaging [108]. A data-driven approach to sustainable SC integration should also be adopted, where technology can strategically enable SC participants to have an end-to-end visibility of the SC, agility, and trust. Furthermore, businesses should adjust and ensure social protection systems and fair remuneration practices as a strategy to all employees along the SC, for sustainability [107]. Firms have to build partnership with local organizations as a strategy and should support vulnerable groups and

Table 2.4 Strategies for enhancing SSCM integration.

Author/ year	Awareness	Top management support	Collaboration/ alignment	Waste optimization and carbon footprint reduction	Concern for stakeholders	Responsiveness	Enshrining sustainability in supplier selection	Government funding projects that promote sustainability	Adopting sustainable technology	Responsiveness/ continuous improvement	Resilience/ reliability
[107]				√	√						
[84]			√								
[104]			√				√	√			
[49]		√									
[101]				√							
[105, 108]							√				
[102]	√										
[39]					√						
[109]						√				√	√
[65]								√	√		

joint ventures [39] for attaining resilient sustainability support. Various strategies can be used to employ success and attaining SSCM. The same are represented in Table 2.4, which gives various views of the researchers regarding strategies that organizations should employ to curb the challenges associated with SSCM integration.

2.3 Methodology

A qualitative multiple–case study approach was used to conduct the study presented in this chapter. A multiple-case study of managers working in the manufacturing and service companies was conducted in order to establish the SSCM practices integrated in the manufacturing and service companies, to identify the challenges encountered during the integration process, and to establish the solutions to the challenges faced by the companies. A multiple-case study is considered as it is a useful qualitative research strategy and can aid in studying about connections between organizational factors, environmental factors, social factors, etc., in order to guide organizational development. Case study approach is used as a methodology in present chapter as it is important for understanding in-depth studies involving the study of participants' outlooks on the problem within its native surroundings.

2.3.1 Data Collection

Structured interviews were conducted to collect information with 18 purposively samples managers from June 2023 to July 2023. A sample size of 18 was used under qualitative survey. The structured interviews were conducted with managers in charge of SC departments of the manufacturing and service companies in Lusaka. In order to gather sufficient data, the structured interviews with the managers lasted for 30 to 45 min. A thorough brainstorming interview session is conducted to collect quality facts.

2.3.2 Data Analysis

Data was afterward analyzed using the thematic approach. Firstly, the data that were collected were transcribed, and then codes were created in order to establish the cross-cutting answers from each question that was asked to each participant. Later, the established codes were reduced in words in order to develop the themes that were cross-cutting. In this case, themes were developed on the basis of what the researcher identified in the data.

2.4 Results

2.4.1 SSCM Practices

The findings from the interviews highlighted a number of themes regarding the practices being implemented by companies to achieve a sustainable SC, and these include green packaging, green productions, stakeholder engagement, supplier collaboration, human rights adherence and workplace health and safety, innovation through digitization, and adherence to international sustainability standards, laws, and regulations.

2.4.1.1 Green Packaging

Most of the interviewees mentioned that green packaging is one of the common practices that their company implements to ensure a sustainable SC. One of the respondents indicated that their firms integrate green packaging by using paper-based packaging in form of boxes, packets, and carrier bags. They indicated that the paper-based packaging has been easy to handle especially paper boxes. It was learnt that organizations are now streamlining their packages to reduce materials usage and lower costs. A SC manager indicated that "We made several commitments regarding sustainable packaging in 2019, including designing our packaging for sustainability and setting targets for use of sustainable packaging, especially for our own brands and in-store packaging".

2.4.1.2 Green Production

Another theme that came up was green production. One interviewee disclosed that green production is done through ensuring that their activities reduce carbon emissions and conserve resources. One respondent indicated that "We are committed to operating efficiently and to minimize our footprint onto the environment, including our carbon emissions, water usage, and waste". From the interviews conducted, it was clear that firms were in favor of eco-design, which supports the greening of the environment in terms of cleaner production. Others indicated that production is done locally and involves the procurement of local raw materials to support production of goods from natural hides and rubber and chemicalized synthetic polymers. Sustainable agriculture, water stewardship, reduced energy usage, reduced waste production, and reduced pollution were cited as sustainable practices being implemented. One respondent indicated that "We are committed to train our employees to become champions in reducing food losses and waste, and setting targets to improve our performance in this regard". Another respondent articulated that "Our circular economic outlook encompasses renewable energy sources, plastic reduction, setting up recycling stations, and resource minimization".

2.4.1.3 Stakeholder Engagement

One SC professional interviewed spoke on stakeholder's engagement as one of the main sustainable SC practices implemented by their firm. This was identified as a key by respondents, and they mentioned that it is done by following through communication with other stakeholders, which are involved from design of products to reuse and recycling. The firm ensures that customers use environmentally friendly disposal methods. One respondent said "stakeholder integration is required to lead to collaboration and joint development of products and processes, which then lead to the development of innovative technologies". Likewise, another respondent revealed that their company's SC strategy bordered on nurturing relationships that are beneficial to them and to all their stakeholders. He said "We have designed an effective supply chain strategy, which is crucial to meet our business objectives, serve the customer better, and stay ahead of competition".

2.4.1.4 Supplier Collaboration

Supplier collaboration was identified as a common practice by the firm, and it was found that a closer relationship with suppliers ensures resource efficiency and waste reduction. One interviewee indicated that "suppliers undergo screening prior to selection through a due diligence process to ascertain the commercial feasibility of collaboration, ensure compliance to all applicable laws and regulations (such as the labor laws), and evaluate the suppliers on parameters such as quality, price competitiveness, and response time".

Moreover, it was indicated by the firm that "all our suppliers are required to have sustainable business practices in place and embraces supplier code of conduct". Similarly, another respondent shared that "our priorities include partnering with suppliers, who provide goods/services at supreme quality and competitive cost, in addition to a short lead time. Furthermore, they must also have a strong focus on innovation, risk mitigation and agility to evolve with the dynamically changing customer needs. Adding that in doing so, we consistently strive to maintain an efficient procurement process operating under sound governance and ensure compliance to all applicable laws and regulations". One respondent indicated that despite the adverse effects brought by the COVID-19, the company managed to stay afloat due to strategic alliances made with local suppliers and effective communication with international suppliers.

2.4.1.5 *Risk Mitigation*

Another common theme among the respondents was risk mitigation through adherence to standards, laws, and regulations and by upholding ethical behavior. Respondents mentioned that they ensure that they follow global sustainability guidelines (ISO 14001, ISO 45001, and ISO 27001) and compliance to laws and regulations. One chief procurement officer disclosed that "We require all our suppliers to have sustainable business practices in place and to embrace our supplier code of conduct, which encompasses various ESG (environmental, social, and governance)-related guidelines. Further, we encourage our suppliers to adhere to international sustainability standards such as ISO 14001, ISO 45001, and ISO 27001. Currently, more than 80% of our procurement by value from critical suppliers comes from those having at least one of these ISO certifications and we target at least 85% procurement by value from such suppliers by 2024". Another firm indicated that their code of conduct also encompasses cooperation to conduct audits and training. Another said that the company has codes that regulate behavior to mitigate risks in terms of data privacy.

2.4.1.6 *Social Sustainability*

On the social aspect, one respondent indicated that their firm works at ensuing workplace health and safety, as well as employee and supplier continuous training, and ensures that they adhere to human rights. One SC manager mentioned that their firm has codes that must be followed, and these include guidelines to ensure supplier adherence to healthy and safe working conditions, so as to maintain a workplace that is free from recognized hazards and prevent any workplace injuries or ill-health. He further indicated that any non-compliances identified are addressed through implementation of corrective action plans, which are then periodically reviewed. The respondents reported that sustainability of groups is supported by continuous training and upliftment of staff and suppliers, with special emphasis on making local purchases, wherever possible. In terms of serving the local communities, one respondent highlighted that "our objectives currently include serving the underserved, promoting healthy living, ensuring local economic growth".

2.4.1.7 *Innovative Infrastructure and Technology Systems*

During the interviews, innovation came out as one of the common themes among respondents. One respondent mentioned that the "adoption of innovative infrastructure and technology systems into the SC processes helps to reduce costs and enhance efficiency and this was done through the introduction of AI, machine learning, and robotic process automation". Another interviewee indicated that "We have implemented a cloud-based sourcing and SC application across all markets to bring transparency and efficiency into its operations for its internal stakeholders and its suppliers". It was revealed that the adoption of e-government procurement in Zambia (e-GP) is marked as a new era in public procurement and is in line with government's overall policy to digitize its operations and to improve the effectiveness and efficiency of the tender process. One respondent indicated that they have continued to invest in innovative information and technology systems and infrastructure to ensure their SC remains world class and efficient.

2.4.2 SSCM Challenges

During the interviews with respondents, the study established that while companies wish to comply with good practices of sustainable SCM, they face numerous challenges. The common challenges identified during the interviews conducted were found as difficult in managing multi-tier suppliers as they do not comply with sustainable SC practices. One respondent indicated that lower-tier suppliers are not as equipped to handle sustainability requirements. In this vein, one respondent alluded to the fact that adopting green packaging sometimes attracts more costs; he said that, most times, a bag made of paper is more expensive than a bag made of plastic, which thereby affects the implementation of the sustainable SC practices. Moreover, another respondent alluded to the fact that a lack of availability of reusable, recyclable, and durable products hinders their firm from implementing sustainability policies. SC professionals interviewed revealed that the other common problem in SC is stock-outs as suppliers sometimes do not have the capacity to meet the demand. The other challenge indicated by the respondents was fluctuations in the exchange rate. A lack of technology adoption was another common theme mentioned by respondents as a challenge. In addition, one respondent said that "The company has invested a lot of money in new machinery; however, some workers especially casual farm workers still have difficulties to adjust to the new technologies".

2.5 Discussion on Results

This chapter revealed that the common sustainable SC practices being implemented include green packaging, green productions, stakeholder engagement, supplier collaboration, human rights adherence and workplace health and safety, and innovation. This agrees with literature as Wandosell *et al.* [110] affirm that, recently, there has been a growing awareness among companies and consumers about the importance of the promotion of sustainable SCM through green packaging alternatives. Oloyede and Lignou [111] added that corrugated cardboard and other forms of paper-based packaging are regarded as being among the most environmentally friendly and sustainable packaging solutions.

It was found that the green manufacturing can result in lower raw material costs, more efficient production, lower costs for occupational safety and the environment, and a better reputation for the company. The use of solar energy, recycling of raw materials, and utilization of biodegradable energy sources in manufacturing operations are indications of an effective implementation of green manufacturing practices [112]. Ahi and Searcy [113] stated that businesses should not only prioritize profits but should also look at their social and environmental responsibilities. As part of SSCM implementation, proper budgeting and fund utilization should be guaranteed. Furthermore, Copacino [114] indicated that firms have a responsibility toward the customer and, as such, have to invest in technologies and working methodologies to make the SC efficient. The study presented by Yu *et al.* [115] indicated that regular communication and information sharing enable the manufacturer to give suppliers comprehensive environmental requirements along with product and process specifications. It also empowers suppliers to enhance their environmental image by applying for ISO 140001 certification. The challenges given by respondents in the implementation of SSCM practices agree with Lechler *et al.* [116], who indicated that lower-tier suppliers do not comply with the sustainability requirements proposed in the focal companies' codes of conduct. It was found that managers mostly do not have visibility of the supply base beyond the first tier of suppliers and of suppliers located in developing economies, where environmental and labor laws are lax or, where laws exist, enforcement is dubious. It was found significant to exploit critical technologies [117, 118] and critical factors for handling suppliers and implement resources to indeed achieve sustainability [119, 120]. Delays in material delivery of raw materials and sole use of foreign suppliers are the factors that may involve an extension in throughput time thus exposing companies to possible disruptions in material flows [121]. The results revealed the other common problem in SC is stock-outs as suppliers sometimes do not have the capacity to meet the demand. A study by Keith [122] indicated that stock-outs can be costly, if demand exceeds available product supply. The other challenge indicated by respondents was fluctuations in the exchange rate. Yang *et al.* [123] indicated that the risks faced by suppliers in the SC include increasing national tax affairs policy, tariff, and fluctuations in exchange rate, and stable exchange rate attracted foreign investment. Technology adoption was found as a crucial challenge that is critically needed toward ensuring sustainability in the SC. Testa and Vigolo [124] indicated that SMEs face barriers toward technology investment due to lack of knowledge, low priority given to technology investment, and low financial resources.

2.6 Conclusion and Recommendations

The chapter highlighted that the companies surveyed were implementing a limited number of sustainable SC practices. These practices include green packaging, green productions, stakeholder engagement, supplier collaboration, human rights adherence and workplace health and safety, innovation through digitization, and adherence to international sustainability standards, laws, and regulations. The chapter advocated the need for the development of more practices that must be adapted for implementation and satisfaction of sustainability dimensions. Lawmakers should do more in enlightening firms on the benefits that come from implementation of SSCM practices, and much support is needed especially to companies that have showed interest in implementing SSCM practices. Top management should be in the forefront for evident implementation of suitable possible practices for true development. A quantitative framework on the topic is recommended to find out the gaps and drivers in the implementation of sustainable SCM practices. The chapter presented sustainability practices, challenges, and strategies, which will help in attaining SSCM. The same are elaborately discussed and their role is individually highlighted to give a good understanding of drivers and barriers to managers and practitioners. The chapter presented findings that can help in adapting SSCM. The chapter discussed SSCM importance based on qualitative marks and cases. The chapter recommends that future studies should focus on applying a quantitative approach in order to complement the qualitative perspectives to explore crucial boundaries of SSCM. It is found that sustainable SC can be fruitful and can be attained by enduring practices related with green packaging, green productions, stakeholder engagement, supplier collaboration, human rights adherence and workplace health and safety, innovation through digitization, and adherence to international sustainability standards, laws, and regulations. The results presented in the chapter can be used as a referential basis for helping practitioners, researchers, and policymakers in preparing quality frameworks and strategic decisions.

References

1. Mwanza, B.G. and Arnesh, T., Supply chains risks: An interpretative structural modelling approach. *Int. J. Supply Chain Oper. Resilience*, *5*, 2, 217–234, 2022.
2. Solingen, E.B.M. and Xu, A., Rising Risks to Global Value Chains, in: *Global Value Chain Development Report*, 2021.
3. Desore, A. and Narula, S.A., An overview on corporate response towards sustainability issues in textile industry. *Environ. Dev. Sustain.*, *20*, 1439–1459, 2018.
4. Gunasekara, K.R.H.L., Yapa, S.T.W.S., Shantha, A.A., Assessing the Sustainability Performance of Social Practices in Manufacturing Supply Chains. *Asian J. Manage. Stud.*, *2*, 2, 2022.
5. Abdul-Rashid, S.H., Sakundarini, N., Ghazilla, R.A.R., Thurasamy, R., The impact of sustainable manufacturing practices on sustainability performance: Empirical evidence from Malaysia. *Int. J. Oper. Prod. Manage.*, *37*, 2, 182–204, 2017.
6. Wolf, J., Sustainable supply chain management integration, a qualitative analysis of the German manufacturing industry. *J. Bus. Ethics*, *102*, 221–235, 2011.
7. Rebs, T., Marcus S. Seuring, B., Stohler, M., Stakeholder influences and risks in sustainable supply chain management: a comparison of qualitative and quantitative studies. *Bus. Res.*, *11*, 197–237, 2018.
8. Nambela, N., The impact of adopting sustainable supply chain management practices on the performance of public sector organizations in Zambia, a case study of Zambia Air Force (ZAF), Masters Dissertation, The University of Zambia Library, Zambia, 2020.
9. Mukuka, K., An Evaluation of The Effectiveness of Sustainable Practices In Supply Chain Management Of Zambian Breweries Plc, PhD Dissertation, Cavendish University Library, Zambia, 2022.

10. Manchisi, R., Sustainability in Supply Chain Management Practices In Least Developed Countries: A Case Of SMEs In Zambian Mining Sector, PhD Dissertation, University of Bradford, United Kingdom, 2019.
11. Chileshe, M.J. and Phiri, J., The impact of supply chain management practices on the performance of small and medium enterprises in developing countries, a case of agro-dealers in Zambia. *Open J. Bus. Manage.*, *10*, 2, 591–605, 2022.
12. Kiekens, A., Dierckx de Casterlé, B., Vandamme, A.M., Qualitative systems mapping for complex public health problems: A practical guide. *PloS One*, *17*, 2, e0264463, 2022.
13. Sony, M., Green supply chain management practices and digital technology: A qualitative study, in: *Technology Optimization and Change Management for Successful Digital Supply Chains*, pp. 233–254, 2019.
14. Dubey, R., Gunasekaran, A., Papadopoulos, T., Childe, S.J., Shibin, K.T., Wamba, S.F., Sustainable supply chain management: Framework and further research directions. *J. Clean. Prod.*, *142*, 1119–1130, 2017.
15. Münch, C. and Benz, L., Collaborative Partnerships-Conceptualization of Strategic Alliances for Sustainable Supply Chains. *Acad. Manage. Proc.*, *1*, 2022.
16. Norfarah, N., Ali, S.M., Siti-Nabiha, A.K., Ebrahim, N.A., Firms' sustainable practice research in developing countries: Mapping the cited literature by bibliometric analysis. *Int. J. Sustain. Strategic Manage.*, *7*, 1–2, 5–26, 2019.
17. Ahmad, W.N.K.W., Rezaei, J., Sadaghiani, S., Tavasszy, L.A., Evaluation of the external forces affecting the sustainability of oil and gas supply chain using Best Worst Method. *J. Clean. Prod.*, *153*, 242–252, 2017.
18. Musonda, M.M.J., Mwanaumo, E.M., Thwala, D.W., Risk management in the supply chain of essential medicines, in: *Proceedings of the 21st International Symposium on Advancement of Construction Management and Real Estate*, Chau, K.W., Chan, I.Y.S., Lu, W., Webster, C. (Eds.), pp. 1275–1287, 2018.
19. Luthra, S., Garg, D., Haleem, A., Empirical analysis of green supply chain management practices in Indian automobile industry. *J. Inst. Eng. (India)*, *95*, 119–126, 2014.
20. Saththasivam, G. and Fernando, Y., Integrated sustainable supply chain management: Current practices and future direction, in: *Green Supply Chain Management for Sustainable Business Practice*, pp. 218–233, 2017.
21. Tsai, F.M., Bui, T.D., Tseng, M.L., Ali, M.H., Lim, M.K., Chiu, A.S., Sustainable supply chain management trends in world regions: A data-driven analysis. *Resour. Conserv. Recycl.*, *167*, 105421, 2021.
22. Soosay, C., Fearne, A., Varsei, M., Extending sustainable practices beyond organizations to supply chains, pp. 71–90, Springer, Netherlands, 2014.

23. Liebetruth, T., Sustainability in performance measurement and management systems for supply chains. *Proc. Eng.*, *192*, 539–544, 2017.
24. Dey, A., LaGuardia, P., Srinivasan, M., Building sustainability in logistics operations: A research agenda. *Manage. Res. Rev.*, *34*, 11, 1237–1259, 2011.
25. Zimmon, D. and Domingues, J., Proposal of a concept for improving the sustainable management of supply chains in the textile industry. *Fibres Text. East. Eur.*, 2, 28, 8–12, 2018.
26. Sajjad, A., Eweje, G., Tappin, D., Sustainable supply chain management: Motivators and barriers. *Bus. Strategy Environ.*, *24*, 7, 643–655, 2015.
27. Zimon, D., Tyan, J., Sroufe, R., Drivers of sustainable supply chain management: Practices to alignment with UN Sustainable Development Goals. *Int. J. Qual. Res.*, *14*, 1, 2020.
28. Mujkic, Z., Qorri, A., Kraslawski, A., Sustainability and optimization of supply chains: A literature review. *Oper. Supply Chain Manage.: Int. J.*, *11*, 4, 186–199, 2018.
29. Hasan, M., Sustainable SCM Practices and Operational Performance. *Am. J. Ind. Bus. Manage.*, *3*, 1, 42–48, 2013.
30. Ghadimi, P., Wang, C., Lim, M.K., Sustainable supply chain modeling and analysis: Past debate, present problems, and future challenges. *Resour. Conserv. Recycl.*, *140*, 72–84, 2019.
31. Zhu, Q., Geng, Y., Fujita, T., Hashimoto, S., Green supply chain management in leading manufacturers: Case studies in Japanese large companies. *Manage. Res. Rev.*, *33*, 4, 380–392, 2020.
32. Vila, C., Abellán-Nebot, J.V., Albiñana, J.C., Hernández, G., An approach to sustainable product lifecycle management (Green PLM). *Proc. Eng.*, *132*, 585–592, 2015.
33. Mani, V., Agarwal, R., Gunasekaran, A., Papadopoulos, T., Dubey, R., Childe, S.J., Social sustainability in the supply chain: Construct development and measurement validation. *Ecol. Indic.*, *71*, 270–279, 2016.
34. Bendul, J.C., Rosca, E., Pivovarova, D., Sustainable supply chain models for the base of the pyramid. *J. Clean. Prod.*, *162*, S107–S120, 2017.
35. Marchi, V.D., Maria, E.D., Micelli, S., Environmental strategies, upgrading, and competitive advantage in global value chains. *Bus. Strategy Environ.*, *22*, 1, 62–72, 2013.
36. Chowdhury, M.H., Dewan, M.N.A., Quaddus, M.A., Resilient sustainable supply chain management: A conceptual framework, in: *International Conference on e-Business*, vol. *2*, pp. 165–173, 2012.
37. Ahmad, N.K.W., de Brito, M.P., Rezaei, J., Tavasszy, L.A., An integrative framework for sustainable supply chain management practices in the oil and gas industry. *J. Environ. Plann. Manage.*, *60*, 4, 577–601, 2017.

38. Zimon, D., Tyan, J., Sroufe, R., Implementing sustainable supply chain management: Reactive, cooperative, and dynamic models. *Sustainability, 11*, 24, 7227, 2019.
39. Fritz, M.M., Sustainable supply chain management. Responsible Consumption and Production, in: *Encyclopedia of the UN Sustainable Development Goals*, Leal, F.W., Azul, A., Brandli, L., Özuyar, P., Wall, T. (Eds.), pp. 1–14, 2019.
40. Sánchez-Flores, R.B., Cruz-Sotelo, S.E., Ojeda-Benitez, S., Ramírez-Barreto, M.E., Sustainable supply chain management—A literature review on emerging economies. *Sustainability, 12*, 17, 6972, 2020.
41. Gunasekara, K.R.H.L., Yapa, S.T.W.S., Shantharachi, A., Impact of Green Supply Chain Management Practices on Triple Bottom Line Performance and Moderating Effect of Institutional Pressure. *Asian J. Manag. Studies (AJMS), 3*, 1, 1–21, 2023.
42. Esfahbodi, A., Zhang, Y., Watson, G., Sustainable supply chain management in emerging economies: Trade-offs between environmental and cost performance. *Int. J. Prod. Econ., 181*, 350–366, 2016.
43. Morais, D.O. and Silvestre, B.S., Advancing social sustainability in supply chain management: Lessons from multiple case studies in an emerging economy. *J. Clean. Prod., 199*, 222–235, 2018.
44. Alghababsheh, M., Butt, A.S., Moktadir, M.A., Business strategy, green supply chain management practices, and financial performance: A nuanced empirical examination. *J. Clean. Prod., 380*, 134865, 2022.
45. Hassini, E., Surti, C., Searcy, C., A literature review and a case study of sustainable supply chains with a focus on metrics. *Int. J. Prod., 140*, 69–82, 2012.
46. Sroufe, R. and Melnyk, S.A., *Developing Sustainable Supply Chains: Management Insights, Issues, and Tools: Volume I Foundations*, pp. 7–8, Business Expert Press, New York, 2017.
47. Elmsalmi, M., Hachicha, W., Aljuaid, A.M., Prioritization of the best sustainable supply chain risk management practices using a structural analysis based-approach. *Sustainability, 13*, 9, 4608, 2021.
48. Alzubi, E. and Akkerman, R., Sustainable supply chain management practices in developing countries: An empirical study of Jordanian manufacturing companies. *Cleaner Prod. Lett.*, 2, 100005, 2022.
49. Mulwa, V.M., Sustainable Supply Chain Management Practices and the Performance of United Nations Agencies in Nairobi, Kenya, Doctoral dissertation, University of Nairobi, Kenya, 2015.
50. Kang, M., Yang, M.G., Park, Y., Huo, B., Supply chain integration and its impact on sustainability. *Ind. Manage. Data Syst.*, 118, 9, 1749–1765, 2018.
51. Nsowah, J. and Phiri, M.A., Factors Influencing Adoption of Sustainable Supply Chain Management Practices Among Manufacturing Firms. *Int. J. Environ. Sustain. Soc. Sci.*, 4, 3, 822–831, 2023.

52. Carter, C.R. and Easton, P.L., Sustainable supply chain management: Evolution and future directions. *Int. J. Phys. Distrib. Logist. Manage.*, 41, 1, 46–62, 2011.
53. Emamisaleh, K. and Rahmani, K., Sustainable supply chain management practices and sustainability performance in the food industry. *South East Asian J. Manage.*, 12, 1, 1–19, 2018.
54. Vural, C.A., Sustainable demand chain management: An alternative perspective for sustainability in the supply chain. *Proc. - Soc. Behav. Sci.*, 207, 262–273, 2015.
55. Fousteris, A.E., Didaskalou, E.A., Tsogas, M.M.H., Georgakellos, D.A., The environmental strategy of businesses as an option under recession in Greece. *Sustainability*, 10, 12, 4399, 2018.
56. Srivastava, S., Green supply-chain management: A state-of-the-art literature review. *Int. J. Manage. Rev.*, 9, 1, 53–80, 2007.
57. Seman, N.A.A., Zakuan, N., Jusoh, A., Arif, M.S.M., Saman, M.Z.M., Green supply chain management: A review and research direction. *I. J. M. V. S. C.*, 3, 1, 1–18, 2012.
58. Awan, U., Impact of social supply chain practices on social sustainability performance in manufacturing firms. *Int. J. Innov. Sustain. Dev.*, 13, 2, 198–219, 2019.
59. Nsowah, J. and Phiri, M.A., Factors Influencing Adoption of Sustainable Supply Chain Management Practices Among Manufacturing Firms. *Int. J. Environ. Sustain. Soc. Sci.*, 4, 3, 822–831, 2023.
60. Zhu, Q. and Sarkis, J., Relationships between operational practices and performance among early adopters of green supply chain management practices in Chinese manufacturing enterprises. *J. Oper. Manage.*, 22, 3, 265–289, 2004.
61. Rehman, S.U., Kraus, S., Shah, S.A., Khanin, D., Mahto, R.V., Analyzing the relationship between green innovation and environmental performance in large manufacturing firms. *Technol. Forecasting Soc. Change, 163*, 120481, 2021.
62. Govindan, K., Azevedo, S.G., Carvalho, H., Cruz-Machado, V., Impact of supply chain management practices on sustainability. *J. Clean. Prod., 85*, 212–225, 2014.
63. Kam-Sing Wong, S., The influence of green product competitiveness on the success of green product innovation: Empirical evidence from the Chinese electrical and electronics industry. *Eur. J. Innov. Manage., 15*, 4, 468–490, 2012.
64. Chin, T.A., Tat, H.H., Sulaiman, Z., Green supply chain management, environmental collaboration, and sustainability performance. *Proc. CIRP, 26*, 695–699, 2015.

65. Baig, S.A., Abrar, M., Batool, A., Hashim, M., Shabbir, R., Barriers to the adoption of sustainable supply chain management practices: Moderating role of firm size. *Cogent Bus. Manage.*, *7*, 1, 1841525, 2020.
66. Allenbacher, J. and Berg, N., How assessment and cooperation practices influence suppliers' adoption of sustainable supply chain practices: An inter-organizational learning perspective. *J. Clean. Prod.*, *403*, 136852, 2023.
67. Rao, P. and Holt, D., Do green supply chains lead to competitiveness and economic performance. *Int. J. Oper. Prod. Manage.*, *25*, 9, 898–916, 2005.
68. De Giovanni, P. and Vincenzo, E.V., The benefits of a monitoring strategy for firms subject to the Emissions Trading System. *Transp. Res. Part D: Transp. Environ.*, *33*, 220–233, 2014.
69. Wu, T., Shen, Q., Xu, M., Peng, T., Ou, X., Development and application of an energy use and CO2 emissions reduction evaluation model for China's online car hailing services. *Energy*, *154*, 298–307, 2018.
70. Seuring, S. and Müller, M., Core issues in sustainable supply chain management–a Delphi study. *Bus. Strategy Environ.*, *17*, 8, 455–466, 2008.
71. Azevedo, S.G., Carvalho, H., Machado, V.C., The influence of green practices on supply chain performance: A case study approach. *Transp. Res. Part E: Logist. Transp. Rev.*, *47*, 6, 850–871, 2011.
72. Tyan, J. and Wee, M.H., Vendor managed inventory: A survey of the Taiwanese grocery industry. *J. Purch. Supply Manage.*, *9*, 1, 11–18, 2003.
73. Vachon, S. and Klassen, R.D., Extending green practices across the supply chain: The impact of upstream and downstream integration. *Int. J. Oper. Prod. Manage.*, *26*, 7, 795–821, 2006.
74. Zailani, S., Jeyaraman, K., Vengadasan, G., Prem kumar, R., Sustainable supply chain management (SSCM) in Malaysia: A survey. *Int. J. Prod. Econ.*, *140*, 1, 330–340, 2012.
75. Lambert, D.M., Cooper, M.C., Pagh, J.D., Supply chain management: Implementation issues and research opportunities. *Int. J. Logist. Manage.*, *9*, 2, 1–20, 1998.
76. Soda, S., Sachdeva, A., Garg, R.K., GSCM: Practices, trends, and prospects in Indian context. *J. Manuf. Technol. Manage.*, *26*, 6, 889–910, 2015.
77. Azmat, F. and Ha, H., Corporate social responsibility, customer trust, and loyalty-perspectives from a developing country. *Thunderbird Int. Bus. Rev.*, *55*, 3, 253–270, 2013.
78. Al Zaabi, S., Al Dhaheri, N., Diabat, A., Analysis of interaction between the barriers for the implementation of sustainable supply chain management. *Int. J. Adv. Manuf. Technol.*, *68*, 895–905, 2013.
79. Jia, F., Zuluaga-Cardona, L., Bailey, A., Rueda, X., Sustainable supply chain management in developing countries: An analysis of the literature. *J. Clean. Prod.*, *189*, 263–278, 2018.

80. Negi, S. and Anand, N., Green and sustainable supply chain management practices- A study of Wal Mart. In A.D. Dubey (IIM Calcutta), *Emerging Business Sustainability* (1st ed.), pp. 141–157, Research India Publication, New Delhi, India, 2014.
81. Li, G., Fan, H., Lee, P.K., Cheng, T.C.E., Joint supply chain risk management: An agency and collaboration perspective. *Int. J. Prod. Econ.*, *164*, 83–94, 2015.
82. Murillo-Luna, J.L., Garcés-Ayerbe, C., Rivera-Torres, P., Barriers to the adoption of proactive environmental strategies. *J. Clean. Prod.*, *19*, 13, 1417–1425, 2011.
83. Abbasi, M. and Nilsson, F., Themes and challenges in making supply chains environmentally sustainable. *Supply Chain Manage.: Int. J.*, *17*, 5, 517–530, 2012.
84. Pereseina, V., Jensen, L.M., Hertz, S., Cui, L., Challenges and conflicts in sustainable supply chain management: Evidence from the heavy vehicle industry. *Supply Chain Forum: Int. J.*, *15*, 1, 22–32, 2014.
85. Paul, A., Shukla, N., Paul, S.K., Trianni, A., Sustainable supply chain management and multi-criteria decision-making methods: A systematic review. *Sustainability*, *13*, 13, 7104, 2021.
86. Moktadir, M.A., Ali, S.M., Rajesh, R., Paul, S.K., Modeling the interrelationships among barriers to sustainable supply chain management in the leather industry. *J. Clean. Prod.*, *181*, 631–651, 2018.
87. Gandhi, A.V., Shaikh, A., Sheorey, P.A., Impact of supply chain management practices on firm performance: Empirical evidence from a developing country. *Int. J. Retail Distrib. Manage.*, *45*, 4, 366–384, 2017.
88. Chen, T.J., Supply Chain Positioning and Innovation: Taiwan's Challenge. *Paper presented at the Mega-Regionalism – New Challenges for Trade and Innovation workshop*, East-West Center, 2016.
89. Kumar, G. and Goswami, M., Sustainable supply chain performance, its practice and impact on barriers to collaboration. *Int. J. Product. Perform. Manage.*, *68*, 8, 1434–1456, 2019.
90. Yun, N.Y. and Ülkü, M.A., Sustainable Supply Chain Risk Management in a Climate-Changed World: Review of Extant Literature, Trend Analysis, and Guiding Framework for Future Research. *Sustainability*, *15*, 17, 13199, 2023.
91. Giunipero, L.C., Hooker, R.E., Denslow, D., Purchasing and supply management sustainability: Drivers and barriers. *J. Purch. Supply Manage.*, *18*, 4, 258–269, 2012.
92. Oelze, N., Sustainable supply chain management implementation – enablers and barriers in the textile industry. *Sustainability*, *9*, 8, 1435, 2017.
93. Charles Jr, O.H., Schmidheiny, S., Watts, P., *Walking the talk: The business case for sustainable development.* Routledge, 2017.
94. Ageron, B., Gunasekaran, A., Spalanzani, A., Sustainable supply management: An empirical study. *Int. J. Prod. Econ.*, *140*, 1, 168–182, 2012.

95. Clarke, T. and Martijn, B., The governance of global value chains: Unresolved human rights, environmental and ethical dilemmas in the apple supply chain. *J. Bus. Ethics*, *143*, 111–131, 2017.
96. Bartley, T., Transnational private regulation in practice: The limits of forest and labor standards certification in Indonesia. *Bus. Politics*, *12*, 3, 1–34, 2010.
97. Huq, F.A., Social sustainability implementation in global supply chains, Lancaster University, United Kingdom, 2014.
98. Alghababsheh, M. and Gallear, D., Social sustainability in the supply chain: a literature review of the adoption, approaches and (un)intended outcomes. *Manage. Sustain.: Arab Rev.*, *1*, 1, 84–109, 2022.
99. Sharma, V.K., Sachdeva, A., Singh, L.P., A meta-analysis of sustainable supply chain management from different aspects. *Int. J. Supply Oper. Manage.*, *8*, 3, 289–313, 2021.
100. Saeed, M.A. and Kersten, W., Supply chain sustainability performance indicators: A content analysis based on published standards and guidelines. *Logist. Res.*, *10*, 12, 1–19, 2021.
101. Ahmad, A., Ikram, A., Rehan, M.F., Ahmad, A., Going green: Impact of green supply chain management practices on sustainability performance. *Front. Psychol.*, *13*, 973676, 2022.
102. Ibrahim, N. and Ahmed, N., Investigating Regenerative Ideation within Sustainable Development Goals. *Sustainability*, *14*, 16, 10137, 2022.
103. Maranesi, C. and De Giovanni, P., Modern circular economy: Corporate strategy, supply chain, and industrial symbiosis. *Sustainability*, *12*, 22, 9383, 2020.
104. Blome, C., Paulraj, A., Schuetz, K., Supply chain collaboration and sustainability: A profile deviation analysis. *Int. J. Oper. Prod. Manage.*, *34*, 5, 639–663, 2014.
105. Hussain, M., Ajmal, M.M., Gunasekaran, A., Khan, M., Exploration of social sustainability in healthcare supply chain. *J. Clean. Prod.*, *203*, 977–989, 2018.
106. Adesanya, A., Yang, B., Bin Iqdara, F.W., Yang, Y., Improving sustainability performance through supplier relationship management in the tobacco industry. *Supply Chain Manag.: Int. J.*, *25*, 4, 413–426, 2020.
107. Massaroni, E., Cozzolino, A., Wankowicz, E., Sustainability in supply chain management – a literature review. *Sinergie Ital. J. Manage.*, *33*, 98, 331–355, 2015.
108. Al-Odeh, M. and Smallwood, J., Sustainable supply chain management: Literature review, trends, and framework. *Int. J. Comput. Eng. Manage.*, *15*, 1, 85–90, 2012.
109. Madhani, P.M., Strategic Supply Chain Management (SSCM): Developing Conceptual Framework and Research Propositions, in: *Facets of Business Excellence in IT*, pp. 389–399, 2022.
110. Wandosell, G., Parra-Meroño, M.C., Alcayde, A., Green packaging from consumer and business perspectives. *Sustainability*, *13*, 3, 1356, 2021.

111. Oloyede, O. and Lignou, S., Sustainable Paper-Based Packaging: A Consumer's Perspective. *Foods, 10*, 5, 1035, 2021.
112. Amemba, C., Nyamboke, P., Osoro, A., Mburu, N., Elements of Green Supply Chain Management. *Eur. J. Bus. Manage.*, *5*, 51–61, 2013.
113. Ahi, P. and Searcy, C., A Comparative Literature Analysis of Definitions for Green and Sustainable Supply Chain Management. *J. Clean. Prod.*, *52*, 329–341, 2013.
114. Copacino, W.C., *Supply Chain Management: The Basics and Beyond*, Routledge, New York, 2019.
115. Yu, Y., Zhang, M., Huo, B., The impact of supply chain quality integration on green supply chain management and environmental performance. *Total Qual. Manage. Bus. Excell.*, *30*, 9–10, 1110–1125, 2019.
116. Lechler, S., Canzaniello, A., Hartmann, E., Assessment sharing intra-industry strategic alliances: Effects on sustainable supplier management within multi-tier supply chains. *Int. J. Prod. Econ.*, *217*, 64–77, 2019.
117. Sahu, A.K., Katyayan, A., Khandey, U., Jangde, P., Sahu, A.K., Sahu, N.K., Adaptation of blockchain technology in SCM for steering managerial strategies: investigative study under Indian context. *J. Int. Logist. Trade*, 21, 1, 18–40, 2023, https://doi.org/10.1108/JILT-06-2022-0020.
118. Ahmed, K.A., Sahu, A.K., Sahu, A.K., Sahu, N.K., Quantify the Behavior Intention of Individuals to Control SC Performance by Exploring Cloud Storage Services: An Extended UTAUT2 Approach. *Int. J. Technol. Hum. Interact. (IJTHI)*, 18, 7, 1–28, 2022, http://doi.org/10.4018/IJTHI.306227.
119. Bag, S., Viktorovich, D.A., Sahu, A.K., Sahu, A.K., Barriers to adoption of blockchain technology in green supply chain management. *J. Glob. Oper. Strateg. Sourc.*, 14, 1, 104–133, 2021.
120. Sahu, A.K., Sharma, M., Raut, R.D., Sahu, A.K., Sahu, N.K., Antony, J., Tortorella, G.L., Decision-making framework for supplier selection using an integrated MCDM approach in a lean-agile-resilient-green environment: evidence from Indian automotive sector. *TQM J.*, 35, 4, 964–1006, 2023, https://doi.org/10.1108/TQM-12-2021-0372.
121. Martino, G., Fera, M., Iannone, R., Miranda, S., Supply chain risk assessment in the fashion retail industry: An analytic network process approach. *Int. J. Appl. Eng. Res.*, *12*, 2, 140–154, 2017.
122. Keith, R.D., John, D.S., Jeroen, S., Supply constraints and waitlists in new product diffusion. *Syst. Dyn. Rev.*, *33*, 3–4, 254–279, 2017.
123. Yang, H.C., Syarifuddin, F., Chang, C.P., Wang, H.J., The impact of exchange rate futures fluctuations on the macroeconomy: Evidence from ten trading markets. *Emerg. Mark. Finance Trade*, *58*, 8, 2300–2313, 2022.
124. Testa, F. and Vigolo, V., Sustainability through energy efficiency: an Italian perspective. *Sinergie Ital. J. Manage.*, *33*, 93–111, 2015.

3

Advocating Lean Practices and Strategies in Decision-Making for Reinforcing Industrial and Manufacturing Designs

Sanju Nishad, Atul Kumar Sahu* and Nitin Kumar Sahu

Department of Industrial and Production Engineering, Guru Ghasidas Central Vishwavidyalaya, Bilaspur, India

Abstract

In the present chapter, lean practices and strategies are advocated for integration in numerous structures of decision-making (DM) for attaining reinforced industrial and manufacturing (INM) designs. The present chapter is reporting the capabilities of lean thinking and refers that lean enrollment can be attain in organizational model based on the collection of qualitative and quantitative information. The remarkable attainments in terms of waste management, cost management, energy depreciation, and manufacturing efficacy can be attained by the enrollment of lean practices and strategies. Nowadays, numerous assets and resources are associated by the INM chains, and, thus, the sustainable designs are needed to be built by the utilization of various practices like lean, agile, green, and resilient by the organizations. The said practices demands collection of qualitative and qualitative information for making sustainable designs, and, after collection; optimization and synchronization is needed for exploiting the capabilities, contentment, satisfaction, and sustainability. Accordingly, the present chapter is advocating the importance of lean management and lean manufacturing to be inducted in DM and demonstrated the need to exploit lean practices for designing INM environment for better results and utilization. A large number of lean practices are available and the same are discussed in the present chapter. The chapter presents a cluster of lean practices, where; the authors have presented 19 momentous lean practices in the first segment, which are established lean practices, and additionally defined five critical lean practices in the second segment that are radical to be

**Corresponding author*: atul85sahu@gmail.com

Atul Kumar Sahu, Rakesh D. Raut, Rohit Raja, Anoop Kumar Sahu and Nitin Kumar Sahu (eds.) Industrial and Manufacturing Designs: Quantitative and Qualitative Analysis, (75–104) © 2024 Scrivener Publishing LLC

utilized in various fields. An evident number of cases linking lean are discussed in this chapter to know about lean practices and the effect of their implementation in manufacturing industries (MIs), the effect of lean on health and safety of workers in manufacturing, lean and agile manufacturing impacts in work-in-progress control, lean importance in small- and medium-sized enterprises to support Industry 4.0, and lean for attaining process innovation performance. Additionally, a conceptual model of lean manufacturing dimensions in various industries, i.e., furniture industry, electronic industry, and water heater MI, is also presented. The lean model in various fields like the manufacturing field, academic institute, management structure, environmental image, social dimension, economical attainments, and automotive industrial application is presented. Furthermore, some strategies related to lean management and lean manufacturing are also presented for realizing the easy realization of lean practices in the INM fields. The chapter will help readers in understanding various lean concepts.

Keywords: Lean practices, decision-making, strategies, industrial and manufacturing designs, qualitative, quantitative, lean concepts

3.1 Introduction

Lean is a systematic approach toward the elimination of wastes, increasing productivity and improving efficiency within manufacturing industries (MIs) and management domain. By applying lean principles, manufacturers can improve their competitiveness, affix customer values, increase customer satisfaction, reduce lead time, augment brand image and efficiency, and reduce induced costs. In this chapter, lean practices are presented that can be utilized by practitioners for uplifting the capabilities of applied industrial and manufacturing (INM) resources under confined boundaries. Lean manufacturing (LM) indicates the elimination of wastes in production systems that are related to human being efforts and inventories at different stages of production. Lean manufacturing's major objective is to support organizations looking toward advance operations, improve quality, and amplified customer happiness while spending less resource. In the present era, there is the need of new production theories in the MI that can aim to create more with fewer undesirable activities, where LM and management principles can be utilized. Lean manufacturing promotes amplified attainments in the total production output and confers toward workers' as well as customers' satisfaction [1]. Lean is completely a customer-based strategy, whose strong focus is on the optimization of applied resources with the help of optimization technique, obsolescence reduction, energy saving, and identification of capable assets for improved productivity. The resources that do not provide worth in the work environment is a target for modification or elimination according to LM principles. It is nowadays affirmed as one of the crucial tool internationally to increase a firm's productivity and efficiency [2]. Accordingly, authors as an objective of this chapter tried to collect a cluster of lean practices to help readers in understanding various lean concepts (LCs). The present chapter is enrolling its existence toward defining the capabilities of lean thinking and embracing that the lean enrollment is required under organizational model for reaching toward sustainable decision-making (DM), where the information collected in both qualitative and quantitative measures can deliver fruitful results. The said practices can be enrolled by the decision-makers based on the collection of qualitative and qualitative information, where optimization and synchronization can be done for pleasing the manufacturing capabilities and societal satisfaction. In the present chapter, the authors attempt to answer the following critical questions (CQ) related to lean implementation in the INM domain:

CQ 1: What are the critical lean practices reported in the published literature and what can be the radical lean practices for enrolling sustainability in current manufacturing industries?
CQ 2: What are the lean strategies followed by the manufacturing industries?
CQ 3: What are the steps followed by the concerns for implementation of lean thinking?
CQ 4: What are the various lean models applied in different industries and what kind of motivation and challenges can be attained by the lean.

In the present stage, it is required by the large number of countries to adopt lean in manufacturing and industrial boundaries due to its huge potential benefits and demonstrable capabilities in quality assurance, cost savings, and attributed response. The majority of manufacturing companies have found difficult to implement LCs in the present production methods and equipments due to unawareness, lack of payback knowledge, low understandability, and low application of competitive marks among industries. Nowadays, organizations have investigated toward adopting some of the prominent manufacturing concepts for receiving effectiveness with the existing resources, where LCs are quite competent. Lean practices stimulate toward cost reduction, setup time reduction, Kaizen, value stream mapping (VSM), 5S, KANBAN, total quality management (TQM), and inventory management as motivating domain for receiving sustainable operations [3]. This chapter aims to present bunch of lean practices, where the authors have presented well-established and critical lean practices to be utilized in various INM fields for shaping future effectiveness. Here in the present chapter, the authors have presented 19 lean practices, which are momentous, and additionally defined five radical lean practices to be utilized in diverse sectors.

3.2 Literature Review

Lean manufacturing enables a company to compete by removing non-value activities. For receiving sustainability, the companies should focus on protecting its resources by eliminating waste. Lean manufacturing skills are needed to be presently applied in number of industries, i.e., automobile, electronics, plastics, job shops customized products, food products, dairy, foundry, steel manufacturing, and maintenance. The same is needed to be introduced individually or in integration with other bottle neck techniques. The reimbursements that can be attained by using individual or combined LM practices are reduction in cycle time, deletion of non-valued operations, recognition of value-added sources, and attainment of clean and competent environment. Lean assists in delivering a smooth production flow, increase in productivity, reduction in production overheads, refined employee participation, organized records, inventory reduction, and low breakdowns with improved intra- and inter-interconnection among decision parties to make quick decisions. Accordingly, the present chapter attempt to review LCs on various manufacturing segments to examine the impact of implementation of LM methods toward enhanced procedures and utility while decreasing waste [4]. Currently, there is an intense competition among manufacturers, which necessitates the delivery of top-notch products cost-effectively from every facility, where LCs are the most recent tool to achieve this. The chapter investigates the lean thinking under the boundaries of lean practices, lean strategies, lean techniques, and lean modeling under various case studies. This chapter evaluates the technical status of LM to demonstrate its practical implementation. This chapter discusses the LM idea, lean waste, LCs, lean hurdles, and lean implementation cycle. This chapter will attempt to give a clear understanding of lean based on the literature review to clarify the condition of LM and its strategies [1].

The purpose of this chapter is to investigate and appraise precedent investigations related to LCs with integrated connection and linkages among lean and sustainable manufacturing with allied development. The chapter identified many paradigms under the said shadow, and, after investigation, correlation mapping among received solutions, overlapping areas, differences in productivity, framework integration, and categorization of factors are found as crucial segments for attaining sustainability, and, accordingly, it is needed to exploit the dimensional factors and related qualitative and qualitative information for preparing DM framework, which can have the capacity to integrated all the aforesaid interconnections. Several pieces of evidence demonstrated that lean is advantageous for receiving sustainable manufacturing, particularly in terms of receiving deprived environmental load and improved economy. Accordingly, it is needed to conduct significant research under the dimensions of integrating lean and sustainable manufacturing to enhance business presentation while employing it as a methodological approach [5].

A preliminary investigation is required under INM boundaries for developing an original conceptual framework for assessing the degree of leanness within the manufacturing sector. A conceptual model for measuring leanness in the manufacturing sector is always needed to be established in the primary phase and designed to be needed at at-least two levels, i.e., dimensional and determinants in the secondary phase. The model is afterward needed to be developed on the basis of the primary and secondary phases and considering eight different sources of waste, which is needed to be benchmark in terms of efficiency and productivity. Production process and equipment, scheduling and planning of production, visual data system, supplier collaboration, customer engagement, workforce management, and product innovation and technology are found as the seven primary factors to be used for measuring leanness under LC [6].

Nowadays, market competition is increased, which has compelled multiple business organizations including small- and medium-sized businesses to reconsider, rethink, and reevaluate their manufacturing methods, outputs, and performance indexes. Nowadays, it is needed by major industries to use lean to maintain their operations. The goal of this chapter is to present a structural study for advocating the significance of LM and lean practices. A bunch of case studies with the established and critical lean practices are produced for ready implementation in this chapter for improvements as a result of modifications. Today, there is need to develop several new lean practices that could be employed to eradicate waste in the production flow line [7]. Today, it is needed to exploit lean thinking under currently open explore information technology (IT) for developing new age of LM principles. Today, it is also needed to implement LCs and methods in a wide range of organizations with a focus on project-based industries to traditional service sectors like hospitals, logistics, and supply chain and in small continuous manufacturing firms to customized product handling. To apply LCs in diversified fields, many well-known methodologies like KANBAN, Kaizen, SMED (single-minute exchange of dies), and 5S are required to be integrated in close umbrella. The use of IT is to create such integration that is a deemed solution in today's scenario, where the role and utilization of IT in the implementation of LCs is a crucial challenge in recent years [8].

3.3 Lean Tools, Motivation, and Methodology

It is found that the lean idea presented several benchmarks in a metalworking company, where, after lean adaptation, various opportunities converted into strength for the manufacturing organization [9]. Here, the study targeted the company's machining area for reducing waste and increasing productivity. Flowcharts and VSM were used to identify and map essential processes for this aim, and, additionally, lean tool SMED is utilized to reduced setup times, which was found remarkable through aforesaid setup implementation. Installation times were found decreased by 40% just on the company's milling machines and 57% on horizontal milling machines [9]. Today, every industry is under pressure to increase quality, output, and customer happiness. The lean tool SMED is applied on a specific production line in the Romanian pharmaceutical industry [10], where the outcomes obtained after using SMED tools are investigated. It is found that product changeovers on a specific packaging machine using the SMED technique allowed to reduce machine downtime while enhancing ultimate productivity. Here, it is found that the SMED along with the utility of other LM tool, the main interchange time just at bottlenecks process was reduced by 30% in 12 months. Quality standardization and teamwork have also found improved in addition to the economic benefits [10]. It is found that lead time is an essential component for improving the image of the firms in the highly competitive global market. By identifying and removing bottlenecks, the manufacturing firms can reduce the overall lead time. Work-in-progress (WIP), layout optimization, and line balancing are crucial production line activities that can assist in the reduction of waste and improvement in multiple segments. Based on a detailed analysis of the product's manufacturing process and existing manufacturing methods, a study was conducted and a total available production duration is determined to investigate that the demand will be met or not. Under existing condition, VSM was sketched, and the bottleneck units were located, and flexibility line management program was used to balance the production line after producing the current state of VSM. The lean approaches under fish bone diagrams, Pareto chart, and layout optimization are utilized to lead further advancements. The aforementioned lean strategies were put into practice, and the future state VSM was drawn, where it was found that the lead time had been cut by roughly 15 min [11].

In the present chapter, the theoretical conception of lean is exploited and established, the critical lean practices along with lean strategies are presented, and a number of models related to implementation of lean theory are also presented on the basis of the literature review. To put lean methodology into practice, it is primarily required to describe the concept, allied parameters, and measures. Firstly, introduce the questions to know about the availability of pace and segments, where the modification and elimination can be done. For the same practitioner and researcher can utilize present chapter for realizing the easy implementation of lean practices in the INM fields. In the present chapter, an invention related to LM in automotive industry is presented to assist and to explain "how the LM implemented in different segments" for sustainability. Additionally, many case studies related to LM practices and its adoption in the different MIs are also presented to reinforce the understanding of the readers about LCs. The various lean models used in different fields like MI, furniture industry, electronic industry, and automotive industry are presented for the readers under single umbrella. The theory about various lean practices and the strategies used for realization of LM are discussed to help in its implementation and adaptation. The allied segments of present chapter are presented in Figure 3.1, where it can be understand that lean theory, established lean practices, lean strategies, modeling, and case studies are demonstrated for defining the robust concept of lean management in the present chapter.

3.4 Lean Theory and Practices

Lean practices refer to the set of principles and methodologies that focus on maximizing value for customers and minimizing wastes in the production and delivery of products. Lean is usually defined as an approach to reduce waste, to increase productivity, and to improve efficiency within INM industries. By applying these principles, manufacturers can improve their competitiveness and can add value, satisfaction, and sustainable

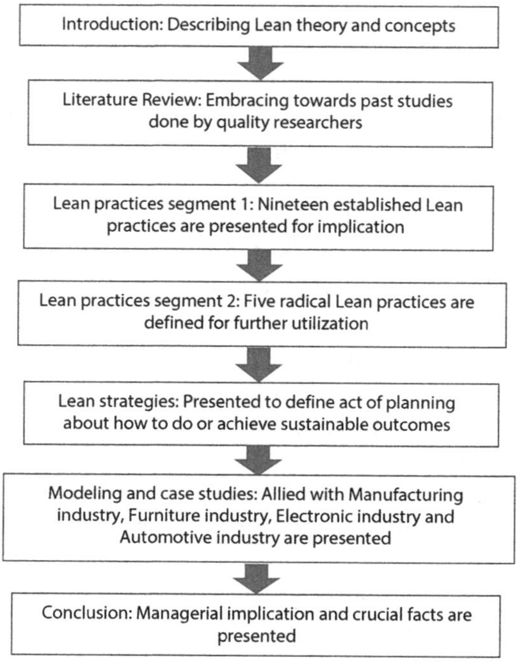

Figure 3.1 Flow diagram representing the allied segments of present chapter.

dimensions with reduce costs. In this chapter, some lean practices are presented in the first segment that are momentously utilized by the practitioners for uplifting the capabilities of INM resources, which are established lean practices. Additionally, few lean practices are also proposed as critical lean practices in the second segment that are radically needed for implication in new fields. This section presents a cluster of lean practices to be used in various fields. The lean practices are shown in Table 3.1, whose description can be understood in Sections 3.4.1 and 3.4.2.

3.4.1 Lean Practices (Segment 1)

The lean practices are defined as a set of managerial actions that are put to into operation to improve effectiveness and efficacy of the system under consideration by eliminating waste. Some lean practices are momentous and have applied by the practitioners for uplifting the capabilities of INM resources, as shown in Sections 3.4.1.1 to 3.4.1.9. The central part of LC is to eliminate non–value-adding actions and wastes, which can be attained by the application of lean practices as depicted in Sections 3.4.1.1 to 3.4.1.19.

Table 3.1 Lean practices for sustainability.

Lean practices	Depiction
Segment 1	Value stream mapping (VSM)
	Kaizen
	5S
	KANBAN
	Six Sigma
	Total productive maintenance (TPM)
	Total quality management (TQM)
	Overall equipment effectiveness (OEE)
	Plan-do-check-act (PDCA)
	Inventory management
	Production leveling
	Zero-defect concept
	Bottleneck analysis
	Root cause analysis (RCA)
	Just in time
	Time and motion study
	Single-minute exchange of dies (SMEDs)
	DMAIC
	Poka-yoke
Segment 2	Redundancy
	Digitalization
	Health, safety, and allowance for continuous flow
	Simplification and standardization
	Teamwork and partnering

3.4.1.1 Value Stream Mapping (VSM)

VSM is a practice that believes in creating a visual channel related to all the components working under a system to be used for evaluation and eradication of un-useful activities. VSM considers the evaluation of each and every component under a value chain with the objective of optimizing the entire process. The same concentrates on visualizing the flow of materials and information from the raw material to delivery to customers. It displays the process mechanism related to current as well as future states in a manner that focuses on areas for improvement.

3.4.1.2 Kaizen

It promotes the process of continuous improvement that involves the organization efforts on making incremental improvements toward processes and products infinitely. It is a principle, a philosophy, or a practice that is at the heart of lean and believes in identifying change in the system for the betterment of future in a continuous way at every time.

3.4.1.3 5S

5S is a visual management system that aids in the maintenance of a clean, organized, and efficient workplace, removing anything that is no longer useful. The 5S are named as Seiri, Seiton, Seiso, Seiketsu, and Shitsuke, whose ultimate aim is to plan cleaning and storage activities. Here, Seiri stands for sorting and separate essential and nonessential items. Seiton main intension is to set the items in order. Seiso main intension is to make cleanness. Seiketsu main intension is to standardize the things. Shitsuke main intension is to attain self-discipline.

3.4.1.4 KANBAN

It is a visual control system that aids in the management of workflow. It involves using signals (such as cards or containers) to indicate when more materials are needed and when production should stop.

3.4.1.5 Six Sigma

It is a process improvement practices that focus on reducing defects and variability in processes through statistical analysis and measurement. It aims to achieve a quality level, where the possibility of a defect arising is less than the opportunities. It focuses on indentifying deviation from the target.

3.4.1.6 *Total Productive Maintenance (TPM)*

TPM aims to increase equipment uptime, decrease maintenance time, optimize equipment reliability, and maximize equipment effectiveness by preventing breakdowns.

3.4.1.7 *Total Quality Management (TQM)*

Organizations can create a culture of continuous improvement by focusing on customer needs by way of making evident strategies that promotes the involvement of each and every employee in the improvement process. The same can be attained, and data-driven decisions can be prepared by the integration of aforesaid aspects, which thereby induces implementation of TQM. This has the potential to significantly improve quality, efficiency, and customer satisfaction.

3.4.1.8 *Overall Equipment Effectiveness (OEE)*

Organizations can significantly increase quality, efficiency, and customer satisfaction by implementing lean thinking related with OEE. Continuous improvement is emphasized in lean thinking, which includes identifying and eliminating waste, improving processes, and increasing efficiency, where OEE can be significant, which concentrates on determination of "how much planned productive time is actually productive". OEE believes in determining the percentage of "acceptable component" created that holds the quality standards. OEE considers multiplication of three parameters named as availability, performance, and quality.

3.4.1.9 *Plan-Do-Check-Act (PDCA)*

PDCA can be utilized to practice to manage change. Between the raw material and the finished product, various stages of production are there, which need to be evaluated on the basis of PDCA for continuous improvement. PDCA is a technical scheme that can be used toward handling change. The same is also known as the Deming cycle and involves on identifying prospect or route that can be improved by testing and analysing results in an iterative way.

3.4.1.10 Inventory Management

Inventory management is an essential component of lean practices, which is a managerial implication that primarily focused on maximizing inventory value while minimizing waste.

3.4.1.11 Production Leveling

Production leveling is known as Heijunka, which is a key concept in lean practices that are used to improve efficiency by smoothing out the manufacturing process. It focuses on volume leveling and leveling of product variants for continuous flow.

3.4.1.12 Zero Defect (ZD) Concept

The ZD concept of LM is a management philosophy that focuses on maximizing value and minimizing waste. The ZD concept emphasizes the importance of producing products that are free of defects, errors, and poor quality.

3.4.1.13 Bottleneck Analysis (BA)

It is a practice that focuses on evaluating the number of times the process has been immovable or stopped, somewhere between receiving raw material to manufacturing of finished products or between development and delivery. BA is a planned approach toward identifying faults at any zone and looking toward the barriers in processes and exertions under a product or service. BA is used toward categorizing and recognizing operational and process challenges.

3.4.1.14 Root Cause Analysis (RCA)

Waste elimination is a key component of lean practices that can assist organizations in improving quality, lowering costs, and increasing customer satisfaction; accordingly, it is feasible to determine the foundation of an issue to remove the occurrence of the said problem. Organizations can improve their efficiency and competitiveness in the marketplace by identifying RCA and eliminating the said sources. RCA can be a valuable tool for increasing operational efficiency and decreasing waste. For manufacturers looking to implement lean practices in their operations, an RCA practice can be a valuable asset. To achieve maximum efficiency and effectiveness, the system must be properly designed and integrated into the overall production process by continuous evaluation through RCA.

3.4.1.15 *Just in Time (JIT)*

The primary goal of JIT is to reduce the amount of WIP inventory and finished inventory, which can aid in cost reduction, quality improvement, and customer satisfaction. JIT is a management strategic practice that believes in purchasing the raw material from suppliers directly in conjunction with production schedules.

3.4.1.16 *Time and Motion Study*

A time and motion study is an important practice that concentrates in attaining the standard time for completing the process. The same is significant for improvement, driving efficiency, and quality improvements in the manufacturing process.

3.4.1.17 *Single-Minute Exchange Dies (SMED)*

It is the practice to shorten the change over time to 1 min, and it is important in discrete or batch production at the time of change of product.

3.4.1.18 *DMAIC*

It is a data-driven quality practice used to improve processes. DMAIC stands for Define-Measure-Analyze-Improve-Control scheme to be used for improvement. It is a five-step continuous activity cycle.

3.4.1.19 *Poka-Yoke*

It is an idea, mechanism, or solution in the primary detection of mistakes. Poka-yoke is an essential component of LM because it promotes a culture of continuous improvement by preventing errors and defects from occurring in the first phase. Poka-yoke is an error-proofing analysis–based practice that is based on prevention. It is feasible toward identifying the foundation of an issue to prevent it rather than the occurrence. It is a mistake-proofing practice, which prevents assembly and connection-related procedural errors like two-component assembly that should be planned by striving on the mechanical component design shape, size, and assembly in one direction.

3.4.2 Lean Practices (Second Segment)

The five critical practices are proposed in this section as a result of exploration and evaluation of the previous work, which are radical and focused on the relationship and links between lean, value addition, academic knowledge, incremental improvements, and responsiveness. The same are explained in Sections 3.4.2.1 to 3.4.2.5.

3.4.2.1 Redundancy

This practice is usually required to provide a supplementary network or means for avoiding failure in network and to keep manufacturing maintained and valuable throughout the network to avoid delay in lead time- and support-dependent networks. This refers to a situation where an operator quits their work because of the no longer availability of resources. The practice believes in keeping backup systems or components in case of breakdown or malfunction that is referred to as redundancy. Redundancy in a system can also refer to the duplication of information or processes, which can lead to inefficiencies or errors. Streamlining or reducing unneeded redundancies in such situations can improve system efficiency and lower expenses.

3.4.2.2 Digitalization

Digitalization is the process of converting traditional analog or manual processes to digital ones through the use of digital technologies. The application of digital technologies such as automation, robotics, and artificial intelligence improves the efficiency, accuracy, and productivity of manufacturing operations.

3.4.2.3 Health, Safety, and Allowance for Continuous Flow

This practice refers the enrollment of an effective dimension for retaining the safety and health of labors. It supports on significantly reducing the onsite harm and ensures that no staff members are harmed onsite and believes on predicting possible risks that can happen and measures to avoid them as well as taking proactive estimates in designing to prevent them. This practice can be beneficial in planning for continuous flow by assessing, identifying, and controlling potential risks to avoid failures.

3.4.2.4 *Simplification and Standardization*

This practice refers to designing a machine components and parts following simplification and standardization principals by the design team. Standardization refers to the use of accepted dimensions and criteria for each function involved in the products that help to develop standard machine, ease assembly, and aid in running the manufacturing process continuously without any delay. With the help of standardization, productivity has increased and all costs have reduced, whereas simplification refers to a variety of reduction.

3.4.2.5 *Teamwork and Partnering*

This practice believes in maintaining collaboration from all parties involved in the manufacturing process, including clients, designers, planners, engineers' contractors, suppliers, and others parties. This helps to decrease the variability of workflow and continuous improvement of manufacturing process and projects.

3.5 Lean Strategy: Discussions and Implications

Lean strategy refers to a management philosophy and a set of principles for streamlining and optimizing business processes and motions, reducing waste, and improving overall efficiency [4, 12]. The goal of lean strategies is to create a value in the system most efficiently and effectively as possible [13]. The authors found some of the strategies related to lean management and LM that are crucial and momentous. Thus, the chapter presented the clusters of lean strategies for realizing the easy implementation of lean thinking in the INM fields. The same are presented in Table 3.2, and the steps for its implementation are shown in Figure 3.2. Figure 3.2 represents the methodological steps that are required for implementing LM practices in any type of MI. Firstly, identify the type of wastes that can be originated and then differentiate them and determine the cause and effect of these wastes in either operators or the environment. Use the most appropriate LM method to eliminate all types of wastes. Determine the wastage ratio by putting the value of total manufactured products to the total estimated wastes. The same will help in determining the use of most appropriate lean measurement strategy to find out the leanness in any MI.

Table 3.2 Lean strategies.

S. no.	Strategies (Implementation)	Definitions
1.	Focus on customer value and need	This lean strategy starts with a thorough understanding of what customer's value and is willing to pay for the consumption of the product. This understanding aids in determining, where the organization's efforts and resources should be directed. The same will help in implicating the extent of efforts for the production.
2.	Continuous improvement	This lean strategy emphasizes on the importance of continuous improvement across the board. This entails a continuous search for better ways to do things, as well as a commitment to ongoing learning and development.
3.	Empowerment of employees	This lean strategy recognizes the value of employees and strives toward empowering them to identify and solve problems. This approach provides employees with the tools and training that they require to improve processes and contribute to the organization's success.
4.	Eliminate seven types of wastes of production	Excess inventory, waiting, overproduction, defects, unnecessary motion, transportation, and over-processing are the seven types of examples of waste that must be identified and eliminated as part of a lean strategy. Organizations can improve efficiency and cut costs by eliminating these wastes.
5.	Create a culture of excellence	This lean strategy strives toward creating an excellence culture in which everyone in the organization is dedicated toward providing the highest quality products or services. This entails a strong emphasis on teamwork, collaboration, and a shared commitment to achieving the organization's objectives.

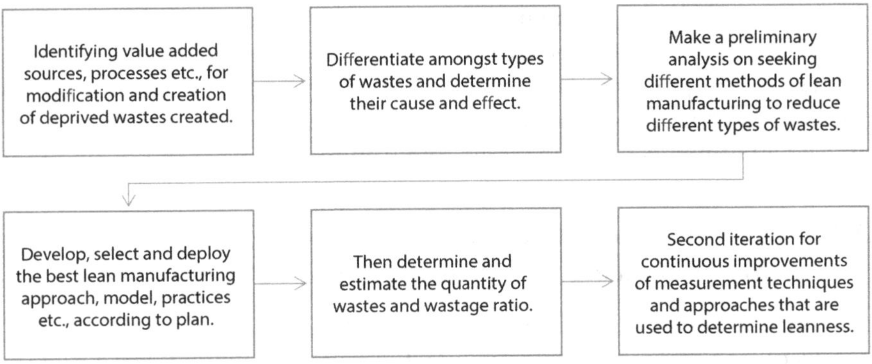

Figure 3.2 Steps for implementation of lean manufacturing practices.

3.6 Lean-Based Case Investigations and Discussions

The present chapter has exploited various cases to show the competencies of lean practices and strategies for reinforcing IBM designs as shown in Sections 3.6.1 to 3.6.11. The 11 cases will help to know about the lean practices and the effect of lean implementation in various sectors like MIs to enhance productivity, sustainability, significance of lean on health in manufacturing, importance of lean and agile simultaneously in WIP control in SMEs to induce Industry 4.0 adaptation in manufacturing, and recognition of lean as a tool for on-process innovation performance. Additionally, the theoretical modeling of LM dimensions in a number of industries considering electronic industry and water heater MI is also presented in this chapter to understand its implementation.

3.6.1 Lean Manufacturing is a Vital Tool to Enhance Productivity in Manufacturing

Chahala and Narwal [1] have applied the LM strategy in the MIs to increase productivity, reduce cost, and eliminate waste. In their study, the review concentrated on various manufacturing sectors to understand the impact of lean practices. The implication is found essential to be implicated in the number of MIs, i.e., automobile, electronics, textiles, paper, job-based customized products, food industries, dairy, foundry, and maintenance. The implementation of individual or combined lean practices including reduced cycle time; elimination of non-value activities; and clean, tidy, and hygienic workplace is found in their study. Additionally, a smooth production flow, a reduction in production costs, employee involvement, appropriate ordering of records, inventory reduction, and a breakdown of job toward understanding connecting parts for making quick decisions are the indeed attainments received by the implication of LCs.

3.6.2 The Linkage Between Lean and Sustainable Manufacturing for Attaining Refined Performance

Hartini and Ciptomulyono [5] have explored and discussed several frameworks of lean and sustainable manufacturing for receiving refined performance. Their study has analyzed and assessed the relationship and links between various factors to transform lean with sustainable manufacturing. Correlation, overlapping area, difference, integration, and classification under factorial dimensions are some of the aspects consider by them for understanding impacts due to their relationships. Their study identifies significant research gaps and found to explore lean and green dimensions simultaneous to be integrated to improve business performance for sustainability as a DM approach. Their paper conducted a quantitative and qualitative analysis to examine the relationship between lean and to receive sustainable manufacturing [5] and found that improved productivity, resources savings, better lead times, and improved environmental image and customer satisfaction can be attained with simultaneous modeling of lean and sustainable manufacturing.

3.6.3 A Conceptual Model of Lean Manufacturing Dimensions for Sustainability

Wahab *et al.* [6] have described a preliminary investigation to define a way to measure leanness in the MI. Their study has given a conceptual model for measuring leanness that has been designed on two levels named as dimensions and factors. Manufacturing process and equipment, manufacturing planning and scheduling, visual information system, supplier collaboration, customer relationship, workforce, and product development and technology are the seven main dimensions defined by them toward measuring leanness. Furthermore, the model demonstrates lean dimensions under eight different types of waste [6].

3.6.4 Lean Practices Align Toward the Health and Safety of Workers in Manufacturing Industries (MIs)

Singh *et al.* [14] have explored literatures to investigate the impact of lean practices on worker health and safety. Their study has provided insights into the relationship between lean and safety in MI. Resources and time spent on injuries and employee compensation are presented as wastes and considered to be reduced as part of a lean approach. The integration of lean principles with safety measures is deemed crucial for the creation of a workplace that not only sees efficiency enhancements but also places the well-being of employees as a top priority. Safety incorporation into lean practices results in accident reduction, fewer injuries, and reduced hazards, all while simultaneously boosting productivity and cutting down on waste. Accordingly, it is suggested by them to maintain a little emphasis for establishing a link between specific lean practices, and occupational safety and health. The study has concluded that supplementary research is needed to determine the connections between particular lean tools and worker safety and health issues in specific MI [14].

3.6.5 The Linkage Between Lean and Agile Manufacturing for Work-In-Progress (WIP) Control

Hemalatha *et al.* [15] have focused on the root cause of the problem, fundamental system problems, KANBAN implementation, inventory cost, reorder point, and so on for identifying improvements in system. Their work provides methodologies for reducing total WIP inventory across all boiler components. Their study has given the concept of lean and agile manufacturing to be applied simultaneously for receiving elevated outcomes. They have applied simultaneous approach it their study related to boiler component fabrication work, intending to determine the factors affecting WIP inventory levels to congregate the requisite demand for each product effectively. They identified decisive variables and examined the effects of those variables, where it is found that LM takes priority in value addition by removing repetitive, low-value processes from the manufacturing cycle and that agile manufacturing promotes process optimization, standardization, and automation. They affirmed the dignified role of lean and agile manufacturing to provide customer satisfaction at a low cost and emphasized that lean and agile are two distinct methodologies in business and manufacturing and are characterized by its own set of principles and objectives. Nevertheless, their effective integration can be observed to generate a more adaptable and efficient organization, when managed appropriately [15].

3.6.6 Adaptations of Lean Practices in SMEs to Support Industry 4.0 in Manufacturing

Mofolasayo *et al.* [16] have found that lean principles can help Industry 4.0 to achieve greater customer value and manufacturing excellence. Their study provided the desire for process efficiencies and waste reduction by small- and medium-sized enterprises (SMEs) and declares that lean practices can ease assist in accomplishing primarily implementation of Industry 4.0 through resource reduction, foundational digitalization, and automatic data processing followed by a stepwise approach to cyber-physical systems [16].

3.6.7 Implementation of Lean Practices in the Water Heater Manufacturing Industry for Value Adding

Barot *et al.* [17] have given a framework to implement lean practices using primary lean tools in water heater MIs. In their study, future state value stream map and string diagram have been produced and implemented to attainment improvements. Their study demonstrated the in-process improvements and potential time reductions as a result of modifications. The waiting time has decreased from 645 min to 462 min, productivity has increased from 50% to 70%, and material movement has minimized from 283.08 m to 153.92 m in their study [17]. It is recommended by them to utilize lean tools to eliminate waste in the production flow line.

3.6.8 Lean Practices in Indian Machine Tool Industries for Receiving Productivity

Eswaramoorthi *et al.* [17] have conducted a survey on lean practices in Indian machine tool industries. They conducted a study to determine low priority toward adapting lean practices and determine lean implementation in machine tool sector under talk time, lead time, value chain, value-added activity, etc. A survey is done to determine the status of the lean methods in machine tool production, which has a component manufacturing of the vehicle value chain. A machine tool–specific questionnaire tool has been built and verified. The survey data from the major machine tool makers have been examined, and the results have been provided. The results suggest that lean implementation in the machine tool sector remains in its early life, and it is required to develop lean thinking and cultures based on the encouragement of employees to take initiative to remove non valuable activities and to craft small improvements derived from their observations and experiences [17].

3.6.9 Lean Manufacturing (LM) Practices for Influencing Process-Based Innovation and Performance

Möldner *et al.* [18] have investigated the impact of LM strategies on developing process innovation and performance strains in MI. Their study suggested a moderately significant favorable effect of both technical and human lean approaches on the input and found the occurrence of incremental innovation and radical process performance as a result in manufacturing organizations. The same, in turn, derived improved businesses' operational performance as a result of process innovation. The findings refute the conception held by academics and managers that it is challenging for LM and innovation to coexist [18].

3.6.10 The Implementation of Lean Manufacturing in the Furniture Industry

Abu *et al.* [19] have given a review and analysis of the adaptation of lean in the furniture sector based on the motives, barriers, challenges, and factors in emerging economies. Their assessment provides a little overview of lean concerns through scenarios carried out in developing countries (the Indian scenarios). A crucial aspect that motivates the furniture industries to implement lean and the critical challenges that prevent companies from using LCs are evaluated. A complete review of the challenges that are faced by the furniture industries in the early stages of lean implementation was presented, where obstructions related with employee issues including lack of labor resources, lack of implementation know-how, and employee resistance to change are found crucial [19].

3.6.11 Implementation of Lean Manufacturing in the Electronics Industry

Venkat Jayanth *et al.* [20] have demonstrated a principle to implement LM in the electronic industry to increase productivity by reducing errors and improving the quality of the electronic product. Their study highlighted the implementation of lean methodology with many tools for continuously improving productivity and quality of electronic devices in the electronic industry. After the implementation in the electronic industry, it is found that the productivity and the level of quality has improved by 23% in comparison to the system previously exists [20].

3.7 Modeling of Lean Under Industrial and Manufacturing Sphere

The term "lean" is primarily originated in the MI and refers to the practice of reducing waste in a production process to improve productivity and effectiveness. However, the conception of lean has recently been applied to other fields, such as academic institutes, automotive industry, social fields, managerial structure, service-related organizations, economical analysis, and environmental science. The present chapter has exploited seven modeling depicted in Sections 3.7.1 to 3.7.7 under different spheres to show the competencies of lean practices and strategies toward reinforcing INM designs in different fields like automotive, manufacturing, academic, management, environmental, social, and economic. The chapter will help readers in understanding various LCs.

3.7.1 Lean Modeling in Manufacturing Industries

The modeling of lean in different types of MIs like cement, oil, leather, and iron, plays a very important role to minimize waste and enhance productivity. Nowadays, many industries have started implementing new tools and technology to produce better quality products for sustainability, where lean can be a positive tool. MIs embrace many challenges such as reducing operating costs, market instability, competitiveness, and demand turbulences. Lean practices support the sustainability of MI by stimulating them toward doing more with limited resources by eliminating non-essential activities to maintain their efficiency and profitability [11, 21]. Lean modeling in manufacturing strives toward deploying a systematic approach for reducing diverse waste under a manufacturing structure based on the recognition of non- critical motions and process segments while retaining convinced margins of control such as productivity and quality [10, 22].

3.7.2 Lean Modeling in Academic Institutes

The modeling of leanness is adopted in academic institutes to eliminate non–value-adding activities. Here, in academic institutes, the lack of skills, lack of practical knowledge, lack of technical capabilities, lack of resources, and insufficient trained and inexperienced teaching staff are evident hurdles, which are obstructions toward implementation of lean properly in the institute. The same are great challenges that occur from the side of students and academic management to handle for lean modeling in academic institutes. Lean practices support the academic institutes by removing various challenges and unskilled teaching staff and improving technology as well as practical knowledge to maintain efficiency and academic record.

3.7.3 Lean Modeling in Managerial Structure and Service-Related Organizations

In any type of managerial structure either industry- or service-based, the modeling of lean plays a very important role to give well-quality management and to remove non–value-adding work. Here, the lack of experienced management staff, lack of commitment, less support from senior managers, lack of empowerment of workers, and due to the non-continuation of managerial services not running continuously hamper the adaptation of lean practices, and, accordingly, models are needed to be designed to remove the aforesaid barriers to improve services and value-added strings and to reduce dispersion [23].

3.7.4 Lean Modeling in Social Fields

The model of leanness in society can give value-added provisions at the stage of handling social platform, social work, funds, frauds, working ethics, and making top class social services. Many social services are inefficient due to lack of knowledge, lack of financial support, and deficient implication of lean practices for removing waste and increasing effectiveness in the delivery of social things to customers. The implication of lean modeling is required to necessarily imply in streamlining procedures, removing useless steps or tasks, and trying to maximize the use of available resources [24].

3.7.5 Lean Modeling in Environmental Science

Lean modeling in the environmental contexts involves the integration of LCs and methodologies into environment to decrease wastes receiving by the environment due to unstrained activities, deprived energy sources, and unevaluated eco-mechanism. Lean principles such as waste identification, removal, and continuous improvement via considering respect of people habits in using eco-resources can be used in modeling environmental practices for waste management, energy efficiency, and resource conservation. Lean approaches, for example, can be used to optimize production processes to reduce waste and emissions and to streamline supply chains to reduce transportation-related carbon emissions for deprived environmental loads. The application of lean principles to environmental management is known as green-lean management or lean environmental management [25].

3.7.6 Lean Modeling in Economics

In economics, lean modeling is the application of lean ideas and methodologies under business chains and economic processes toward improving efficiency and reducing invaluable activities and profitability. Lean principles strives toward identification and establishment of analytical methods to apply a holistic view toward reducing different types of waste, continuous improvement, and ethical integrity under variety of operations for receiving economic dimensions in manufacturing, supply chain management, and customer services. Overall, applying LCs to economics can help businesses become more aggressive and competent to retain economical sustainability while also enhancing the community's economic and social outcomes [24].

3.7.7 Lean Modeling in the Automobile Industry

Lean manufacturing is a production-based approach that focuses on smart manufacturing based on reduction of unutilized streams and dimensions for maximizing efficiency and quality. Lean manufacturing is widely employed in the automotive sector, where JIT manufacturing, quality improvement, reducing defects, identification of obsolescence, implication of new process, equipments, and integration of resources for maximum their utility are some of the main principles of the lean model. Implementing LM practices in the automobile sector requires a strong focus on teamwork, communication, and participation. The results of lean implementation in the automotive industry increases production, reduce cost, and reduce product delivery lead time. For modeling lean, it is needed to discover areas of wastes and create methods to remove them by cooperating and exchanging knowledge and ideas [26].

It is always significant to model the characteristics of the system based on the exploitation of critical factors and drivers [27, 28]. Accordingly, to make modeling successful, it is required to explore wide dimensions of research practices and their hierarchical measures [29, 30]. The same is required to attain sustainability and the same can aids in receiving outputs crucially and effectually [31, 32]. Identification of circular measures for modeling always remains a prime focus [33, 34], where deficient and unacceptable results can be received due to wrong information and implication of insignificant practices [35, 36]. It is essential to expand modeling and tools for receiving superior outputs for resources and grabbing more productivity from manufacturing and industrial applications. Additionally, it is productive to appraise numerous factors for assuming effectiveness from involved resources.

3.8 Conclusions

The current chapter investigated the technical domain of lean literature to demonstrate the importance of lean practices and strategies for developing INM efficiencies. In the present chapter, 19 momentous lean practices are described in the first segment and defined five critical lean practices in the second segment that are radical to be utilized in various fields. Additionally, five lean strategies are presented for its easy implementation in the MI. A total of 11 attainments are discussed after realizing its implementation in MI, furniture industry, electronics industry, and Indian machine tool industry. Here, a total of seven lean models are presented to highlight the lean modeling in various sectors like in MI, academic institutes, management, social, environment, ecology, and automotive industry.

After understanding the attainments of the 11 cases, it is found that LCs needed momentous measures and parameters for conducting an empirical review and for defining prominent lean strategies. The authors found indeed necessary to investigate various LCs in the context of various lean strategies. A wide extent of exploratory research that can aid in determining the current state of LM and its methods of implementation is needed to be explored crucially. An evident research that can discuss the LM concept, lean waste, lean strategies, lean barriers, and the lean implementation cycle is needed. The collection and development of relevant research is needed thoroughly based on the literature review and expert judgments to clarify the status of LM and its strategies. The present chapter finally concluded that the implementation of LM is prominent to eliminate waste, increase productivity, and reduce the cost of production in the many dimensional industries and other sectors like management and academic institutes.

References

1. Chahala, V. and Narwal, M.S., An empirical review of lean manufacturing and their strategies. *Manage. Sci. Lett.*, 7, 7, 321–336, 2017, doi: 10.5267/j.msl.2017.4.004.
2. Sharma, M.K. and Lata, S., Effectuation of Lean Tool '5S' on Materials and Work Space Efficiency in a Copper Wire Drawing Micro-Scale Industry in India. *Mater. Today Proc.*, 5, 2, 4678–4683, 2018, doi: 10.1016/j.matpr.2017.12.039.
3. Kafuku, J.M., Factors for effective implementation of lean manufacturing practice in selected industries in Tanzania. *Proc. Manuf.*, 33, 351–358, 2019, doi: 10.1016/j.promfg.2019.04.043.
4. Palange, A. and Dhatrak, P., Lean manufacturing a vital tool to enhance productivity in manufacturing. *Mater. Today Proc.*, 46, 729–736, 2021, doi: 10.1016/j.matpr.2020.12.193.
5. Hartini, S. and Ciptomulyono, U., The relationship between lean and sustainable manufacturing on performance : literature review. *Proc. Manuf.*, 4, 38–45, 2015, doi: 10.1016/j.promfg.2015.11.012.
6. Wahab, A.N.A., Mukhtar, M., Sulaiman, R., A Conceptual Model of Lean Manufacturing Dimensions. *Proc. Technol.*, 11, 1292–1298, 2013, doi: 10.1016/j.protcy.2013.12.327.
7. Barot, R.S., Raval, K., Berawala, H.S., Patel, A., Implementation of lean practices in water heater manufacturing industry. *Mater. Today Proc.*, 38, 5, 2227–2234, 2020, doi: 10.1016/j.matpr.2020.06.304.
8. Riezebos, J. and Klingenberg, W., Advancing lean manufacturing, the role of IT. *Comput. Ind.*, 60, 4, 235–236, May 2009, doi: 10.1016/j.compind.2009.01.005.
9. Monteiro, C., Ferreira, L.P., Fernandes, N.O., Sá, J.C., Ribeiro, M.T., Silva, F.J.G., Improving the machining process of the metalworking industry using the lean tool SMED. *Proc. Manuf.*, 41, 555–562, 2019, doi: 10.1016/j.promfg.2019.09.043.
10. Karam, A.A., Liviu, M., Cristina, V., Radu, H., The contribution of lean manufacturing tools to changeover time decrease in the pharmaceutical industry. A SMED project. *Proc. Manuf.*, 22, 886–892, 2018, doi: 10.1016/j.promfg.2018.03.125.

11. Nallusamy, S., Execution of lean and industrial techniques for productivity enhancement in a manufacturing industry. *Mater. Today: Proc.*, 37, 568–575, 2020, doi: 10.1016/j.matpr.2020.05.590.
12. Sahu, A.K., Raut, R.D., Gedam, V.V., Cheikhrouhou, N., Sahu, A.K., Lean–agile–resilience–green practices adoption challenges in sustainable agri-food supply chains. *Bus. Strategy Environ.*, 32, 6, 3272–3291, 2022. https://doi.org/10.1002/bse.3299.
13. Sahu, A.K., Sharma, M., Raut, R.D., Sahu, A.K., Sahu, N.K., Antony, J., Tortorella, G.L., Decision-making framework for supplier selection using an integrated MCDM approach in a lean-agile-resilient-green environment: evidence from Indian automotive sector. *TQM J.*, 35, 4, 964–1006, 2023.
14. Singh, C., Singh, D., Khamba, J.S., Exploring an alignment of lean practices on the health and safety of workers in manufacturing industries. *Mater. Today: Proc.*, 47, 6696–6700, 2020, doi: 10.1016/j.matpr.2021.05.116.
15. Hemalatha, C., Sankaranarayanasamy, K., Durairaaj, N., Lean and agile manufacturing for work-in-process (WIP) control. *Mater. Today: Proc.*, 46, 10334–10338, 2021, doi: 10.1016/j.matpr.2020.12.473.
16. Mofolasayo, A., Young, S., Martinez, P., Ahmad, R., How to adapt lean practices in SMEs to support Industry 4.0 in manufacturing. *Proc. Comput. Sci.*, 200, 934–943, 2022, doi: 10.1016/j.procs.2022.01.291.
17. Eswaramoorthi, M., Kathiresan, G.R., Prasad, P.S.S., Mohanram, P.V., A survey on lean practices in Indian machine tool industries. *Int. J. Adv. Manuf. Technol.*, 52, 9–12, 1091–1101, 2011, doi: 10.1007/s00170-010-2788-y.
18. Möldner, A.K., Garza-Reyes, J.A., Kumar, V., Exploring lean manufacturing practices' influence on process innovation performance. *J. Bus. Res.*, 106, 233–249, 2020, doi: 10.1016/j.jbusres.2018.09.002.
19. Abu, F., Gholami, H., Mat Saman, M.Z., Zakuan, N., Streimikiene, D., The implementation of lean manufacturing in the furniture industry: A review and analysis on the motives, barriers, challenges, and the applications. *J. Clean. Prod.*, 234, 660–680, 2019, doi: 10.1016/j.jclepro.2019.06.279.
20. Venkat Jayanth, B., Prathap, P., Sivaraman, P., Yogesh, S., Madhu, S., Implementation of lean manufacturing in electronics industry. *Mater. Today: Proc.*, 33, 23–28, 2020, doi: 10.1016/j.matpr.2020.02.718.
21. Babalola, O., Ibem, E.O., Ezema, I.C., Implementation of lean practices in the construction industry: A systematic review. *Build. Environ.*, 148, 34–43, 2019, doi: 10.1016/j.buildenv.2018.10.051.
22. Sahu, A.K., Kottala, S.Y., Narang, H.K., Rajput, M.S., Intertwining green SCM- and agile SCM-based decision-making framework for sustainability using GIVTFNs. *J. Glob. Oper. Strateg. Sourc.*, ahead-of-print, 2022, https://doi.org/10.1108/JGOSS-06-2022-0060.

23. Mayr, A., Weigelt, M., Kühl, A., Grimm, S., Erll, A., Potzel, M., Franke, J., Lean 4.0-A conceptual conjunction of lean management and Industry 4.0. *Proc. CIRP*, 72, 622–628, 2018, doi: 10.1016/j.procir.2018.03.292.
24. Mohaghegh, M., Blasi, S., Größler, A., Dynamic capabilities linking lean practices and sustainable business performance. *J. Clean. Prod.*, 322, 129073, Nov. 2021, doi: 10.1016/J.JCLEPRO.2021.129073.
25. Chen, X., Kurdve, M., Johansson, B., Despeisse, M., Enabling the twin transitions: Digital technologies support environmental sustainability through lean principles. *Sustain. Prod. Consum.*, 38, 13–27, 2023, doi: 10.1016/j.spc.2023.03.020.
26. Prakash, D. and Kumar, C., Implementation of Lean Manufacturing Principles in Auto Industry. *Ind. Eng. Lett.*, 1, 1, 56–61, 2011, http://iiste.org/Journals/index.php/IEL/article/view/640.
27. Sahu, A.K., Sahu, N.K., Sahu, A.K., Laminating STRATH block chain technology-SWOT architectures to endure business strategy between digital transformation, firms and supply chains capabilities for sustainability. *J. Clean. Prod.*, 383, 1–20, 2023, https://doi.org/10.1016/j.jclepro.2022.135531.
28. He, Z., Ma, X., Luo, J., Sahu, A.K., Sahu, A.K., Sahu, N.K., Exploitation of the advanced manufacturing machine tool evaluation model under objective-grey information: a knowledge-based cluster with the grey relational analysis approach. *Grey Syst.: Theory Appl.*, 11, 3, 394–417, 2021.
29. Sahu, A.K., Sahu, A.K., Sahu, N.K., Benchmarking of Advanced Manufacturing Machines Based on Fuzzy-TOPSIS Method, in: *Theoretical and Practical Advancements for Fuzzy System Integration*, pp. 309–350, 2017.
30. Bag, S., Sahu, A.K., Kilbourn, P., Pisa, N., Dhamija, P., Sahu, A.K., Modeling barriers of digital manufacturing in a circular economy for enhancing sustainability. *Int. J. Product. Perform. Manage.*, 71, 3, 833–869, 2021, https://doi.org/10.1108/IJPPM-12-2020-0637.
31. Kang, D., Prabhu, M., Ahmed, R.R., Zhang, Z., Sahu, A.K., Digital-IIoTs spheres approach toward public development: an exploiting fuzzy-grey mathematical modeling of IIoTs spheres. *Grey Syst.: Theory Appl.*, 12, 2, 389–416, 2022.
32. Sahu, A.K., Sahu, N.K., Sahu, A.K., Fuzziness: A Mathematical Tool, in: *Theoretical and Practical Advancements for Fuzzy System Integration*, pp. 1–30, 2017.
33. Bag, S., Viktorovich, D.A., Sahu, A.K., Sahu, A.K., Barriers to adoption of blockchain technology in green supply chain management. *J. Glob. Oper. Strateg. Sourc.*, 14, 1, 104–133, 2021.
34. Sahu, A.K., Sahu, N.K., Sahu, A.K., Benchmarking CNC machine tool using hybrid fuzzy methodology a multi indices decision making approach. *Int. J. Fuzzy Syst. Appl.*, 4, 2, 28–46, 2015.

35. Sahu, A.K., Sahu, N.K., Sahu, A.K., Rajput, M.S., Narang, H.K., An Investigation Tool for Mounting Sustainable Practice: Modeling Using GIVTFNs in an Indian Context. *Int. J. Decis. Support Syst. Technol.*, 12, 2, 25–49, 2020.
36. Wang, W., Huang, L., Zhu, Y., Jiang, L., Sahu, A.K., Sahu, A.K., Sahu, N.K., Decision support system toward evaluation of resilient supplier: A novel fuzzy gain-loss computational approach. *Kybernetes*, 49, 6, 1741–1765, 2019.

4

A Qualitative Study to Rank Non-Conventional Energy Sources for Industrial Sustainability and Energy Management Decisions Using MoSCoW Prioritization Method

Pankaj Singh[1*], Ruchi Kushwaha[2], Jyoti Kushwaha[3] and Ajit Kumar Singh[4]

[1]Institute of Business Management, GLA University, Mathura, India
[2]School of Studies in Management, Jiwaji University, Gwalior, India
[3]Department of Home Management, Govt. KRG PG College, Jiwaji University, Gwalior, India
[4]Jaipuria Institute of Management, Jaipur, India

Abstract

Energy is pre-essential to financial improvement for any country. Power generation and energy utilization are vital to the monetary turn of events. The Earth has no more ability to retain and process the immense waste created from the utilization of fossil energy as poisonous gases are upsetting the biological system. An endeavor has been made in the current chapter to look at the job of the board of energy and the allied directories in the Indian setting. The essential goal of this chapter is to examine the extent of non-traditional energy sources in satisfying the future energy necessity in the context of non-conventional energy sources. The examination procedure adapted for this chapter is a qualitative study based on the expert interview analysis by utilizing Must have, Should have, Could have and Would not have (MoSCoW) method-based requirement prioritization. Consequently, the conversation of this chapter is connected with the capability of the board of energy to see that the ongoing energy needs of the nation are palatably met at least expense as of now and in the future to broaden conceivably. This chapter aims to explore and rank the non-conventional energy requirements in

**Corresponding author*: pankajabviiitm@gmail.com

Atul Kumar Sahu, Rakesh D. Raut, Rohit Raja, Anoop Kumar Sahu and Nitin Kumar Sahu (eds.) Industrial and Manufacturing Designs: Quantitative and Qualitative Analysis, (105–138) © 2024 Scrivener Publishing LLC

industrial systems using the MoSCoW prioritization–based qualitative method. In this regard, a focused expert interview has been accomplished among the non-conventional energy experts in industrial area of Delhi-NCR region of India. The outcomes of the MoSCoW method reported that non-conventional energy source–related factors like resource availability, financial-economic viability, and institutional preparedness are prioritized under the "must have" category, whereas factors like appropriateness of the technology, energetic feasibility, awareness, and users' training are placed under the "should have" category. Subsequently, factors like environmental sustainability and availability of the after sales and service come under the "could have" category. The findings are helpful for policymakers, engineers, and energy managers in order to prioritize the non-conventional energy sources as per their industrial needs.

***Keywords*:** Energy management, energy policy, non-conventional energy sources, industrial sustainability, qualitative study, requirement prioritization

4.1 Introduction

For any nation to develop economically, energy is a must. Energy is a crucial component of economic growth, and the future development of a country is closely related to the energy supply. Economic growth depends on both the production of electricity and the consumption of energy. Energy management is becoming a bigger issue for both developed and developing countries [1]. It is concerned that the world is using up the large but finite supply of fossil fuel extraction and production, and many businesses and countries are paying more attention to the problem of energy management. The idea of energy management is neither novel nor difficult. Energy should be viewed as a business expense on par with other expenses like labor and raw materials. India is extremely reliant on fossil fuels to meet its energy needs [2]. The main source of greenhouse gases, which cause pollution, is fossil fuels. India's need for crude oil is largely satisfied by imports. Future economic burdens from increasing fossil fuel imports would also cause an energy crisis in the power sector. Dependence on fossil fuels and excessive resource usage will exacerbate climatic changes, which already pose a threat to many ecosystems throughout the world, raise sea levels, and perhaps have an impact on food production, resulting in instability and conflict driven by resource scarcity. Investment in the alternative energy sector is the only realistic way to improve energy management, reduce coal usage, reduce oil imports, and safeguard the environment [3]. The non-conventional sources are able to solve the decentralized energy supply issue and contribute to maintaining a cleaner environment. Due to the recent global energy crisis, widespread electricity theft, and privatization of universal energy supplies, energy management has become increasingly important [4].

For any nation to develop economically, energy is a must. Energy management should always be accessible because blackouts can increase manufacturing costs and negatively impact human well-being and livelihood. The wealthy and industrialized nations have a good understanding of the necessity for long-term energy security planning. Energy is a crucial component of economic growth and industrial sustainability, and the future development of a country is closely related to the energy supply. Any nation's development is inversely correlated with its energy usage. India is experiencing an energy crisis, which is slowing down its rate of industrial growth and economic expansion. The construction of new power plants depends heavily on the importation of extremely volatile fossil fuels, which is unsuitable for a sustainable environment. Therefore, it is crucial for this country's policymakers to address the energy crisis by making the most use of the plentiful renewable energy supplies. Non-conventional energy sources will aid India in combating climate change in addition to enhancing a steady energy supply [5].

In terms of energy consumption, India is one of the fastest expanding economies. It currently consumes the fifth largest amount of energy globally, and, by 2030, it will consume the third largest amount. India uses roughly 580 W of energy per person on average, which is far less than developed nations like the United States, Europe, Russia, and Japan. However, because of the strong economic growth and extensive industrialization, this number is anticipated to rise quickly [6].

Energy is a critical contribution to financial development, and there is a nearby connection between the accessibility of energy and the future development of a country. Energy is a growing concern for developing worry for rich and arising countries. Many companies and nations are increasing attention to the issue of energy management because they dread that the world is rapidly spending the immense yet limited measure of petroleum derivative extraction and creation. The board of energy is neither a new concept nor confounded. Energy ought to be viewed as a business cost like another business cost including unrefined substance and work. India intensely relies upon petroleum products toward fulfilling its power need. Petroleum products are the greatest wellspring of greenhouse gases, which makes contamination. India relies intensely upon imports for meeting its unrefined petroleum prerequisite. Developing import of petroleum products could make ever more prominent monetary weight and furthermore made an energy emergency in the power area [7].

Non-renewable energy source reliance and inefficient utilization of assets will deteriorate environmental changes, which, as of now, take steps to jeopardize large numbers of the world's biological systems, raise ocean levels, and influence food creation. The same, in turn, will conceive prompting asset shortage-driven shakiness and struggle. The main down-to-earth choices for upgrading the board of energy and decreasing coal utilization as well as oil import and safeguarding climate with interest in the elective energy area. The non-customary sources are equipped for taking care of the issue of energy supply in a decentralized way and aid in supporting a cleaner climate. Energy executives have accepted impressive significance in the upcoming years attributable to the worldwide energy emergency, high pilferage of power in the nation, and privatization of general energy assets. The cutting edge intricate and quickest developing industrialized world with an extremely enormous populace requires expanding the contribution of energy to make a big difference for it.

The non-conventional energy sources, also known as renewable energy sources or green energy, are essential components of a sustainable and environmentally friendly energy future. Unlike conventional energy sources such as fossil fuels (coal, oil, and natural gas), non-conventional energy sources are replenishable and have a minimal impact on the environment. These sources harness natural processes to generate power and are considered crucial in mitigating climate change and reducing our dependence on finite fossil fuels [8].

4.1.1 Major Non-Conventional Energy Sources

Non-conventional energy sources are renewable energy sources because they are found naturally, are limitless, and can be utilized repeatedly to

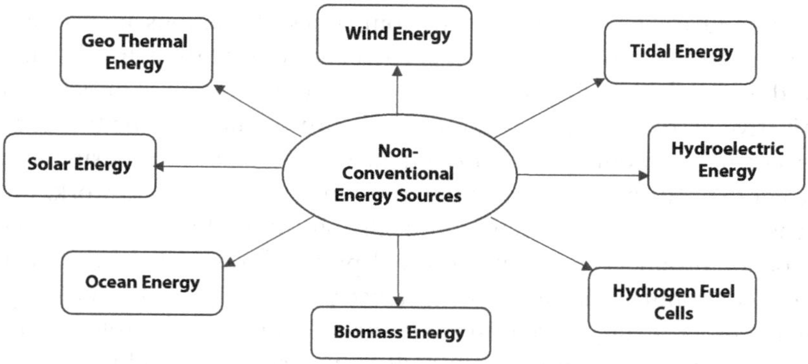

Figure 4.1 Major non-conventional energy sources.

generate energy. These occur naturally in an illimitable quantity and take a little time to mature. Renewable energy has the potential to provide a decentralized solution to the issue of reliable electricity supply and support a cleaner environment. These include biomass, tidal, ocean, wind, and water energy. Renewable energy is a natural gift that is readily accessible to all people [9]. The key non-conventional energy sources are depicted in Figure 4.1.

4.1.1.1 Solar Energy

Solar power harnesses energy from the sun's radiation using photovoltaic (PV) cells or solar thermal systems. It is abundant, is clean, and can be harnessed both at the utility scale and in distributed applications like rooftop solar panels.

4.1.1.2 Wind Energy

Wind turbines capture kinetic energy from the wind and convert it into electricity. Wind farms can be found onshore and offshore, making use of the Earth's constant air movement.

4.1.1.3 Hydroelectric Power

This source relies on the gravitational potential energy of water. Dams or waterfalls are used to generate electricity as water flows downhill, turning turbines in the process.

4.1.1.4 Biomass Energy

Biomass includes organic materials such as wood, agricultural residues, and even municipal waste. It can be converted into biofuels like ethanol or used to generate heat and electricity through processes like combustion or gasification.

4.1.1.5 Geothermal Energy

Geothermal power taps into the Earth's internal heat by harnessing steam or hot water from underground reservoirs. This heat can be used directly for heating buildings or for electricity generation.

4.1.1.6 *Tidal and Wave Energy*

Tidal energy harnesses the gravitational forces between the Earth, the moon, and the sun to generate power as tides rise and fall. Wave energy, on the other hand, captures energy from ocean waves.

4.1.1.7 *Hydrogen Fuel Cells*

This is not a direct energy source; hydrogen can be considered as a non-conventional energy carrier. It can be produced using renewable energy and used in fuel cells to generate electricity with water as the only byproduct.

4.1.2 Significance of Non-Conventional Energy Source

Non-conventional energy sources offer numerous advantages, including reducing greenhouse gas emissions, enhancing energy security, and decreasing our reliance on fossil fuels, which are finite and contribute to climate change. However, they also present challenges such as intermittency (in the case of wind and solar), high upfront costs, and the need for efficient energy storage solutions. As technology advances and economies of scale are realized, non-conventional energy sources are becoming increasingly vital in our efforts to create a more sustainable and resilient energy system. The non-conventional energy sources are significant because they contribute to a cleaner environment, enhance energy security, stimulate economic growth, promote resource sustainability, mitigate climate change, and foster technological advancement [10]. Their adoption and expansion are essential for building a more resilient and sustainable energy system

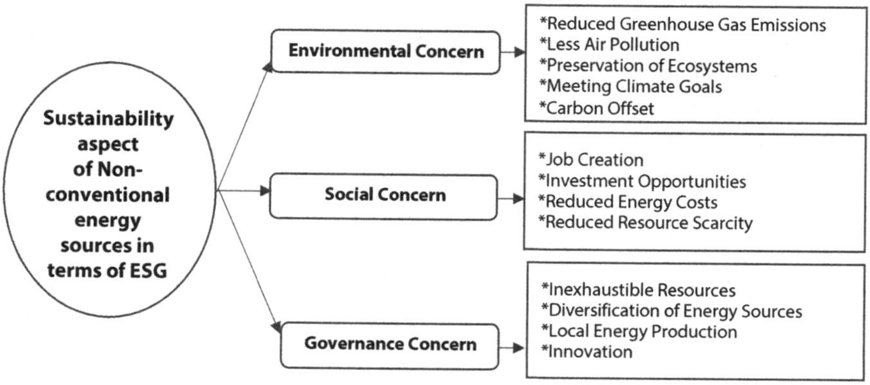

Figure 4.2 Consequence of non-conventional energy sources on industrial sustainability.

globally as depicted in Figure 4.2. The significance of non-conventional energy sources, also known as renewable energy sources or green energy, is profound and far-reaching. These sources play a crucial role in addressing a wide range of global challenges and are vital for a sustainable and environmentally responsible future. Some of the key significances of non-conventional energy sources are mentioned in the following.

4.1.2.1 *Environmental Benefits*

- **Reduced Greenhouse Gas Emissions:** Non-conventional energy sources produce little to no greenhouse gas emissions, when generating electricity or heat. This helps to combat climate change by reducing the carbon footprint of energy production.
- **Less Air Pollution:** Unlike fossil fuels, renewable energy sources do not release harmful pollutants such as sulfur dioxide, nitrogen oxides, or particulate matter, which have adverse effects on air quality and public health.
- **Preservation of Ecosystems:** Renewable energy projects typically have a smaller impact on natural habitats compared to conventional energy sources like coal mining or oil drilling. This helps protect biodiversity and ecosystems.

4.1.2.2 *Energy Security*

- **Diversification of Energy Sources:** Relying on a variety of non-conventional energy sources reduces dependence on finite fossil fuels, making energy supplies more secure and less vulnerable to price fluctuations and geopolitical tensions.
- **Local Energy Production:** Many renewable energy systems can be deployed at the local level, reducing the need for long-distance transportation of energy resources and enhancing energy resilience during emergencies.

4.1.2.3 *Economic Benefits*

- **Job Creation:** The renewable energy sector has been a significant source of employment, from manufacturing and installation to research and development.

- **Investment Opportunities:** Non-conventional energy sources have attracted substantial investment, fostering innovation and economic growth in related industries.
- **Reduced Energy Costs:** As technology advances and economies of scale are realized, the cost of renewable energy generation has decreased, making it increasingly competitive with fossil fuels.

4.1.2.4 *Resource Sustainability*

- **Inexhaustible Resources:** Renewable energy sources are inherently sustainable as they rely on naturally occurring processes like sunlight, wind, and geothermal heat that are not depleted over time.
- **Reduced Resource Scarcity:** Non-conventional energy sources can help alleviate the strain on finite resources such as coal, oil, and natural gas, which are subject to depletion.

4.1.2.5 *Climate Change Mitigation*

- **Meeting Climate Goals:** The transition to non-conventional energy sources is a critical component of global efforts to limit global warming to safe levels, as outlined in international agreements like the Paris Agreement.
- **Carbon Offset:** Utilization of renewable energy for electricity generation and transportation can significantly reduce carbon emissions and help countries and regions to meet their carbon reduction targets.

4.1.2.6 *Technological Advancement*

- **Innovation:** The development and deployment of non-conventional energy technologies drive innovation in energy storage, grid management, and energy efficiency, benefiting various sectors beyond energy production.

4.1.3 Scope of Non-Conventional Energy in Industrial Sustainability

Non-conventional energy sources play a critical role in enhancing industrial sustainability and influencing energy management decisions within the industrial sector [11]. Their crucial role in this context is multifaceted and can be summarized as follows:

- **Cost Reduction:** Non-conventional energy sources, particularly solar and wind power, can help industries to reduce their energy costs over the long term. By generating electricity on-site or purchasing it from renewable sources, industries can hedge against volatile fossil fuel prices and reduce their operational expenses.
- **Energy Security:** Incorporating renewable energy into industrial operations provides greater energy security. Industries can become less reliant on external energy suppliers, mitigating the risks associated with supply disruptions or price fluctuations in traditional energy markets.
- **Environmental Compliance:** Many industries are subject to strict environmental regulations and emissions targets. Non-conventional energy sources produce minimal to no greenhouse gas emissions and air pollutants, helping industries to meet regulatory requirements and reduce their environmental impact.
- **Resource Efficiency:** Renewable energy technologies, such as solar panels and wind turbines, can be integrated into industrial processes to improve resource efficiency. They can generate power on-site, reducing transmission and distribution losses associated with centralized power generation.
- **Corporate Social Responsibility (CSR):** Utilizing non-conventional energy sources aligns with CSR goals. Many consumers and investors are increasingly conscious of a company's environmental footprint, and a commitment to green energy can enhance a company's reputation and attractiveness.
- **Energy Independence:** By generating their own renewable energy, industries can reduce their dependence on external energy suppliers. This independence allows for greater control over energy production and consumption patterns.
- **Resilience:** Renewable energy sources, when combined with energy storage solutions like batteries, provide industrial facilities with greater resilience during grid outages or emergencies. These systems can ensure critical operations to continue even when the grid is down.
- **Long-Term Planning:** Non-conventional energy sources offer long-term energy solutions. Industries can make more

informed energy management decisions by investing in technologies with predictable lifespans and operating costs.

- **Reduced Peak Demand Charges:** Many industrial facilities incur peak demand charges based on their highest electricity usage during specific periods. By using renewable energy to offset peak demand, industries can reduce these charges, resulting in significant cost savings.
- **Market Competitiveness:** Industries that demonstrate a commitment to sustainability and clean energy can gain a competitive advantage in the market. It can open up opportunities to access green supply chains, attract environmentally conscious customers, and stay ahead of regulatory changes.
- **Carbon Emissions Reduction:** Non-conventional energy sources are instrumental in reducing carbon emissions from industrial operations. This is particularly important as industries are under increasing pressure to decarbonize and to achieve net-zero emissions targets.

4.1.4 Problem Formulation

It is recognized from review of existing literatures on the boundaries of non-conventional energy management decision and industrial sustainability that almost there is no concrete qualitative analysis that is previously carried out by scholars. It provides a sufficient analysis gap to perform a qualitative analysis on chosen topic. Furthermore, this analysis is significant in futuristic scenario of forthcoming energy crisis for industrial sustainability. Therefore, in this chapter; the authors have articulated their analysis problem as "A qualitative study to rank non-conventional energy sources for industrial sustainability and energy management decisions".

4.1.5 Objectives of Chapter

In order to address the aforementioned queries on non-conventional power–based energy management decision and industrial sustainability, the following objectives are formulated:

- to visualize the significance, challenges, and impact of non-conventional energy sources in perspective of industrial sustainability;

- to analyze the role of non-conventional energy sources in industry related energy management decision; to prioritize and rank the non-conventional energy sources through industrial experts as per the need of Industrial sustainability.

4.1.6 Methodology of Chapter

In the present chapter, the authors have employed a requirement prioritization–based qualitative methodology to carry out analysis, i.e., the MoSCoW method [12]. In the initial phase, a pertinent literature review has been conducted to detect the relevant non-conventional energy management decision based on existing scale that will further utilize to perform an expert interview. In this regard, a nine-dimensional renewable energy technology–based scale has been adopted as developed by Chandrasekar and Kandpal [13]. In the next phase, a focused interview with energy management experts from both industry and academia has been carried out in order to prioritize and rank the different non-conventional energy sources as per the needs of industries.

4.1.7 Organization of Chapter

The chapter is structured in such a way to easily highlight the rank, importance, allied challenges, and impact of non-conventional energy sources in perspective of industrial sustainability. Here, in the first segment, the authors presented a brief overview of the need and significance of non-conventional energy in industrial sustainability. In the second segment, a review of associated studies in the area of non-conventional energy management decisions is presented. In the third segment, current scenario of responsible leadership is discussed, and, in the fourth segment, the qualitative data analysis is presented. Findings and discussion part of chapter is deliberated in the fifth segment. At last, the conclusion is portrayed and further extensions are determined.

4.2 Review of Literature

A review of the literature is conducted on non-conventional energy sources to provide a comprehensive understanding of the current state of research, developments, and trends in the field of renewable and sustainable energy as presented in the following.

4.2.1 Solar Energy

Solar energy is one of the most abundant and widely studied non-conventional energy sources. Solar PV technology, concentrating solar power, and solar thermal systems have been extensively researched for their efficiency and practical applications. Various researchers have made significant contributions to the understanding of solar cell technology, where solar PV technology advancements are efficiency improvements and cost reductions, solar thermal systems for electricity generation and thermal applications, integration of solar energy into urban environments and building designs, and solar resource assessment and forecasting for optimal system design and operation. This review provides an overview of the development, technologies, and applications of PV solar energy [14].

4.2.2 Wind Energy

Wind energy is a mature and reliable source of renewable energy. Many studies have explored the advancements in wind turbine technology and their integration into the existing power grid. Many studies have stressed on developing turbines design and appropriate site selection to receive maximum energy and to provide clean electricity. In wind energy, turbines translate the kinetic energy of wind into mechanical or electrical energy that can be used in the form of power. Thus, it is found imperative to develop sustainable practices related to wind turbine technology, wind energy potential, and environmental impacts of wind farms for green objectives [15].

4.2.3 Hydropower

Hydropower has been a traditional source of renewable energy. Recent research has focused on the environmental impact assessment of hydropower projects and the optimization of hydropower generation systems. It is found significant to explore the challenges and opportunities associated with hydropower development, including environmental and social considerations [16].

4.2.4 Biomass and Bioenergy

Biomass and bioenergy technologies have gained attention due to their potential to reduce greenhouse gas emissions. Several researches have explored the production methods of biofuels and their environmental implications. It is significant to examine the status and future prospects of biomass energy in China, focusing on feedstock availability and conversion technologies [17].

4.2.5 Geothermal Energy

Geothermal energy is usually derived from the Earth's heat and is a consistent and sustainable source of power. Various studies reported the technical aspects of geothermal energy production. It is significant to provide a global overview of geothermal resources and reserves for emphasizing the importance of this renewable energy source [18].

4.2.6 Tidal and Wave Energy

Tidal and wave energy are emerging fields of study. Notable studies investigated the potential and challenges associated with harnessing energy from ocean tides and waves. Focusing on wave energy, it is required to discuss recent developments, challenges, and future prospects for harnessing energy from ocean waves [19].

The literature on non-conventional energy sources reflects a growing global interest in transitioning to sustainable and clean energy systems. Researchers explore technical innovations, environmental considerations, economic implications, and policy measures to accelerate the adoption of renewable energy technologies and to reduce our reliance on fossil fuels. This body of knowledge is crucial for informing energy policies, investment decisions, and sustainable development efforts worldwide.

4.3 Current Scenario of Non-Conventional Sources in Industrial Sustainability

The non-conventional energy sources are highly significant for industrial sustainability and energy management decisions. They offer cost savings, energy security, environmental compliance, and other strategic advantages that enable industries to operate more efficiently, responsibly, and competitively in an evolving energy landscape. Incorporating renewable energy into industrial processes and energy management strategies is an essential step toward a sustainable and resilient industrial sector.

Energy management and security is a strategically important for industrial sustainability. Achieving goals for national economic development and raising people's standard of living are significantly impacted by energy security. It is in India's long-term interest to increase the proportion of new and renewable energy in the fuel mix. Due to demographic pressure in densely populated locations across the nation, energy crisis issues are critical. One should always take into account non-fossil renewable energy sources due to their ability to regenerate for sustainability [20]. The below-mentioned sections represent the current scenario, industrial sustainability, and prospects of non-conventional sources.

4.3.1 Wind Energy

- Wind turbine technology and design enhancements are required for higher efficiency and reliability.
- Offshore wind farms and their potential development are needed for generating large amounts of clean energy.
- Wind resource assessment, wind modeling, and wind energy forecasting models are needed to enhance efficiency.
- Social and environmental impacts of wind energy projects, including community engagement, are needed to be evaluated.

4.3.2 Hydroelectric Power

- Small-scale and micro-hydro projects are needed for rural electrification and off-grid applications.
- Environmental effects of large-scale hydroelectric dams and efforts to mitigate these impacts need evaluation.
- Pumped hydro storage for grid stability and energy storage requires sustainable development.

4.3.3 Biomass Energy

- Biomass feedstock sources, availability, and sustainability dimensions are needed to be assessed.
- Bioenergy conversion technologies, including combustion, gasification, and pyrolysis, are needed to be modeled for optimization.
- Biogas production from organic wastes and agricultural residues should be developed.
- Challenges related to land use, food security, and the competition for biomass resources should be identified.

4.3.4 Geothermal Energy

- Geothermal resource exploration and assessment techniques are needed to be developed.
- Enhanced geothermal systems (EGSs) and their potential for expanding geothermal power production should be identified.
- Geothermal heat pumps for space heating and cooling applications need development and adaptation.
- Geothermal energy as a base load power source with minimal emissions should be developed.

4.3.5 Tidal and Wave Energy

- Developments in marine energy technologies, including tidal stream and wave energy converters, are needed.
- Challenges associated with deploying and maintaining marine energy systems in harsh ocean environments should be identified.
- Environmental impacts and mitigation strategies for marine energy projects need assessments.

4.3.6 Hydrogen Fuel Cells

- Hydrogen production methods, including electrolysis, reforming, and biological processes, need sustainable development.
- Fuel cell technology advancements for various applications, such as transportation and stationary power, are needed.
- Hydrogen storage, distribution challenges, and solutions need persistent evaluation.

4.3.7 Energy Storage

- Advances in energy storage technologies, including batteries, supercapacitors, and thermal storage, are needed.
- Grid-scale energy storage systems and their role in integrating variable renewable energy sources need development.
- Economic and technical aspects of energy storage deployment should be optimized.

4.3.8 Policy and Regulation

- Regulatory frameworks and market mechanisms for promoting renewable energy adoption are needed.

4.3.9 Integration and Grid Management

- Strategies for integrating high levels of renewable energy into existing power grids should be devised.
- Demand-side management, smart grid technologies, and grid resilience need development.
- Modeling and simulation tools for optimizing renewable energy integration need assessment.

4.4 Overview of Indian Non-Conventional Energy Sector

India has an abundance of renewable energy resources, and numerous technologies have recently been created to maximize their potential. The nation already has several industrial bases for a range of renewable energy sources, including biomass, solar, thermal, wind, and tidal energy. In order to address the nation's increasing energy demands, the rapidly diminishing non-renewable conventional energy sources, and the energy requirements of rural areas, the Indian government is promoting the use of new and non-conventional energy sources [21]. The potential and scope of non-conventional energy sources are presented in Table 4.1.

The Ministry of Non-Renewable Energy formerly known as the department of non-conventional energy sources was established in 1982 and has since grown into a full-fledged ministry responsible for the development of new and renewable technology. The provision of facilities for the well-being of the weakest segments of society has shown this energy to be especially helpful for industrial sustainability specifically in isolated and hilly locations [22]. The expected usage of electricity generation from non-conventional energy sources is mentioned in Table 4.2.

Table 4.1 Scenario of non-conventional energy management in India.

S. no.	Type of non-conventional energy sources	Status	Potential	Remarks
1	Solar energy	Close to feasibility; marginally economic in some areas	Large in favorable areas	Useful only in areas of high direct sunlight
2	Wind energy	Feasible, but still scope for further development	Limited by site availability	Likely market for small turbines, role in large scale is not clear
3	Geothermal energy	Economic now in some area: room for further technological development	Ultimate large	Also a very useful source of low grad heat
4	Ocean energy	Close to feasibility; economics uncertain	Large in favorable areas	Most useful in tropical areas.
5	Tidal energy	Feasible now, but scope for further development	Limited to favorable areas	Could become economic in medium term
6	Wave energy	Principles demonstrated but many practical problems remain	Fairly large in favorable areas	Could become economic in medium term

Table 4.2 Electricity generation from non-conventional energy sources.

		2002		2030	
Renewable sources		**Electricity generation**	**Share in total renewable %**	**Electricity generation**	**Share in total renewable %**
1.	Hydropower	2,610	89	4,248	69
2.	Biomass	207	7	627	10
3.	Wind	52	2	929	15
4.	Geothermal	57	2	167	3
5.	Solar	1	0	119	2
6.	Tide/wave	1	0	35	1
	Total	**2,927**	**100**	**6,126**	**100**

Source: Energy Statistics [23]

4.4.1 SWOT Analysis of Non-Conventional Energy Sources

The SWOT (strengths, weaknesses, opportunities, and threats) analysis is the comprehensive assessment of an organization's opportunities, threats, weaknesses, and strengths. The business' strengths and weaknesses can be revealed by the internal environment analysis using SWOT. Any firm can identify its opportunities and challenges by doing an external environment study [24]. It is a method for assessing the overall strategic position of an organization. The SWOT analysis of non-conventional energy management is performed as mentioned in the following.

4.4.1.1 Strength

- It has unlimited potential.
- It is naturally available and recycled.
- It can help in sustainable development.
- It can help in green economy.
- It is environment friendly.

4.4.1.2 Weaknesses

- Its availability is seasonal in nature.

- It initially requires high cost for implementation.
- It has low supply of power.
- It lacks customer awareness.
- There is an absence of appropriate policy planning.

4.4.1.3 *Opportunity*

- Social benefits can be availed through increased production.
- It can help in environment protection and climate change.
- It can generate employment in rural areas.
- It can provide alternative for electrification in rural areas.

4.4.1.4 *Threats*

- Enough research and development is needed.
- It has new and advanced technologies.
- It has subsidy or grants issues.
- It has transmission problems.

4.4.2 Energy Management Decision in Indian Context

Energy management is a growing concern for rich and emerging nations. Many companies and nations are increasing their attention to the issue of energy management because they worried that the world is rapidly depleting the enormous but limited supply of fossil fuels extraction and production. Energy management is not a novel or difficult idea. Energy prices should be considered as business expenses on par with those of labor and raw materials. Energy management has assumed considerable importance in the recent years owing to the global energy crisis, high pilferage of electricity in country, and privatization of universal energy resources due to increased rapid industrialization and urbanization [25].

Energy management is neither a novel idea nor a difficult procedure. Ensuring that the nation's energy needs are adequately supplied at the lowest feasible cost for current demand and future demand is the primary goal of the energy management function. Energy management calls for a methodical approach, which encompassing management and managerial decision-making abilities founded in energy data analysis. Energy management is required to propose a new stage of total cost reduction through energy-saving techniques that boost profitability for commercial utilities and industry. Planning, directing, and regulating the supply and consumption of energy is known as energy management. The goal of energy management is to use energy wisely, efficiently, and with consciousness in order to maximize productivity while minimizing costs and pollution [26].

Energy efficiency and conservation from energy inputs are required to underline a way toward the efficient use of energy and energy services for all economic sectors in terms of policy, goals, and concepts of energy management. The industrial sector, which consumes a large amount of energy in the nation, has a great deal of opportunity to increase end-use energy efficiency, which would minimize costs and reduce energy consumption. This is especially important for energy-intensive industries like steel, iron, cement, and fertilizers. The goal of energy management is to reduce energy waste in an organization's activities in order to use energy more effectively and efficiently. The goal of energy management is to minimize energy costs and usage while achieving predetermined organizational objectives [27]. Three essential energy management tenets are following:

- to identify and exploit the best available technology for utility;
- to manage the utility of energy to the highest possible degree of efficiency;
- to buy energy at the lowest possible price.

4.5 Qualitative Analysis Using MoSCoW Method

In the direction to prioritize the industrial requirements of non-conventional energy sources, a study is presented in this chapter, where a focused and in-depth expert interview has been conducted as mentioned in the following.

4.5.1 Research Design

The primary data were gathered by the expert opinions through focused interview based on purposive sampling method from the Delhi-NCR region. The expert interview has been performed in August 2023 among the five specific experts from industry and five experts from academia. The chosen experts were highly experienced and knowledgeable in the domain of non-conventional energy management. This analysis intends to prioritize the need of non-conventional energy sources by applying the MoSCoW technique. The authors have employed the MoSCoW prioritization tool as suggested by Ahmad *et al.* [28]. The authors adopted a nine-dimensional renewable energy technology–based scale that has been adopted and developed by Chandrasekar and Kandpal [13].

4.5.2 Renewable Energy Technology Dimensions Based on Industrial Sustainability

The industrial sustainability is a prime importance of concern for industries in terms of energy management in contemporary scenario. The availability of limited stock of conventional energy sources marks a question on future industrial sustainability. In this regard, non-conventional energy sources are one of the best alternative solutions to the future energy crisis.

This chapter adopted a nine-dimensional scale based on industrial requirement of non-conventional energy sources. The first dimension of the adopted scale is resource availability (RA) that is related to the accessibility and convenience of non-conventional energy sources to industries. The second dimension is appropriateness of the technology (AT) that is associated with the relevance of particular non-conventional energy technology as per the requirement of industries. The third dimension is financial and economic viability (FEV) that is related to the economic feasibility of required renewable energy sources to fulfill the energy needs of industries. The fourth dimension is energetic feasibility (EF) that is related to viability of non-conventional energy sources in terms of their robustness and power capacity. The fifth dimension is socio-cultural acceptability (SA) that is associated with suitability and adequacy of particular renewable energy source to industrial energy management decision. The sixth dimension is environmental sustainability (ES) that is related to environmental consideration of specific renewable energy sources. The seventh dimension is institutional preparedness (IP) that is related to industrial readiness and preparation to adopt the specific renewable energy sources. The eighth dimension is related to the awareness and users' training (AUT) that is associated with the need of training to industrial persons regarding the better operation of renewable energy sources at employee level. The last and ninth dimensions is related to availability of after sales and service (ASS) is related to convenience of after sale service in order to ensure the smooth functioning of the adopted renewable energy sources [13].

4.5.3 MoSCoW Prioritization Approach

MoSCoW approach is a requirement-based prioritization technique that is broadly employed in different domains [29]. In MoSCoW-based questionnaire, experts have to elect merely one alternative from the following:

- must have
- should have
- could have
- will not have

The above four elements as presented in Table 4.3 are required for evaluating questionnaire in the MoSCoW technique. Questionnaires are framed in the present chapter with the motive to prioritize the non-conventional energy sources requirement of industries expectations into four categories, i.e., must have, should have, could have, and will not have, based on the MoSCoW method.

For the purpose of explaining the structure of the question, the question related to "resource availability" dimension as "the fulfillment of renewable energy technology dimension", i.e., RA as an energy management decision for industrial sustainability is depicted in Table 4.3. The MoSCoW technique is an acronym that is based on a four-phase requirement prioritization method known as must-have (M), should have (S), could have (C), and will not have (W) as depicted in Figure 4.3.

With due employment of MoSCoW questionnaire standards, expert answers were collected with the help of structured questionnaires comprising a total of nine questions from the five non-conventional energy management experts from industry as professional and five experts from academia. The description of expert profile is mentioned in Table 4.4. The responses are collected, and the collected responses are evaluated to get one specific prioritized category based on the MoSCoW assessment criteria.

Table 4.3 MoSCoW questionnaire used for expert opinion.

Questions	Prioritized moscow class
How do you rank the fulfillment of renewable energy technology dimension, i.e., resource availability (RA) as an energy management decision for industrial sustainability?	❒ Must have ❒ Should have ❒ Could have ❒ Will not have this time

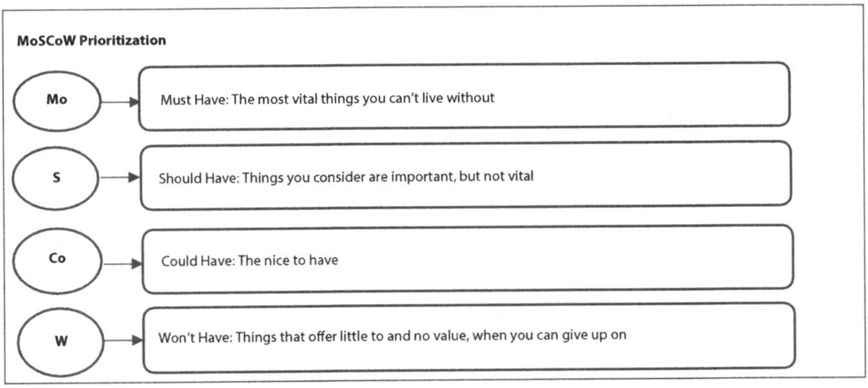

Figure 4.3 MoSCoW framework.

Table 4.4 Expert profile.

Expert	Area	Designation	Experience
Industry	Production	Production manager	16 years
	Operations	Deputy director operations	19 years
	Manufacturing	Techno manager	14 years
	Energy supply	Executive engineer	12 years
	Power transmission	Director renewable energy	27 years
Academia	Energy research and development	Department head	30 years
	Non-conventional energy	Professor	25 years
	Energy studies	Associate professor	22 years
	Electrical engineering	Assistant professor	18 years
	Renewable energy	Principle scientist	15 years

4.5.4 Results

The intention of this chapter is to explore and prioritize the non-conventional energy sources as energy management decisions in industries through deploying the MoSCoW method. This analysis has been accomplished among non-conventional energy management experts from industry and academia from Delhi-NCR region of India. The MoSCoW analysis is performed through the focused interview with chosen experts with the motive to prioritize ranking information that was collected from the experts by adopting a nine-dimensional renewable energy technology scale based on the MoSCoW questionnaire. The categorization of requirement is done to prioritize class to each dimension of non-conventional energy based on frequency analysis scale [30]. Subsequently, an ultimate prioritized rank is attained to each required element on the basis of the highest frequency of MoSCoW responses in every class as presented in Table 4.5.

Out of the nine non-conventional energy–based prioritized adoption elements, three elements were placed under the "must have" category and three elements fall under the "should have" category. Whereas, two elements are found to belong to the "could have" category and one element is found under the "will not have" category as depicted in Table 4.6.

Table 4.5 MoSCoW prioritization analysis.

S. no.	Renewable energy technology–based adoption Dimensions for industrial sustainability (Code)	Frequency (N = 10)				Ultimate prioritized class
		Mo	S	Co	W	
1	How do you rank the resource availability (RA) dimension in the context of non-conventional energy sources?	7	1	1	1	Must Have (Mo)
2	How do you rank the appropriateness of the technology (AT) dimension in the context of non-conventional energy sources?	1	6	1	1	Should have (S)

(Continued)

Table 4.5 MoSCoW prioritization analysis. (*Continued*)

S. no.	Renewable energy technology–based adoption Dimensions for industrial sustainability (Code)	Frequency (N = 10)				Ultimate prioritized class
		Mo	S	Co	W	
3	How do you rank the financial and economic viability (FEV) dimension in the context of non-conventional energy sources?	5	1	2	1	Must have (Mo)
4	How do you rank the energetic feasibility (EF) dimension in the context of non-conventional energy sources?	1	8	1	0	Should have (S)
5	How do you rank the socio-cultural acceptability (SA) dimension in the context of non-conventional energy sources?	1	1	1	6	Will not have (W)
6	How do you rank the environmental sustainability (ES) dimension in the context of non-conventional energy sources?	2	1	7	0	Could have (C)
7	How do you rank the institutional preparedness (IP) dimension in the context of non-conventional energy sources?	8	1	1	0	Must have (Mo)

(*Continued*)

Table 4.5 MoSCoW prioritization analysis. (*Continued*)

S. no.	Renewable energy technology–based adoption Dimensions for industrial sustainability (Code)	Frequency (N = 10) Mo	S	Co	W	Ultimate prioritized class
8	How do you rank the awareness and users' training (AUT) dimension in the context of non-conventional energy sources?	1	7	0	1	Should have (So)
9	How do you rank the availability of after sales and service (ASS) in dimension context of non-conventional energy sources?	1	1	7	0	Could have (Co)

Table 4.6 MoSCoW requirement prioritization graph.

3-Must have requirements * Resource availability (RA) * Financial and economic viability (FEV) * Institutional preparedness (IP)	**3-Should have requirements** * Appropriateness of the technology (AT) * Energetic feasibility (EF) * Awareness and users' training (AUT)
2-Could have requirements * Environmental sustainability (ES) * Availability of after sales and service (ASS)	**1-Won't have requirements** * Socio-cultural acceptability (SA)

4.6 Discussion

In this chapter, a requirement prioritization–based MoSCoW method has been employed to rank the non-conventional energy adoption factors for industrial sustainability. Focused interviews were conducted among the industrial and academic experts in order to prioritize the non-conventional energy management decision in Delhi-NCR region of India. The study presented in this chapter aims to report the prioritization of non-conventional energy sources by adopting a renewable energy technology adoption scale. Results revealed that, out of the nine non-conventional energy–based prioritized adoption dimensions, three factors were placed under the "must have" category and three factors fall under the "should have" category. While two factors found to belong to the "could have" category and one factor is found under the "will not have" category. This analysis is performed with numerous limitations like other studies.

To sustain in the current state of complexity and rapid industrialization, the world's population demands an ever-increasing amount of energy. The massive amounts of waste produced by the burning of fossil fuels are harmful gases that disrupt ecosystems and are in large quantity for the Earth to absorb and process. Toxic waste accumulation is manifested everywhere. The inherent dynamic equilibrium of many cycles including those of nitrogen and carbon is constantly distressed and is in the state of the deteriorating environment. Coral reefs are dying, and the oceans are becoming more acidic, which are two of the main effects of global warming. Today, the need of energy is meeting by the readily available dirtier alternatives, which probably increasing the threat to the issue of climate change. It is imperative to comprehend in a clear and rational manner to address the issue of climate change seriously and accordingly, fossil fuels must be replaced with renewable alternative energy sources [31].

The most crucial metrics for determining economic success in the renewable alternative are productivity gains and energy cost reductions, which are being driven by the WTO's regulations, as well as increasing globalization and liberalization of trade. Today, a formal framework is employed for energy management, but that is not sufficient for energy management [32]. Accordingly, it is needed by the business organizations to increase their energy performance by adhering and adaptation of a straight forward procedure as follows:

- obtain support from senior management;
- evaluate the current state of affairs;
- set goals and objectives;
- create an action plan, assign resources, and carry out the plan; and
- review and assess the results.

Worldwide, the amount of electricity consumption is rising. Every type of production process or economic activity requires energy, and sustainable renewable energy is a key component of India's industrialization and socio-economic growth. In the present scenario, conventional energy resources are shrinking throughout the world in a rapid pace. Consequently, there will be a situation for future generations for facing a choice between, either using the renewable energy resources and wars for resources in the near future. For the sake of the next generation's survival, India must use renewable energy sources. Decentralized electricity generation using renewable energy sources, such as modest wind, hydro, solar, biomass, biofuels, and energy from wastes is best suited to give the much-needed options and is evidently required. However, grid extension in rural and distant areas is very frequently and not cost-effective [33].

Due to exponential industrial growth, rising living standards, and the increased usage of electronic devices, traditional "conventional" electricity production is no longer adequate. Micro-generation is required, which needs significant alteration on customers' attitudes and behaviors related with energy production, management, conservation, and usage. The same can also serve as a catalyst for cultural shifts in consumer attitudes. Small wind turbines, biomass, solar power, micro-hydropower, and/or a mix of these technologies are examples of micro-generation technologies, which need sustainable development. Renewable energy may initially seem more expensive than conventional energy, but when you consider its benefits related with constant power availability, environmental friendliness, and significant reduction in global warming, it is well worth the investment [34].

4.7 Conclusion

The present chapter reciprocated in the direction to examine the extent of non-traditional energy sources, where the rank of the non-conventional energy requirements is presented using MoSCoW prioritization–based qualitative method. The outcome analysis of the MoSCoW method reported that non-conventional energy sources factors like "resource availability", "financial-economic viability", and "institutional preparedness" are found prioritized under the "must have" category, whereas factors like "appropriateness of the technology", "energetic feasibility", and "awareness and users' training" are found under the "should have" category. Subsequently, factors like "environmental sustainability" and "availability of after sales and service" come under the "could have" category, whereas the factor "socio-cultural acceptability" falls under the "won't have" category.

In context of energy-related industrial decisions, the consideration of business procedures that are used in finance, production, marketing, and administration also needs to be used in energy management. It is highlighted that, when there will be a defined direction that is ingrained in the company's long-term business plan, the energy management system will only yield sustained benefits after gaining the attention of top management. In the current scenario, majority of businesses and allied top management is considering energy management as a low priority. Nevertheless, businesses can cut their energy expenses up to 20% by using sound energy management practices. Therefore, it is imperative that businesses should take a strategic approach to implementing energy management in the modern world [33].

The majority of corporate entities views energy as an expense that is beyond their control and do not have any plans in this area. Energy management will lessen the organization's environmental effect while enduring money-saving. As a result, it addresses issues such as rising productivity, raising living standards, and boosting profits in addition to energy conservation. The development of renewable energy sources requires increased research and efforts. Encouraging renewable energy technology is one of the strategies to solve the issues of economic growth and energy security. In order to lessen the effects of climate change and to reduce reliance on imported gas and oil, which takes up a significant portion of foreign exchange reserves, India needs to prioritize research and development into non-conventional energy sources. Additionally, it is evident that an integrated energy management system with two or more renewable energy sources is stable, dependable, and profitable. Finally and importantly, it is critical and challenging to create responsiveness, so that people in both urban and rural areas comprehend the potential of renewable energy sources and their role in environmental sustainability [34].

The present chapter provides valuable insights into the current state of non-conventional energy sources, including their technologies, challenges, opportunities, and the policy and economic considerations surrounding their adoption. Researchers and policymakers can use the above findings to prepare decisions related to non-conventional energy development and sustainability. In addition, the outcomes of this chapter contributed to the systematic theory building of non-conventional energy management literature. The findings of the chapter would be helpful to other scholars to conduct further research on the inclusion on non-conventional energy sources for industrial sustainability [35].

4.7.1 Limitations

As any other study, this chapter also possesses certain limitations [36]. This study is only based on the qualitative analysis of non-conventional energy management decision through requirement prioritization method, i.e., the MoSCoW method. This study provides insights from the existing non-conventional energy literature from the perspective of industrial sustainability. This entire study mainly focuses its attention on the prioritization of non-conventional energy management decision through the expert opinion method [37].

4.7.2 Further Avenues

The major issues related with the further exploration for the utility of non-conventional power related energy management decision in terms of industrial sustainability is a big concern. Thus, future energy management decision–based studies needed careful consideration, more qualitative criteria, quantitative criteria, and focused research for enhancing the inclusiveness of non-conventional energy sources. Further exploration is required to explore the role of non-conventional energy in sustainability of industries in terms of environmental, social, and governance (ESG) concerns [38].

Additional investigation is needed to identify the relevance of non-conventional energy sources in future energy needs of various industrial sectors [39]. Also, there is scope for research to measure the efficacy of non-conventional energy sources as per their cost-effectiveness in the long run [40] in direction to attain the targets of energy planning.

References

1. Bhetalu, A.D., Patil, S.S., Ingole, N.W., Studies on generation of power alcohol as a non-conventional energy source from aquatic macrophytes—a critical review. *J. Eng. Res. Stud.*, 3, 1, 9–17, 2012.
2. Goetzberger, A., Hebling, C., Schock, H.W., Photovoltaic materials, history, status, and outlook. *Mater. Sci. Eng.: R: Rep.*, *40*, 1, 1–46, 2003.
3. Kushwaha, J., Singh, P., Kushwaha, R., Predicting working sole Indian mothers' satisfaction towards work-family balance integrating Kano model and weighted average method. *Int. J. Soc Econ.*, ahead-of-print, 2023, https://doi.org/10.1108/IJSE-02-2023-0074.

4. Jain, P. and Singh, P., Lethargic approach in expediting the financial reporting system: a case of a construction firm. *I. J. T. C. S.*, *14*, 1, 1–10, 2023, doi: 10.1504/IJTCS.2023.10055878.
5. Miyan, M. and Shukla, M.K., Review on Non-Conventional Energy Resources in India. *SAMRIDDHI: J. Phys. Sci. Eng. Technol.*, *10*, 02, 87–94, 2018.
6. Sen, S., Ganguly, S., Das, A., Sen, J., Dey, S., Renewable energy scenario in India: Opportunities and challenges. *J. Afr. Earth. Sci.*, *122*, 25–31, 2016.
7. Sharma, S., Non-Conventional Energy's Overview Sources of India. *Int. J. Eng. Res. Gen. Sci.*, 2, 4, 12–20, 2014.
8. Singal, S.K., Review of augmentation of energy needs using renewable energy sources in India. *Renew. Sustain. Energy Rev.*, *11*, 7, 1607–1615, 2007.
9. Gaurav, N., Sivasankari, S., Kiran, G.S., Ninawe, A., Selvin, J., Utilization of bioresources for sustainable biofuels: a review. *Renew. Sustain. Energy Rev.*, *73*, 205–214, 2017.
10. Crini, G., Non-conventional low-cost adsorbents for dye removal: a review. *Bioresour. Technol.*, *97*, 9, 1061–1085, 2006.
11. Amata, I.A., The use of non-conventional feed resources (NCFR) for livestock feeding in the tropics: a review. *J. Glob. Biosci.*, *3*, 2, 604–613, 2014.
12. Kushwaha, J., Sharma, A., Singh, P., Exploration and prioritization of enablers to organization work-family balance planning for working sole Indian mothers integrating Fuzzy-Delphi and AHP. *Int. J. Soc Econ.*, 50, 3, 398–418, 2023, https://doi.org/10.1108/IJSE-05-2022-0348.
13. Chandrasekar, B. and Kandpal, T.C., An opinion survey based assessment of renewable energy technology development in India. *Renew. Sustain. Energy Rev.*, *11*, 4, 688–701, 2007.
14. Jamel, M.S., Abd Rahman, A., Shamsuddin, A.H., Advances in the integration of solar thermal energy with conventional and non-conventional power plants. *Renew. Sustain. Energy Rev.*, *20*, 71–81, 2013.
15. Sharma, A., Srivastava, J., Kar, S.K., Kumar, A., Wind energy status in India: A short review. *Renew. Sustain. Energy Rev.*, *16*, 2, 1157–1164, 2012.
16. Azad, A.S., Rahaman, M.S.A., Watada, J., Vasant, P., Vintaned, J.A.G., Optimization of the hydropower energy generation using Meta-Heuristic approaches: A review. *Energy Rep.*, *6*, 2230–2248, 2020.
17. Kumar, A., Kumar, N., Baredar, P., Shukla, A., A review on biomass energy resources, potential, conversion, and policy in India. *Renew. Sustain. Energy Rev.*, *45*, 530–539, 2015.
18. Razdan, P.N., Agarwal, R.K., Singh, R., Geothermal energy resources and their potential in India. *Earth Sci. India*, *1*, 1, 30–42, 2008.
19. Chowdhury, M.S., Rahman, K.S., Selvanathan, V., Nuthammachot, N., Suklueng, M., Mostafaeipour, A., Habib, A., Akhtaruzzaman, M., Amin, N., Techato, K., Current trends and prospects of tidal energy technology. *Environ. Dev. Sustain.*, *23*, 8179–8194, 2021.

20. Prashar, A., Towards sustainable development in industrial small and medium-sized enterprises: An energy sustainability approach. *J. Clean. Prod.*, *235*, 977–996, 2019.
21. Saroha, J., Nonconventional energy resources in India, in: *Spatial Diversity and Dynamics in Resources and Urban Development: Volume 1: Regional Resources*, pp. 431–447, 2015.
22. Wang, Q. and Li, R., Drivers for energy consumption: A comparative analysis of China and India. *Renew. Sustain. Energy Rev.*, *62*, 954–962, 2016.
23. Energy Statistics, *Energy Year Book*, Ministry of Statistics and Programme Implementation, Government of India, 2015, https://mospi.gov.in/statistical-year-book-india/2015/185.
24. Singh, P. and Agrawal, G., Development, present status, and performance analysis of agriculture insurance schemes in India: Review of evidence. *Int. J. Soc Econ.*, *47*, 4, 461–481, 2020, https://doi.org/10.1108/IJSE-02-2019-0119.
25. Dey, S., Sreenivasulu, A., Veerendra, G.T.N., Rao, K.V., Babu, P.A., Renewable energy present status and future potentials in India: An overview. *Innov. Green Dev.*, 1, 1, 100006, 2022.
26. Upadhyay, S. and Sharma, M.P., Selection of a suitable energy management strategy for a hybrid energy system in a remote rural area of India. *Energy*, *94*, 352–366, 2016.
27. Moriarty, P. and Honnery, D., Energy efficiency or conservation for mitigating climate change? *Energies*, *12*, 18, 3543, 2019.
28. Ahmad, K.S., Ahmad, N., Tahir, H., Khan, S., Fuzzy_MoSCoW: A fuzzy-based MoSCoW method for the prioritization of software requirements, in: *2017 International Conference on Intelligent Computing, Instrumentation and Control Technologies (ICICICT)*, IEEE, pp. 433–437, 2017, July.
29. Singh, P. and Agrawal, G., Mapping the customer-centric weather index insurance service design using quality function deployment. *TQM J.*, *34*, 6, 1800–1822, 2022, https://doi.org/10.1108/TQM-08-2021-0236.
30. Singh, P. and Agrawal, G., Efficacy of weather index insurance for mitigation of weather risks in agriculture: an integrative review. *Int. J. Ethics Syst.*, *35*, 4, 584–616, 2019.
31. Singh, P., Unorganized sector and India's informal economy: challenges and prospects. *New Man Int. J. Multidiscip. Stud.*, *1*, 2, 209–218, 2014.
32. Singh, P., Agrawal, G., Kushwaha, R., Evaluating Customer Satisfaction of Weather Index Insurance Service Quality Using Kano Model, in: *2021 IEEE Bombay Section Signature Conference (IBSSC)*, IEEE, pp. 1–6, 2021, November.
33. Eras, J.J.C., Morejón, M.B., Gutiérrez, A.S., García, A.P., Ulloa, M.C., Martínez, F.J.R., Rueda-Bayona, J.G., A look at the electricity generation from non-conventional renewable energy sources in Colombia. *Int. J. Energy Econ. Policy*, *9*, 1, 15–25, 2019.

34. Singh, P., Weather index insurance viability in mitigation of climate change impact risk: A systematic review and future agenda. *J. Sci. Technol. Policy Manage*, 15, 1, 142–163, 2024, https://doi.org/10.1108/JSTPM-07-2021-0102.
35. Singh, P. and Agrawal, G., Modelling the barriers of weather index insurance service adoption integrating expert mining and ISM Fuzzy-MICMAC. *Benchmarking: Int. J.*, *29*, 8, 2527–2554, 2021, doi: 10.1108/BIJ-04-2021-0183.
36. Singh, P., Kushwaha, R., Kushwaha, J., Assessing visitors' satisfaction about museum service quality using Kano model, in: *2021 IEEE 9th Region 10 Humanitarian Technology Conference (R10-HTC)*, IEEE, pp. 1–6, 2021.
37. Singh, P., Kushwaha, R., Kushwaha, J., Evaluating Farmer Participation in Weather Index Insurance in a Socio-Demographic Perspective: Evidence From the Crop Insurance Data of India, in: *Handbook of Research on Managing the Urban-Rural Divide Through an Inclusive Framework*, A. Popoola, H. Magidimisha-Chipungu, L. Chipungu (Eds.), pp. 69–88, IGI Global, 2023, https://doi.org/10.4018/978-1-6684-6258-4.ch005.
38. Rajput, S., Jain, E., Mehra, A., Thakur, K.S., Gupta, O., Singh, P., Technology Driven Tourism: A PLS SEM Model of its Antecedents in Tourist Behavior Intention, in: *2022 IEEE Conference on Interdisciplinary Approaches in Technology and Management for Social Innovation (IATMSI)*, Gwalior, India, pp. 1–6, 2022, doi: 10.1109/IATMSI56455.2022.10119265.
39. Singh, P., Kushwaha, R., Kushwaha, J., Analyzing Accountability of Weather Index Insurance Service in Attainment of Sustainable Development Goals: A Sustainable Accounting Perspective Review, in: *Sustainable International Business: Smart Strategies for Business and Society*, pp. 221–250, Springer International Publishing, Cham, 2023, doi: 978-3-031-43784-7, 531340.
40. Kushwaha, J., Singh, P., Sharma, A., Modelling the enablers of work-family balance among working single mothers integrating expert-mining and TISM-MICMAC. *J. Organ. Eff.: People Perform.*, 10, 4, 501–526, 2023, https://doi.org/10.1108/JOEPP-05-2022-0106.

5

Response Surface Methodology: A Statistical Tool to Optimize Process Parameters (Quantitative Data) to Maximize the Microbial Biomass and Their Bioactive Metabolites

Vijayalakshmi Ghosh* **and Aachal Lonhare**

GD Rungta College of Science and Technology, Kohka, Bhilai, Chhattisgarh, India

Abstract
An effective experimental design is a crucial element in the development of any product or process. It hinges on a comprehensive understanding of the system being studied. Response surface methodology (RSM) is a statistical approach, which can be employed for modeling and optimization of the processes that involve multiple variables. RSM helps to determine the ideal process conditions by merging experimental designs with interpolation, often employing first- or second-order polynomial equations in a systematic testing process. Conventional approaches that are designed to attain multi-factor experimental design typically entail significant time investment and encounter challenges in pinpointing the genuine optimal conditions, particularly when factor interactions are involved. RSM, however, has the capability to overcome these limitations. In the field of biological sciences, RSM is widely utilized for optimizing tasks such as preparing culture media, enhancing product formation, assessing their practical applications, and improving wastewater treatment processes. Additionally, RSM holds the great capacity to optimize both the components of growth media and various physicochemical parameters and represents a promising avenue for enhancing the production of both novel and existing bioactive metabolites (enzymes, antibiotics, probiotics, bioethanol, etc.) from various microbes. The utilization of the RSM methodology demands for quantitative information and proves highly effective

**Corresponding author*: vijayalakshmi.ghosh@gmail.com

Atul Kumar Sahu, Rakesh D. Raut, Rohit Raja, Anoop Kumar Sahu and Nitin Kumar Sahu (eds.) Industrial and Manufacturing Designs: Quantitative and Qualitative Analysis, (139–172) © 2024 Scrivener Publishing LLC

in constructing predictive models and optimizing experimental approaches for substrates and energy utilization. RSM offers a means to systematically assess the influence of various parameters in the form of qualitative or quantitative data on a given process toward enabling the determination of optimal values of these process variables in order to enhance the desired outcome. RSM also finds application in optimization of microbial and polymeric adsorption process to eliminate heavy metal pollution. Formulation (nanoemulsion, liposome, etc.) and development of microbial products using secondary metabolites and essential oils can also be optimized by employing RSM. This chapter will present various cases of quantitative data analysis by RSM pertaining to optimization of process parameters such as culture media composition, substrate concentration, pH, and temperature for increased microbial biomass productions, which will culminate in enhanced production of bioethanol, enzymes, antibiotics, probiotics, etc. Furthermore, the chapter will deal with optimization of quantitative data by RSM in development of formulation using different metabolites and essential oil. This chapter will also deal with optimization of quantitative data related to microbial culture conditions for the applications such as waste water treatment and heavy metal sorption using RSM approach.

***Keywords*:** Response surface methodology, optimization, microbial products, waste water treatment, heavy metal sorption, nanoemulsion

5.1 Introduction

Experimental design is a structured approach toward planning, conducting, and analyzing experiments in a systematic and efficient manner. It is widely used in scientific research, engineering, product development, quality control, and various other fields to gather data, test hypotheses, and make informed decisions. The key components and principles, which are needed in experimental design, include objective, factors, level of factors, number of experiments, replication, randomization, control variables, data collection, analysis, interpretation, and optimization (Figure 5.1). While performing response surface methodology (RSM), it is required to define the objectives of the experiment. Many questions are related to "what are you trying to learn, prove, or optimize?, and understanding the goals is the first step that is required in designing a meaningful experiment. Identification of the independent variables or factors that may influence the outcome or response variable must be done prior to the conduction of experiments physically. Factors can be in the form of qualitative data or quantitative data and can be both controlled (manipulated) and uncontrolled (observed), where optimized values of outputs can be attained with the modeling of controlled factors. Determination of the different levels or settings at which each factor will be tested to understand their capacities toward influencing output can be done by RSM. This demands the exploration of the effects of varying factors across a range of conditions. An appropriate experimental design based on the research question and the number of factors and levels is demonstrated in present chapter. Common experimental designs are as follows:

1. Completely randomized design, where factors are randomly assigned to experimental units.
2. Randomized block design, where, experimental units are divided into blocks, and factors are randomized within each block.
3. Factorial design, which examines the effects of multiple factors simultaneously and uncover interactions between them.
4. Response surface design, which is used for optimizing responses influenced by multiple factors
5. Split-plot design, which is used, when some factors are more challenging or expensive to change during the experiment.

Multiple runs or replicates for each combination of factor levels are needed to be conducted. Replication helps to reduce the impact of random variability and provides more robust results. Experimental units are required to be randomly assigned to treatment groups or factor levels to minimize bias and to account four uncontrolled sources of variation. Variables that may influence the response but are not the part of the experimental run should be identified and controlled during RSM modeling. Keeping multiple quantitative variables in modeling is significant and helps toward understanding the effects of the various key factors. Accordingly, data are needed to be collected systematically and accurately from system characteristics. It must be ensured that data collection methods are consistent across all experimental runs.

Appropriate statistical methods to analyze the collected data play a crucial role in understating various key factors. This includes techniques like analysis of variance (ANOVA), regression analysis, and hypothesis testing. The said techniques assist in the interpretation of the results in the context of research objectives for drawing conclusions based on the data and analysis. The goal of optimization is to find the best conditions for a process and results to identify optimal factor settings, where effective experimental design enables researchers and practitioners to draw meaningful conclusions from their experiments, minimize bias and variability, and make

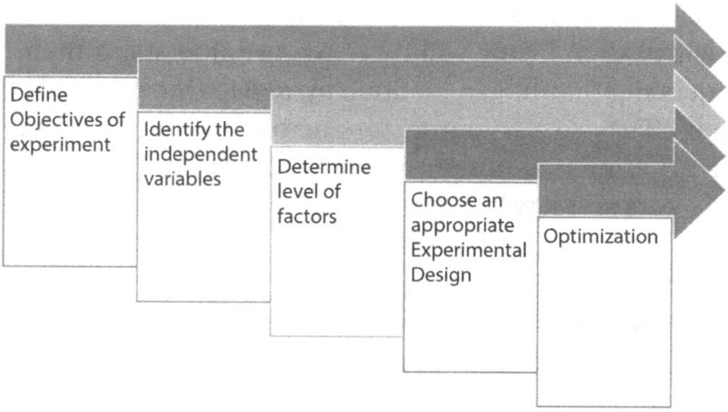

Figure 5.1 The key components and principles of experimental design.

informed decisions. It is a critical component of the scientific method and plays a key role in advancing knowledge and solving practical problems.

Experimental design plays a vital and foundational role in the development of every product or process across a wide range of industries and disciplines. Experimental design provides a structured approach to investigating and understanding complex systems, whether they involve physical products, chemical processes, biological systems, or software algorithms. By following a well-designed experiment, one can collect high-quality data that inform your decisions. This data-driven approach reduces guesswork and subjectivity.

Experimental design allows for the systematic exploration of factors and conditions to find the best combination for product or process optimization. This can lead to improved efficiency, performance, and cost-effectiveness. In manufacturing and production, experimental design is used to ensure consistent product quality by identifying critical process parameters and their acceptable ranges. For research and development, experimental design fosters innovation by helping researchers to explore novel ideas and technologies while systematically evaluating their feasibility and effectiveness. Well-designed experiments reduce resource waste by eliminating unnecessary trials and ensuring that the information gained is maximized. By thoroughly testing hypotheses and potential failure points in a controlled environment, experimental design can help mitigate risks associated with product development and process scale-up. Experimental results generated through well-designed experiments contribute to a deeper understanding of the underlying principles governing a product or process. In highly competitive markets, efficient experimental design can accelerate time to market and give companies an edge by delivering innovative products or processes faster. In fields such as pharmaceuticals and food production, experimental design ensures that products meet regulatory standards and safety requirements. Experimental design supports continuous improvement efforts by providing a structured way to assess and refine products and processes over time. When issues or challenges arise in product development or process operation, experimental design offers a systematic way to diagnose problems and find effective solutions. Experimental design is used in a wide range of disciplines, from engineering and chemistry to biology, social sciences, and beyond, demonstrating its universal utility. Experimental design is an indispensable tool that underpins scientific research, innovation, product development, and process improvement in nearly every field. It ensures that decisions are based on data and that resources are used efficiently, ultimately leading to better products and processes. Thus, one should be very careful about selecting experimental design.

Before designing an experiment, researchers formulate hypotheses about the system's behavior. A comprehensive understanding of the system can be done by generating meaningful hypotheses that can be tested through experimentation. Knowing the system characteristics enables researchers to identify and select the most relevant factors, variables, and qualitative or quantitative data that might influence the system's responses. This reduces the risk of overlooking important factors. An understanding of the system helps in identifying potential confounding variables that might affect the response but are not of primary interest. Researchers can then take steps to control or account for these variables. Understanding the system's limitations and capabilities allows researchers to set realistic and achievable objectives for the experiment. Unrealistic goals can lead to inefficient experimental designs and inaccurate conclusions. Knowing the system's behavior helps in selecting appropriate levels for each factor. This ensures that the experimental conditions are both relevant and feasible. A deep understanding of the system is crucial for interpreting the results of the experiment accurately, where the researchers can contextualize attained findings within the broader system dynamics. In case of unexpected outcomes or issues during the experiment, a strong understanding of the system allows researchers to troubleshoot problems effectively and make necessary adjustments. Understanding the system characteristics aids in validating the experimental results and determining the extent to which the findings can be generalized to real-world situations. Researchers can anticipate potential risks and challenges associated with the experiment and develop contingency plans or mitigation strategies accordingly. A thorough understanding of the system supports ongoing learning and continuous improvement. Researchers can refine their experimental designs based on previous findings and adjust their hypotheses and approaches. An excellent experimental design is built upon a foundation of knowledge and understanding of the system under investigation. This understanding not only guides the design process but also enhances the reliability and relevance of the experiment's outcomes. It allows researchers to make meaningful contributions to their field and draw accurate conclusions that advance scientific understanding to solve real-world problems. The present chapter discusses quantitative data analysis, which is done by RSM pertaining to the optimization of process parameters such as culture media composition, substrate concentration, pH, and temperature for increased microbial biomass productions, which will culminate in enhanced production of bioethanol, enzymes, antibiotics, probiotics, etc.

5.2 Conventional Methods for Multifactor Experimental Design

Conventional methods that are developed for multifactor experimental design involve techniques that have been used for decades and are still valuable for designing experiments, when you have multiple factors or variables for evaluation. These methods help you systematically vary and analyze factors to understand their effects on a response variable of interest. Some commonly used conventional methods for multifactor experimental design that are developed include full factorial design, fractional factorial design, one-factor-at-a-time (OFAT) design, central composite design (CCD), Box-Behnken design, Taguchi method, and Latin square design; a brief representation of aforesaid experimental design is discussed in Sections 5.2.1 to 5.2.7.

5.2.1 Full Factorial Design

In full factorial design, all possible combinations of the levels of each factor are tested. This method provides a complete picture of "how each factor and their interactions influence the response variable". However, it can be resource-intensive when dealing with many factors or wide ranges of factor levels.

5.2.2 Fractional Factorial Design

Fractional factorial designs are used when testing all possible combinations is impractical due to resource constraints. In this approach, only a fraction of the possible combinations is tested. Fractional factorial designs are particularly useful for identifying the factors, which have the most significant impact on the response.

5.2.3 One-Factor-at-a-Time (OFAT) Design

OFAT is a straightforward and easy-to-implement method, where you vary one factor at a time while keeping others constant. This method is quick but may miss interactions between factors, and it does not provide a comprehensive understanding of how all factors interact.

5.2.4 Central Composite Design (CCD)

CCD is an RSM that combines factorial and axial points with center points. It allows for the exploration of both linear and quadratic effects of factors and provides a good compromise between a full factorial and fractional factorial design. CCD is often used when you want to optimize a process or find the optimal conditions for a response.

5.2.5 Box-Behnken Design

Box-Behnken design is another RSM design that is suitable for three-level factors. It minimizes the number of experimental runs that are needed to construct a quadratic model. It is efficient for finding the optimal conditions within a limited range of factors.

5.2.6 Taguchi Method

The Taguchi method focuses on robust design and optimization by systematically varying factors and levels. It is particularly useful when you want to find settings that are less sensitive to variations or noise. Taguchi designs often include orthogonal arrays to efficiently explore factor interactions.

5.2.7 Latin Square Design

Latin square designs are used in situations where two factors, each with several levels, need to be tested simultaneously. They help reduce confounding effects and identify interactions between two factors.

The choice toward determining the use of conventional experimental design method depends on factors such as the number of factors, resource availability, the desired level of precision, and the specific goals of your experiment. In many cases, a combination of these methods may be employed to thoroughly explore the effects of multiple factors on a response variable.

5.3 Response Surface Methodology (RSM)

Conventional methods for multifactor experimental design, such as OFAT experiments, are often time-consuming and less effective toward detecting the true optimum, especially when interactions among factors play a significant role. RSM addresses these limitations and offers several advantages in optimizing processes and systems such as efficiency, interactions detection, optimization, visual representation, statistical rigor, robustness testing, savings in resources, data-driven decision-making, iterative improvement, and applicability to complex systems. RSM is more efficient than OFAT experiments because it allows for the simultaneous variation of multiple factors at different levels. This significantly reduces the number of experimental runs required to explore the entire factor space. RSM is particularly effective in detecting interactions among factors. By fitting mathematical models to experimental data, RSM can reveal complex relationships and synergistic or antagonistic effects that OFAT experiments might miss. RSM is designed for optimization. It systematically explores the factor space to identify the conditions that lead to the best outcome, whether it is maximizing yield, minimizing cost, or achieving another specific goal.

RSM provides visual representations of the response surface, such as contour plots or 3D surface plots. These plots allow researchers to intuitively understand how factors interact and where the optimal conditions lie. RSM uses statistical techniques, such as ANOVA and regression analysis, to rigorously analyze experimental data and validate the mathematical models. This adds a level of confidence in the results. After optimization, RSM can guide robustness testing to assess how well the optimized conditions perform under variations or uncertainties, ensuring that the process remains effective in real-world scenarios. RSM often requires fewer experimental runs compared to traditional methods, which translates savings in time, materials, and resources. RSM is a data-driven approach, allowing researchers to make informed decisions based on empirical evidence rather than assumptions or guesswork. RSM can be used iteratively, where initial experiments can inform subsequent rounds of optimization, allowing for continual process improvement. RSM is suitable for complex systems with multiple factors and interactions, making it a valuable tool in various fields, including engineering, chemistry, biology, and more. In summary, RSM overcomes the limitations of conventional experimental design methods by offering a systematic, efficient, and statistically rigorous approach toward optimizing processes and systems. It is particularly valuable, when interactions among factors are significant, as it helps to uncover the true optimum conditions and provides a data-driven basis for decision-making and process improvement.

RSM is a comprehensive approach that combines various experimental designs and statistical techniques to optimize processes involving multiple factors. It uses mathematical models to describe the relationship between factors and responses and is especially useful for process optimization. RSM is a collection of statistical and mathematical techniques used for modeling and optimizing processes and systems. It is particularly valuable in fields where multiple variables interact to influence a specific outcome or response. RSM focuses on modeling the relationship between one or more independent variables (factors) and a response variable (the outcome of interest). The response surface equation represents how changes in the factors impact the response. RSM employs a factorial experimental design, where factors are varied simultaneously at multiple levels. This allows for efficient exploration of factor interactions and their effects on the response. RSM uses mathematical models, typically polynomials, to describe the relationship between factors and the response variable. These models can be linear, quadratic, or higher-order, depending on the complexity of the system. RSM guides the design and execution of experiments based on the chosen experimental design. Experiments are conducted systematically to collect data for model development. The collected data are used to fit the mathematical model to the experimental results through regression analysis, where the model parameters are estimated from the data. The model is validated to ensure its accurate representation toward the real-world system. Techniques like ANOVA are often used for validation in RSM.

RSM involves visualizing the response surface through contour plots, 3D surface plots, or other graphical representations. These plots help to understand the relationships between factors and the response variables that may be qualitative or quantitative data. Once the model is validated, optimization techniques are applied to identify the optimal conditions that maximize or minimize the response variable. This is often done using calculus or numerical optimization algorithms. RSM can perform sensitivity analysis to determine which factors have the most significant impact on the response. This information helps to prioritize factors for further study or optimization. After optimization, the system's robustness is tested to assess its performance under varying conditions or uncertainties. RSM has a wide range of applications, including process optimization, product formulation, quality improvement, experimental design, and parameter estimation. RSM is a powerful tool for modeling and optimizing complex systems, where multiple factors influence a response variable. It provides a structured approach toward experimentation and data analysis, enabling efficient process improvement, resource optimization, and better understanding of the relationships between variables and outcomes.

RSM is a powerful statistical and mathematical tool used in experimental design and optimization. It can be employed in the field of microbiology to optimize process parameters and maximize desired outcomes, such as microbial biomass and bioactive metabolite production. RSM can be applied to optimize process parameters for microbial biomass and bioactive metabolite production. The same can assist in clearly attaining the objectives of the study, which related toward maximizing microbial biomass and bioactive metabolite production. Here, one should specify the quantitative metrics that will be used to measure microbial outcomes. Here, RSM can be used to identify the process parameters (factors) that may influence microbial biomass and bioactive metabolite production. These factors could include temperature, pH, nutrient concentration, agitation speed, and incubation time.

Conduction of experiments in RSM requires to define the range or levels for each experimental factor that will be investigated during the experiment. The levels should be decided to cover a wide enough range to capture potential variations in the response variables. Here, one should use a statistical experimental design, i.e., CCD or a Box-Behnken design to plan a series of experiments. These designs help in efficiently exploring the factor space with a limited number of experiments. For RSM modeling, conduct the experiments according to the selected experiments matrix, systematically based on varying process parameters (input parameters) at the specified levels. Afterwards, measure and record the microbial biomass and bioactive metabolite production for each experimental run. Ensure that you collect quantitative data accurately to make a precise fit of a response surface model to the experimental data. This model in the field of microbiology will help in relating the process parameters to the response variables (microbial biomass and bioactive metabolite production), where common models including quadratic or polynomial models can be built. The built response surface model is then required to be validated to ensure its adequacy in representing the system. This can be done by checking the goodness of fit and conducting statistical tests. The present chapter demonstrates the RSM model in microbiology and employed response surface model to identify the optimal combination of process parameters that maximize microbial biomass and bioactive metabolite production. Here, optimization is performed using mathematical algorithms to find the factor levels that yield the highest predicted response, and confirmation experiments are conducted with the recommended optimal predictive conditions by the built model to validate the predicted results. This step helps to ensure that the optimized conditions that are practically achievable and yield the desired outcomes. It is to be remembered that while RSM is a powerful tool, its success relies on accurate experimental design, appropriate model selection, and careful interpretation of results. Also, biological systems, such as microbial growth and metabolite production, can be complex and subject to various factors beyond the parameters you control, but one can implement RSM in said field. Response surface methodology (RSM) can be a valuable approach to optimize process parameters for maximizing microbial biomass and bioactive metabolite production in said field, where RSM can be used to provide a systematic way to explore the parameter space, build predictive models, and find optimal conditions for your desired outcomes.

5.4 RSM in Bioprocessing/Fermentation

Microbial fermentation is a biological process, where microorganisms, such as bacteria, yeast, or fungi, are utilized to convert organic compounds into various useful products through the action of enzymes. This process has been used for centuries in various industries, including food, beverage production, pharmaceuticals, biofuels, and more. Various microorganisms can be used in fermentation, including bacteria (e.g., *Lactobacillus* for yogurt), yeast (e.g., *Saccharomyces cerevisiae* for bread and alcohol production), and fungi (e.g., *Aspergillus* for citric acid production). Microbial fermentation typically starts with a substrate or raw material, which serves as a source of carbon and energy for the microorganisms. The choice of substrate can significantly impact the final product. During fermentation, microorganisms metabolize substrates through specific pathways, often producing energy and metabolic intermediates. These pathways vary depending on the microorganism and the desired product. The end products of fermentation can vary widely include alcohols (e.g., ethanol), organic acids (e.g., lactic acid), gases (e.g., carbon dioxide), enzymes, antibiotics, and a range of other chemicals, depending on the microorganism and the process. The success of microbial fermentation depends on several factors, including nutrient availability, pH, incubation time, temperature, agitation, and oxygen levels. Optimizing these conditions is critical for maximizing product yield and quality.

There are different types of fermentation. Aerobic fermentation occurs in the presence of oxygen. Some microorganisms, like certain bacteria, require oxygen for growth and product formation. Anaerobic fermentation occurs in the absence of oxygen. Many fermentation processes, such as ethanol production, are anaerobic. In batch fermentation, the substrate is introduced in the bioreactor/fermenter at the beginning of the fermentation, and products are harvested at the end. In continuous fermentation, substrate is continuously added, and products are continuously removed, allowing for more extended production periods. In fed-batch fermentation, substrate is added in portions throughout the fermentation process, allowing for better control over product formation.

Applications of fermentation are significant in food and beverage industry, in pharmaceuticals, for biofuels production, and in chemical industry. Fermentation is used in the production of various foods and drinks, including cheese, yogurt, bread, wine, beer, and sauerkraut. Fermentation is used to produce antibiotics, vaccines, and therapeutic proteins. Ethanol and other biofuels are produced through microbial fermentation of sugars and starches. Fermentation can be used to produce organic acids, solvents, enzymes, and other chemicals. After fermentation, products often require downstream processing steps like separation, purification, and concentration to obtain the final product in its desired form and purity. The design and optimization of microbial fermentation processes fall under the domain of bioprocess engineering, which involves the integration of microbiology, chemistry, and engineering principles. Microbial fermentation is a versatile and widely used process that harnesses the metabolic capabilities of microorganisms to produce a wide range of valuable products. Its applications span various industries and continue to evolve with advances in biotechnology and bioprocess engineering.

Metabolites that play a vital role in supporting cellular growth and maintenance are synthesized during the exponential growth phase. These essential metabolites encompass enzymes, vitamins, amino acids, organic acids, and nucleosides, and, sometimes, even biomass itself is considered a product. The production of these growth associated metabolites is influenced by many qualitative and quantitative factors such as the type and concentration of nutrients, inoculum size, incubation time, and environmental conditions like oxygen availability, pH, and temperature. It is imperative to precisely modify, evaluate, synchronize, or adjust these influencing factors to stimulate microorganism growth in a manner that maximizes the yield of the desired metabolite. Here, RSM can be used in this regard to evaluate influencing factors and through the utilization of experimental design techniques in RSM, a specialized decision-making can be devised to optimize these factors, ultimately leading to the highest possible production of the target metabolic products.

On the other hand, metabolites generated during the stationary phase are considered as non-essential for the microbial growth, including steroids, antibiotics, toxins, alkaloids, and gibberellins. All these metabolites hold significant importance in the field of industrial microbiology for various human applications. However, the production of these secondary metabolites is greatly influenced by various quantitative parameters, such as the composition of the growth medium and operating conditions. It is crucial to finely tune these parameters to enhance the production of these secondary metabolites, where statistical optimization serves as an important tool in achieving this goal.

When optimizing media components and physicochemical parameters using RSM technique for optimizing microbial growth, metabolite production, or any bioprocess, it is essential to carefully select the qualitative and quantitative factors to be investigated. The choice of factors should be based on specific goals and the characteristics of the microorganism or process under study. Some common factors to consider are media components and physicochemical parameters (Figure 5.2).

RSM can be utilized for optimizing various media components such as carbon source concentration, nitrogen source concentration, trace elements, vitamins and growth factors, salts, and minerals. The type and concentration of carbon source such as glucose, sucrose, glycerol, etc. can significantly impact microbial growth and metabolite production. Nitrogen sources like ammonium salts, peptides, or organic nitrogen compounds play a crucial role in protein synthesis and can influence growth and product formation. The presence and concentration of trace elements, e.g., iron, zinc, and manganese, are essential for the function of various enzymes

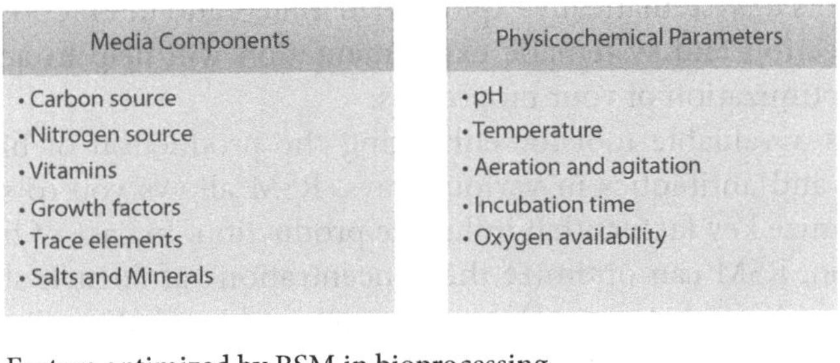

Figure 5.2 Factors optimized by RSM in bioprocessing.

and cofactors. Depending on the microorganism, the addition of specific vitamins or growth factors is necessary for optimal growth and metabolism. The concentration of salts, e.g., NaCl, and minerals, e.g., $MgSO_4$, in the medium can affect osmotic pressure and cellular ion balance.

RSM can optimize various physicochemical parameters such as pH, temperature, aeration and agitation, incubation time, and oxygen availability. The pH of the medium can have a significant impact on microbial growth and metabolism. Different microorganisms have different pH optima. Temperature affects microbial growth rates and enzyme activities. It is crucial to optimize temperature conditions for the specific microorganism. Proper aeration and agitation are vital for oxygen transfer and mixing in fermentation processes, especially for aerobic microorganisms. The duration of incubation or fermentation can influence both biomass and metabolite production. Accordingly, determination of optimized values such as pH, temperature, aeration and agitation, incubation time, and oxygen availability can be significant for receiving elevated outcomes. In aerobic processes, the concentration of dissolved oxygen is critical. Insufficient oxygen can lead to reduced growth and product formation.

RSM holds the capacity to optimize interaction factors. Sometimes, interactions between factors deliver good responses; for example, the interaction between carbon and nitrogen sources can be important. Also, some factors may have a synergistic effect, where the combined impact on the response variable is greater than the sum of their individual effects. RSM can be used in microbiology to optimize response variables, such as microbial biomass, which is often measured as cell dry weight or optical density for microbial growth optimization, metabolite production, and the yield of the product per unit of substrate consumed, which can be an important parameter for optimization. RSM can be used to optimize certain constraints in bioprocessing such as cost and availability and toxicity. It is crucial to perform a preliminary screening to identify and understand which factors are most likely to have a significant impact on the response variables. Once you have identified these factors, you can design your RSM experiments to systematically explore their effects and interactions. Proper factor selection and systematic experimentation will help to achieve the desired optimization of your bioprocess.

RSM is a valuable tool for enhancing the production of bioethanol, enzymes, and antibiotics in various ways. RSM allows you to systematically optimize key factors that influence production. In case of bioethanol production, RSM can optimize the concentrations of fermentable sugars (e.g., glucose), yeast strain, pH, temperature, and incubation time to maximize ethanol yield. During enzyme production, it helps to determine the

Table 5.1 Applications of RSM in production of microbial primary and secondary metabolites.

S. no.	Application	Microorganism used	Experimental design	Factors	Results	References
1	Production of acetoin	*Serratia marcescens* H32	RSM (CCD)	Sucrose	Production of acetoin increased to 41.5 g/L.	Sun *et al.* [1]
			Factorial (PB)	Sucrose, CSLP, citrate, ammonium, $MgSO_4$, $MnSO_4{\cdot}H_2O$, KH_2PO_4, and $FeSO_4{\cdot}_7H_2O$	Sucrose has a significant effect on acetoin production.	
2	Production of avermectin B1	*Streptomyces avermitilis* 14-12A	RSM (CCD)	Corn starch and yeast extract	Optimum concentrations of yeast extract and corn starch were 8.92 g/L and 149.57 g/L, respectively, on the production of Avermectin B1.	Gao *et al.* [2]
			Fact/PB	Corn starch, yeast extract, soya flour, a-amylase, $Na_2MoO_4{\cdot}_2H_2O$, $MnSO_4{\cdot}H_2O$, $CoCl_2$, $(NH_4)_2S_4$, and $CaCO_3$	Corn starch and yeast extract have a significant effect on the production of Avermectin B1.	

(*Continued*)

Table 5.1 Applications of RSM in production of microbial primary and secondary metabolites. (*Continued*)

S. no.	Application	Microorganism used	Experimental design	Factors	Results	References
3	Production of clavulanic acid	*Streptomyces clavuligerus*	RSM (CCD)	Soy meal powder, ornithine, and $FeSO_{4}\cdot_{7}H_{2}O$	Clavulanic acid production increased by 50%.	Oehlert *et al.* [3]
			Fact/FFD	Glycerol, ornithine, soy meal powder, $FeSO_{4}\cdot_{7}H_{2}O$, $KH_{2}PO_{4}$, and $MgSO_{4}\cdot_{7}H_{2}O$	Soy meal powder, ornithine, and $FeSO_{4}\cdot_{7}H_{2}O$ had a significant effect.	
4	Production of neomycin	*Streptomyces marinensis*	RSM (CCD)	Mineral medium, dextrin, and raspberry seed powder	The optimum concentrations were raspberry seed powder of 64.91 g/1 kg rawa, dextrin of 14.1 g/1 kg rawa, and mineral salt solution of 172.6 mL/1 kg rawa	Adinarayana *et al.* [4]

(*Continued*)

Table 5.1 Applications of RSM in production of microbial primary and secondary metabolites. (*Continued*)

S. no.	Application	Microorganism used	Experimental design	Factors	Results	References
5	Production of prodigiosin	*Streptomyces marinensis*	RSM (BB)	$MnSO_4 \cdot 4H_2O$ and $FeSO_4 \cdot 4H_2O$	Prodigiosin production was 7.07 g/L at $MnSO_4 \cdot 4H_2O$ and $FeSO_4 \cdot 4H_2O$ concentration of 3.25 mM and 0.56 mM, respectively.	Chen *et al.* [5]
			Fact/PB	$FeSO_4 \cdot 4H_2O$, $CaCl_2 \cdot 2H_2O$, $MnSO_4 \cdot 4H_2O$, and $MgSO_4 \cdot 7H_2O$	$MnSO_4 \cdot 4H_2O$ and $FeSO_4 \cdot 4H_2O$ had a significant effect on the production of prodigiosin.	

optimal growth conditions for microorganisms or the production medium for enzymes, considering factors like substrate concentration, pH, temperature, and aeration. In antibiotics production, RSM can identify optimal nutrient and culture conditions for antibiotic-producing microorganisms based on factors like carbon and nitrogen sources, pH, and temperature. RSM can be used to analyze and understand "how different factors interact with one another". It can reveal synergistic or antagonistic effects that may not be apparent through single-factor experiments. RSM uses response surface plots to visualize "how changes in multiple factors affect production". These plots can guide to determine the optimal region, where the highest production is achieved. RSM designs experiments are more efficiently than traditional OFAT approaches and reduces the number of experiments needed. This saves time and resources. RSM creates mathematical models describing the relationship between factors and production outcomes. These models can be used for predictive purposes and can be continually refined as more data become available. RSM can optimize the entire production process by identifying conditions that maximize yields while minimizing resource usage and costs.

After optimization, RSM can guide robustness testing to determine the tolerance of the process toward variations in factors, ensuring consistent production under different conditions. RSM results can help in scaling up production processes from the lab to pilot and industrial scales, ensuring that optimized conditions are maintained during larger-scale production. RSM helps identify and control factors contributing to variability in production, leading to consistent product quality. Optimized processes through RSM often result in resource savings (e.g., reduced energy and raw materials) and less waste generation, aligning with sustainability goals. RSM provides a systematic and data-driven approach to optimizing the production of bioethanol, enzymes, antibiotics, and probiotics. It aids in understanding complex interactions, maximizing yields, reducing costs, and ensuring consistent and sustainable production processes. Table 5.1 shows the application of RSM in production of microbial primary and secondary metabolites.

5.4.1 RSM for Antibiotic Production from Microorganisms

AL-Saeedi and Luti [6] conducted a study with the intension to maximize the production of effective bacteriocin from Streptococcus salivarius KA101 by employing RSM with CCD design. The study involved the evaluation of optimal factors such as $CaCO_3$, yeast extract, pH, and incubation time. The optimized concentrations for $CaCO_3$, yeast extract, pH, and incubation time to achieve enhanced bacteriocin production of 80.2 AU/ml were found to be 1.4 g/L, 13 g/L, 7, and 24 hours, correspondingly.

Mohammed *et al.* [7] conducted a study to maximize prodigiosin production using different strategies, including classical optimization and statistical experimental design. They observed peptone and sucrose as the optimal carbon and nitrogen sources for production of prodigiosin yielding approximately 135 mg/L. Productivity of prodigiosin was further enhanced to 331.6 mg/L by optimizing the carbon/nitrogen ratio (C/N) ratio to 60:40 at pH 6 and using a 2% inoculum size. Statistical optimization based on RSM technique was then employed to optimize the composition of trace elements. The design indicated that a predicted prodigiosin productivity of 406 mg/L could be achieved with specific concentrations of trace elements, i.e., $CaCl_2\cdot{}_2H_2O$ (9.22 g/L), $FeSO_4\cdot{}_4H_2O$ (0.32 g/L), $MgSO_4\cdot{}_7H_2O$ (0.67 g/L), and $MnSO_4\cdot{}_4H_2O$ (2.48 g/L).

Luti [8] explored the simultaneous production of undecylprodigiosin and actinorhodin using *Streptomyces coelicolor* by modifying Hobs medium. This medium originally contained glucose, which stimulated only actinorhodin production. They optimized fructose and sucrose concentration for enhance production of undecylprodigiosin and actinorhodin, which demonstrated the role of optimization in attaining desired outputs.

Dyaee and Luti [9] undertook a redesign of the MRS medium to enhance both the biomass yield and bacteriocin production from the bacteria *Lactobacillus plantarum* NH40. They initially employed classical optimization using OFAT method to identify the best sources of carbon and nitrogen. The results demonstrated that glucose and yeast extract were the optimal choices for carbon and nitrogen sources. Additionally, their findings suggested the possibility of removing peptone and meat extract from the original medium composition without negatively impacting the growth of *Lactobacillus plantarum* NH40 and its bacteriocin production.

Mohammed and Luti [10] applied the principles of RSM to study the experimental interaction between Serratia marcescens growth and prodigiosin production in the presence of different carbon sources. Table 5.2 shows some studies on RSM application in antibiotic production, where it can be understand that RSM can be used to mediate optimization for the enhanced production of antibiotics.

It is found that optimizing the media ingredients and all the physicochemical parameters is crucial in various fields and processes for enhancing efficiency, cost reduction, environmental impact, consistency and quality, higher yield, faster development, reducing trial and errors, process stability, resource conservation, customization, scientific understanding, regulatory

Table 5.2 RSM mediated optimization for the enhanced production of antibiotics by various *Streptomyces* spp.

S. no.	Design used	Organism	Product	Yield	References
1	2^3 factorial CCD	*S. marinensis* NUV-5	Neomycin	Maximum production of 17,150 mg/kg	Adinarayana *et al.* [4]
2	2^4 full factorial CCD	*S. coelicolor* A3(2)	Actinorhodin	32% increase	Elibol [11]
3	2^{5-1} central composite factorial design	*S. clavuligerus* NT4	Cephamycin C	2.06-fold increase in production	Bussari *et al.* [12]
4	FFD and BBD	*S. viridochromogenes* Tu57-1	Avilamycin	2.80-fold increasein production	Zhu *et al.* [13]
5	FFD and CCD	*S. clavuligerus*	Clavulanic acid	50% increase in production	Wang *et al.* [14]
6	PBD and 2^{3-1} fractional CCD	*S. olivaceus* MTCC 6820	Olivanic acid	8.00-fold increase in production	Singh *et al.* [15]
7	PBD and 2^{3-1} fractional Factorial CCD	*S. rimosus* MTCC 10792	Oxytetracycline	10 times increase in production	Singh *et al.* [15]
8	PBD and 2^{3-1} fractional factorial CCD	*S. clavuligerus* MTCC 1142	Clavulanic acid	19.37 mg/gds of Maximum production	Vastrad and Neelagund [16]
9	PBD and 2^4 full factorial CCD	*S. fradiae* NCIM 2418	Neomycin	2.70-fold increase in production	Vastrad and Neelagund [17]
10	PBD and 5-level-4-factorial CCD	*Streptomyces sp.* MAPS15	Pyrrolidone	62.41 mg/L of Maximum production	Sathiyanarayanan *et al.* [18]
11	PBD and BBD	*S. aurantiacus* AAA5	Resistomycin	3.70-fold increase in production	Vijayabharathi *et al.* [19]

(*Continued*)

Table 5.2 RSM mediated optimization for the enhanced production of antibiotics by various *Streptomyces* spp. (*Continued*)

S. no.	Design used	Organism	Product	Yield	References
12	PBD and CCD	*S. alboflavus* 313	Cyclic hexapeptide antibiotic NW-G01	2.73-fold increase in production	Guo *et al.* [20]
13	PBD and CCD	*S. aureofaciens* NCIM 2417	Tetracycline	2.87-fold increase in production	Vastrad and Neelagund [21]
14	PBD and CCD	*S. venezuelae* ATCC 15439	Pikromycin	1.45-fold increase in production	Yi *et al.* [22]

compliance, etc. Optimization can be used to maximize the desired outcome (e.g., microbial growth, product yield, and pollutant removal) while minimizing resource consumption (e.g., nutrients, energy, and time). This will lead toward more efficient processes and better resource utilization. By identifying the optimal combination of media components and physicochemical parameters, practitioners can reduce costs associated with raw materials, energy, and waste disposal. This is particularly important in industrial processes, where cost-effectiveness is a primary concern. Optimized processes often produce fewer by-products and waste, leading to a reduced environmental footprint. This aligns with sustainability goals and regulatory requirements. Optimization is necessary to help in maintaining consistent product quality by minimizing variations caused by fluctuations in raw materials or process conditions. This is critical in industries such as pharmaceuticals and food production.

In bioprocessing and biotechnology, optimizing media and conditions can significantly increase yields of valuable products such as enzymes, antibiotics, and biofuels. This, in turn, improves the economic viability of such processes. Optimization is necessary to expedites product and process development by identifying optimal conditions more quickly. This is especially important in competitive industries where time to market is a key factor. Without optimization, development often relies on trial-and-error approaches, which are time-consuming and costly. Optimization is needed to guide the process and to reduce the need for extensive experimentation. Optimization can help to identify robust conditions that are less sensitive to variations in raw materials or external factors. This increases process stability and reduces the risk of production disruptions. In fields like agriculture and aquaculture, optimizing nutrient and water use can conserve scarce resources and minimize environmental impacts, which help to meet global sustainability goals. Optimization allows processes to be tailored for specific objectives. For instance, in biotechnology, media optimization can be customized to maximize cell growth or product formation based on the desired end product. Here, optimization is required to provide insights into the relationships between contributing factors and outcomes for embracing a deeper understanding of complex processes. This knowledge can lead toward innovations and improvements. In industries with strict regulatory requirements, optimization ensures that processes consistently meet quality and safety standards. Thus, one can assume that the optimization of media ingredients and all the physico-chemical parameters is essential in a wide range of applications, including biotechnology, manufacturing, environmental remediation, and agriculture. It not only improves process efficiency but also supports economic, environmental, and quality objectives.

5.4.2 RSM in Enzyme Production

RSM can be applied to various enzyme production processes to optimize conditions and enhance enzyme yield. In the production of cellulase enzymes for the hydrolysis of cellulosic biomass to produce biofuels like ethanol, RSM can be used to optimize factors such as temperature, pH, carbon source concentration, and nitrogen source concentration to maximize cellulase activity and yield. Lipases are enzymes used in the production of biodiesel. RSM has been applied to optimize the fermentation conditions, including factors such as temperature, pH, agitation rate, and substrate concentration, to achieve the highest lipase activity and productivity. Enzymatic detergents often contain proteases to help break down protein stains. RSM can be used to optimize the production of proteases, considering many quantitative factors like temperature, pH, carbon and nitrogen sources, and agitation speed, to maximize protease yield and activity. Amylases are enzymes used in various industries, including the food and beverage industry. RSM can be employed to optimize the production of amylases by adjusting quantitative factors such as temperature, pH, substrate concentration (starch), and inoculum size to achieve the highest amylase activity. Pectinases are used in fruit juice processing to clarify the juice. RSM can be applied to improve the production of pectinase enzymes, optimizing conditions such as temperature, pH, substrate (pectin) concentration, and enzyme induction methods. Xylanases are essential in the pulp and paper industry for the degradation of hemicellulose. RSM can be utilized to optimize xylanase production by adjusting parameters such as temperature, pH, carbon and nitrogen sources, and agitation rate. In the brewing and starch industries, alpha-amylases are used for saccharification. RSM has been employed to optimize the production of alpha-amylase enzymes by varying factors like temperature, pH, substrate (starch) concentration, and inoculum size. In all these cases, RSM helps in systematically exploring the effects of multiple factors on enzyme production and constructing mathematical models to predict enzyme yield under different conditions. This enables the identification of optimal process parameters, leading to improved enzyme production efficiency and cost-effectiveness. Yonis *et al.* [23] focused on optimizing the chitin deacetylase enzyme production from *Aspergillus flavus* F1 using solid-state fermentation. They employed two distinct optimization methods named as statistical experimental design and OFAT approach to show the importance of optimized decision-making in attaining elevated outcomes. Table 5.3 shows some studies on RSM application in enzyme production.

Table 5.3 Production of microbial enzymes enhanced by RSM optimization.

S. no.	Design used	Organism	Product	Yield	References
1	Full factorial CCD	*S. erumpens* MTCC 7317	a-Amylase	3,500 U/mL of maximum production	Kar *et al.* [24]
2	PBD and BBD	*S. griseoloalbus*	a-Galactosidase	2.94-fold increase in production	Anisha *et al.* [25]
3	2^{6-1} fractional factorial CCD	*S. pulveraceus* MTCC 8374	Alkaline protease	2298.16 U/mL of maximum production	Jayasree *et al.* [26]
4	2-level FFD and CCD	*Streptomyces* sp. NK528	Chitinase	9.3% increase in production	Nawani and Kapadnis [27]
		Streptomyces sp. NK1057		29% increasein production	
		Streptomyces sp. NK951		28% increasein production	
5	L_{12}-orthogonal array and CCD	*S. lavendulae* NCIM 2499	Cholesterol oxidase	2.48-fold increase in production	Chauhan *et al.* [28]

(*Continued*)

Table 5.3 Production of microbial enzymes enhanced by RSM optimization. (*Continued*)

S. no.	Design used	Organism	Product	Yield	References
6	PBD and CCD	*Streptomyces* sp. MTCC 3119	Inulinase	Maximum production of 131 U/gds	Dilipkumar *et al.* [29]
7	CCD	*Streptomyces* sp. OM 09	Endoglucanase	3.04-fold increasein production	Karmakar and Ray [30]
8	CCD	*S. malaysiensis* TMS-1a	Keratinase	2.65-fold increase in production	Meetakoti *et al.* [31]
		S. matensis PLS-1		1.95-fold increase in production	
9	PBD and BBD	*S. psammoticus* MTCC 7334	Laccase	3.00-fold increase in production	Niladevi *et al.* [32]
10	CCD	*S. griseoluteus* WS3/1	L-asparaginase	24.56 IU/mL of maximum production	Kumari *et al.* [33]
11	PBD and face-centered CCD	*S. griseus*	L-glutaminase	171.01 IU/mL of maximum production	Suresh Kumar *et al.* [34]
12	BBD	*S. antibioticus*	Phenoxazinone synthase	4052.00 U/L of maximum production	Jia *et al.* [35]
13	2^3 full factorial CCD	*Streptomyces* sp. M49-1	RNase	2.25-fold increase in production	Demir *et al.* [36]
14	CCD	*S. hydrogenans* MGS13	Serratiopeptidase	254.65 IU/mL of maximum production	Vanama *et al.* [37]
15	2^{6-2} FFD and CCD	*Streptomyces* sp. P20	Transglutaminase	1.4 U/mL of maximum production	Macedo *et al.* [38]
16	CCD	*Streptomyces* sp. RCK-2010	Xylanase	3.00-fold increase in production	Kumar *et al.* [39]
17	CCD	*Bacillus subtilis*	Nattokinase	Increased yield at 45°C and pH 5	Eldeen *et al.* [40]

5.4.3 RSM for Bioethanol Production

RSM is commonly used in the bioethanol production process to optimize various factors and maximize ethanol yield. In the production of ethanol from corn, RSM can be used to optimize the conditions of enzymatic saccharification and fermentation. Many factors such as enzyme dosage, temperature, pH, and substrate (corn starch) concentration can be systematically varied, studied, and optimized to maximize ethanol yield. RSM can be employed in sugarcane-based ethanol production to optimize the pre-treatment, hydrolysis, and fermentation steps. Parameters like temperature, pH, enzyme loading, and reaction time can be adjusted to achieve the highest ethanol yield from sugarcane biomass. Ethanol production from lignocellulosic feedstocks like wheat straw involves complex processes, where RSM can help to optimize factors such as pre-treatment conditions (acid concentration, temperature, and time), enzymatic hydrolysis parameters, and fermentation conditions to maximize ethanol production. Switchgrass is a potential feedstock for cellulosic ethanol production, where RSM can be used to optimize the enzymatic hydrolysis of switchgrass by adjusting factors like enzyme concentration, temperature, pH, and reaction time to enhance ethanol yield. RSM can be applied in sorghum-based ethanol production to optimize quantitative factors such as grain size, moisture content, and fermentation conditions. By systematically varying identified quantitative parameters, researchers can determine the optimal conditions for ethanol production from sorghum. RSM can be used to optimize the entire process of ethanol production, including the sequential steps of pre-treatment, enzymatic hydrolysis, and fermentation. By studying the interplay of various factors in each step, RSM helps to identify conditions that lead to maximum ethanol yield. RSM can be employed to optimize the choice of yeast strains and fermentation conditions (e.g., temperature, pH, and agitation) to maximize ethanol production efficiency. The same help in studying the interactions between different yeast strains and fermentation parameters. Beyond ethanol yield, RSM can also be applied to optimize factors related to energy consumption, water usage, and overall production costs in the bioethanol industry. By reducing resource consumption and increasing efficiency, RSM contributes to more sustainable bioethanol production. RSM allows researchers and engineers to systematically investigate the influence of multiple process variables on ethanol yield, construct mathematical models to predict ethanol production under various conditions, and ultimately identify the optimal set of parameters for efficient and cost-effective bioethanol production [41, 42].

5.4.4 RSM in Biosurfactant Production

RSM can be applied to optimize various quantitative factors in biosurfactant production processes, with the goal of improving biosurfactant yield and quality. RSM can be employed to optimize the growth and biosurfactant production of microorganisms, such as bacteria or yeast. Factors like nutrient composition, carbon source concentration, nitrogen source concentration, pH, and temperature can be systematically varied and studied to maximize biosurfactant yield. RSM can help in selecting the most efficient biosurfactant-producing strains. Different strains of microorganisms may produce biosurfactants with varying properties. RSM can be used to optimize conditions for screening and selecting strains that yield biosurfactants with desirable characteristics, such as high emulsification or foaming capacity. RSM can be applied to optimize the utilization of renewable substrates (e.g., agricultural residues and waste materials) for biosurfactant production. Researchers can investigate the effects of factors like substrate type, concentration, and pre-treatment methods on biosurfactant yield. The fermentation process plays a crucial role in biosurfactant production. RSM can be used to optimize fermentation conditions, including agitation rate, aeration, temperature, and pH, to maximize biosurfactant production efficiency. RSM can be applied to optimize the extraction and purification processes for biosurfactants. Factors such as solvent type, extraction time, and purification techniques can be studied to enhance the recovery and purity of biosurfactants. RSM can also be employed to optimize the characterization of biosurfactants, including their physicochemical properties, stability under various conditions, and performance in specific applications. This optimization ensures that biosurfactants meet the required specifications for different industries, such as oil spill remediation or enhanced oil recovery. RSM plays a crucial role in optimizing biosurfactant production processes by systematically exploring the effects of various factors on biosurfactant yield, quality, and efficiency. By employing RSM techniques; researchers and industry professionals can enhance the cost-effectiveness and sustainability of biosurfactant production for a wide range of applications [43].

5.4.5 RSM in Heavy Metal Pollution Elimination

RSM can be applied to optimize processes aimed at eliminating or reducing heavy metal pollution from various environmental sources. RSM can be employed to optimize the design and operation of wastewater treatment processes for the removal of heavy metals. Many parameters such as pH, coagulant dosage, contact time, and the choice of adsorbents or precipitants can be systematically varied to maximize heavy metal removal efficiency. Phytoremediation is the use of plants to remove or stabilize heavy metals in contaminated soil or water. RSM can be used to optimize plant growth conditions, such as the type of plants, nutrient levels, and irrigation methods, to enhance the efficiency of phytoremediation. RSM can be applied in bioremediation processes that utilize microorganisms to detoxify or immobilize heavy metals. Factors such as temperature, pH, microbial inoculum size, and nutrient supplementation can be optimized to improve the bioremediation performance. Adsorption is a common method for removing heavy metals from aqueous solutions. RSM can be used to optimize adsorption parameters, including adsorbent type, dosage, contact time, and initial metal concentration, to maximize adsorption efficiency. Electrochemical methods can be used to remove heavy metals from contaminated water. RSM can optimize the design and operation of electrochemical cells by varying qualitative and quantitative factors like electrode materials, applied voltage, and electrode spacing to achieve efficient heavy metal removal. RSM can be employed to optimize the application of soil amendments or stabilizers to immobilize heavy metals in contaminated soils. Factors like amendment type, application rate, and mixing methods can be optimized to reduce the mobility of heavy metals in the soil. Researchers can use RSM to develop and optimize novel materials, such as nanoparticles or nanocomposites, for the removal of heavy metals from contaminated environments. RSM can help to identify the most effective synthesis conditions and material properties. RSM can be applied to optimize the design of monitoring networks for heavy metal pollution. It can help to determine the optimal placement of monitoring stations and sampling frequencies to detect and respond to pollution events effectively. RSM helps researchers and environmental engineers to systematically investigate the impact of multiple factors on heavy metal pollution elimination processes. By optimizing these processes, RSM contributes to more efficient and cost-effective strategies for addressing heavy metal (Pb) contamination in the environment. Tangahu *et al.* [44] reported removal of lead by using *Bacillus pumilus, Bacillus cereus, Brevibacillus choshinensis, Bacillus subtilis*, and *Rhodococcus rhodochrous*. Copper was also found treated with *Acidithiobacillus ferrooxidans* by a process known as microbial leaching [45] for elimination from printed circuit boards.

5.5 Role of Quantitative Data in RSM

Quantitative data play a fundamental role in making a RSM model, which is a statistical and mathematical technique commonly used in experimental design and optimization. RSM is often applied in various fields, including engineering, chemistry, and manufacturing, to optimize processes, activities, and products. Quantitative data are essential in RSM in model building, experimental design, model variation, optimization, visualization by plots, and error analysis. RSM aims to create mathematical models that describe the relationship between input variables (factors) and the response variables of interest. Quantitative data are necessary to build these models because it provides precise measurements and numerical values that can be used to quantify the relationships. In RSM, experiments are designed to collect quantitative data on the response variables while systematically varying the input variables within a defined experimental region. These data points are used to fit mathematical models that describe the process or system being studied. Quantitative data are essential for validating the accuracy and reliability of the mathematical models created in RSM. Researchers compare the predicted responses from the models with actual experimental results to assess how well the models represent the system under investigation. RSM is primarily used for optimization, whether it is optimizing a manufacturing process for maximum yield or finding the optimal conditions for drug formulation. Quantitative data allow researchers to determine the optimal factor settings that lead to desired response outcomes. Quantitative data help to assess the sensitivity of the response to changes in the input variables. Sensitivity analysis is crucial for understanding how different factors affect the process and for identifying critical variables that require attention. Quantitative data allow for the application of statistical tests and hypothesis testing to draw meaningful conclusions about the relationships between factors and responses. This helps researchers to make informed decisions and draw valid conclusions about the system. Quantitative data are essential for analyzing and quantifying the sources of variation and error within the experimental data. This analysis helps researchers to understand the reliability of their results and can guide improvements in the experimental design. Quantitative data are often visualized in RSM through plots, graphs, contour plots, and surface plots to help researchers to gain insights into the relationships between variables and responses. Here, visualization aids in the interpretation of the results.

5.6 Conclusion

A well-structured experimental design plays a crucial role in the development of any product or process. An effective experimental design necessitates a deep understanding of the system under investigation. RSM is a statistical technique employed for modeling and optimizing multiple variables and enables the determination of optimal process conditions. In the field of biological sciences, RSM is widely used for optimizing tasks such as the preparation of culture media, product formation, applicability studies, and wastewater treatment. Furthermore, RSM offers a valuable approach for optimizing both the composition of media components and physicochemical parameters to enhance the production of novel and existing bioactive metabolites, including enzymes, antibiotics, probiotics, and bioethanol, from various microbial sources. The RSM approach is a proved high efficient evaluation tool that can be used for building predictive models and optimizing experimental methods involving substrates and energy. It can be used for the consideration of factors and evaluation of the effects of different parameters on a process and assists in identifying the optimal values for these variables to optimize the desired response. Additionally, RSM finds applications in optimizing microbial and polymeric adsorption processes aimed at reducing heavy metal pollution. Also, quantitative data are collected, and the identification of qualitative measures is the foundation of RSM. It enables the creation of mathematical models, the design and validation of experiments, the optimization of processes, and the extraction of meaningful insights from experimental data. RSM leverages quantitative data to improve processes, products, and decision-making in various fields.

References

1. Sun, J., Zhang, L., Rao, B., Han, Y., Chu, J., Zhu, J., Wei, D., Enhanced acetoin production by Serratia marcescens H32 using statistical optimization and a two-stage agitation speed control strategy. *Biotechnol. Bioprocess Eng.*, 17, 598–605, 2012.
2. Gao, H., Liu, M., Liu, J., Dai, H., Zhou, X., Liu, X., Zhuo, Y., Zhang, W., Zhang, L., Medium optimization for the production of avermectin B1a by Streptomyces avermitilis 14-12A using response surface methodology. *Bioresour. Technol.*, 100, 4012–4016, 2009.
3. Oehlert, W.G., *A First Course in Design and Analysis of Experiments*, W.H. Freeman and Company, New York, 2000.
4. Adinarayana, K., Ellaiah, P., Srinivasulu, B., Devi, R.B., Adinarayana, G., Response surface methodological approach to optimize the nutritional parameters for neomycin production by Streptomyces marinensis under solid-state fermentation. *Process Biochem.*, 38, 11, 1565–1572, 2003.

5. Chen, W.C., Yu, W.J., Chang, C.C., Chang, J.S., Huang, S.H., Chang, C.H., Wei, Y.H., Enhancing production of prodigiosin from Serratia marcescens C3 by statistical experimental design and porous carrier addition strategy. *Biochem. Eng. J.*, 78, 93–100, 2013.
6. AL-Saeedi, B.S.M. and Luti, K.J.K., Bacteriocin from Streptococcus salivarius optimized statistically by response surface methodology active against different clinical oral pathogenic Streptococci. *Iraqi J. Sci.*, 59, 1C, 463–475, 2018.
7. Mohammed, S.J.B. and Luti, K.J.K., Response surface methodology: A review on its applications and challenges in microbial cultures. *Mater. Today: Proc.*, 42, 5, 2277–2284, 2021, https://doi.org/10.1016/j.matpr.2020.12.316.
8. Luti, K.J.K., Mixture Design of Experiments for The Optimization of Carbon Source for Promoting Undecylprodigiosin and Actinorhodin Production. *J. Pure Appl. Microbiol.*, 12, 4, 1783–1793, 2018.
9. Dyaee, N. and Luti, K.J.K., Classical and Statistical Optimization by Response Surface Methodology for Enhancing Biomass and Bacteriocin Production by Lactobacillus Plantarum. *Iraqi J. Sci.*, 60, 3, 494–508, 2019.
10. Mohammed, S.J. and Luti, K.J.K., Kinetic model for prodigiosin production by Serratia marcescens as a bio-colorant in bioreactor. *AIP Conf. Proc.*, 2213, 020027, 2020, https://doi.org/10.1063/5.0000146.
11. Elibol, M., Optimization of medium composition for actinorhodin production by Streptomyces coelicolor A3 (2) with response surface methodology. *Process Biochem.*, 39, 9, 1057–1062, 2004.
12. Bussari, B., Saudagar, P.S., Shaligram, N.S., Survase, S.A., Singhal, R.S., Production of cephamycin C by Streptomyces clavuligerus NT4 using solid-state fermentation. *J. Ind. Microbiol. Biotechnol.*, 35, 1, 49–58, 2008.
13. Zhu, C.H., Lu, F.P., He, Y.N., Zhang, J.K., Du, L.X., Statistical optimization of medium components for avilamycin production by Streptomyces viridochromogenes Tü57-1 using response surface methodology. *J. Ind. Microbiol. Biotechnol.*, 34, 271–278, 2007.
14. Wang, Y.H., Yang, B., Ren, J., Dong, M.L., Liang, D., Xu, A.L., Optimization of medium composition for the production of clavulanic acid by Streptomyces clavuligerus. *Process Biochem.*, 40, 3–4, 1161–1166, 2005.
15. Singh, N., Rai, V., Tripathi, C.K.M., Production and optimization of oxytetracycline by a new isolate Streptomyces rimosus using response surface methodology. *Med. Chem. Res.*, 21, 3140–3145, 2012.
16. Vastrad, B.M. and Neelagund, S.E., Statistical based media optimization and production of clavulanic acid by solid state fermentation in jackfruit seed powder as novel substrate using Streptomyces clavuligerus MTCC 1142. *Biosci. Eng.: Int. J.*, 1, 1, 59–77, 2014b.
17. Vastrad, B.M. and Neelagund, S.E., Optimization of medium composition for the production of neomycin by Streptomyces fradiae NCIM 2418 in solid state fermentation. *Biotechnol. Res. Int.*, 2014, Article 674286, 1–11, 2014c, doi: DOI: 10.1155/2014/674286.

18. Sathiyanarayanan, G., Gandhimathi, R., Sabarathnam, B. *et al.*, Optimization and production of pyrrolidone antimicrobial agent from marine sponge-associated Streptomyces sp. MAPS15. *Bioprocess. Biosyst. Eng.*, 37, 561–73, 2014, DOI: 10.1007/s00449-013-1023-2.
19. Vijayabharathi, R., Brunthadevi, P., Sathyabama, S., Bruheim, P., Priyadarisini, V.B., Optimization of resistomycin production purified from Streptomyces aurantiacusAAA5 using response surface methodology. *J. Biochem. Technol.*, 3, 4, 402–408, 2012.
20. Guo, Z., Shen, L., Ji, Z., Wu, W., Enhanced production of a novel cyclic hexapeptide antibiotic (NW-G01) by Streptomyces alboflavus 313 using response surface methodology. *Int. J. Mol. Sci.*, 13, 4, 5230–5241, 2012.
21. Vastrad, B.M. and Neelagund, S.E., Optimizing the medium conditions for production of tetracycline by solid state fermentation of Streptomyces aureofaciens NCIM 2417 using statistical experimental methods. *Biosci. Eng.: Int. J.*, 1, 1, 29–44, 2018.
22. Yi, J.S., Kim, M., Kim, S.J., Kim, B.G., Effects of sucrose, phosphate, and calcium carbonate on the production of pikromycin from Streptomyces venezuelae. *J. Microbiol. Biotechnol.*, 25, 4, 496–502, 2015.
23. Yonis, R.W., Luti, K.J.K., Aziz, G.M., Statistical optimization of chitin bioconversion to produce an effective chitosan in solid state fermentation by Asperigellus flavus. *Iraqi J. Agric. Sci.*, 50, 3, 916–927, 2019.
24. Kar, S., Ray, R.C., Mohapatra, U.B., α-amylase production by Streptomyces erumpens MTCC 7317 in solid state fermentation using response surface methodology (RSM). *Pol. J. Microbiol.*, 57, 4, 289–296, 2008.
25. Anisha, G.S., Sukumaran, R.K., Prema, P., Statistical optimization of α-galactosidase production in submerged fermentation by Streptomyces griseoloalbus using response surface methodology. *Food Technol. Biotechnol.*, 46, 2, 171–177, 2008.
26. Jayasree, D., Tadikamal, S., Rao, C.H.S., Rao, J.V., Narasu, M.L., Enhancement of alkali protease production isolated from Streptomyces pulveraceus using response surface methodology. *Int. J. Pharm. Pharm. Sci.*, 4, 4, 1–6, 2012.
27. Nawani, N.N. and Kapadnis, B.P., Optimization of chitinase production using statistics based experimental designs. *Process Biochem.*, 40, 2, 651–660, 2005.
28. Chauhan, A.K., Survase, S.A., Kishenkumar, J., Annapure, U.S., Medium optimization by orthogonal array and response surface methodology for cholesterol oxidase production by Streptomyces lavendulae NCIM 2499. *J. Gen. Appl. Microbiol.*, 55, 3, 171–180, 2009.
29. Dilipkumar, M., Rajasimman, M., Rajamohan, N., Enhanced inulinase production by Streptomyces sp. in solid state fermentation through statistical designs. *3 Biotech.*, 3, 6, 509–515, 2013.
30. Karmakar, M. and Ray, R.R., Endoglucanase synthesis from submerged fermentation of dried grass by Streptomyces sp. *Am. J. Bioeng. Biotechnol.*, 1, 1–10, 2013.

31. Meetakoti, P., Duddu, M.K., Guntuku, G. *et al.*, Keratinolytic activity and statistical optimization of keratinase production by Streptomyces matensis PLS-1 and Streptomyces malaysiensis TMS-1a. *Res. J. Pharm. Biol. Chem. Sci.*, 6, 1, 1575–86, 2015.
32. Niladevi, K.N., Sukumaran, R.K., Jacob, N., Anisha, G.S., Prema, P., Optimization of laccase production from a novel strain—Streptomyces psammoticus using response surface methodology. *Microbiol. Res.*, 164, 1, 105–113, 2009.
33. Kumari, P.V.K., Girija Sankarand, G., Prabhakar, T., Optimization of L-asparaginase production by Streptomyces griseoluteus WS3/1 using experimental methods. *J. Pharm. Biomed. Sci.*, 10, 1–6, 2011.
34. Suresh Kumar, S., Muthuvelayudham, R., Viruthagiri, T., Optimization of L-glutaminase (EC. 3.5. 1.2) enzyme production by Streptomyces griseus using rice bran through statistical designs. *J. Chem. Pharm. Res.*, 5, 386–94, 2013.
35. Jia, H., Gao, Z., Ma, Y., Zhong, C., Xie, Y., Zhou, H., Wei, P., Optimization of phenoxazinone synthase production by response surface methodology and its application in Congo red decolourization. *Electron. J. Biotechnol.*, 16, 5, 2–2, 2013.
36. Demir, T., Gübe, Ö., Yücel, M., Hameş-Kocabaş, E.E., Increased alkalotolerant and thermostable ribonuclease (RNase) production from alkaliphilic Streptomyces sp. M49-1 by optimizing the growth conditions using response surface methodology. *World J. Microbiol. Biotechnol.*, 29, 1625–1633, 2013.
37. Vanama, J., Chintada, H., Guntuku, G., Tadimalla, P., Application of Response Surface Methodology in Medium Components Optimization to Enhance Serratiopeptidase Production by Streptomyces Hydrogenansmgs13. *Eur. Sci. J.*, 10, 12, 196–209, 2014.
38. Macedo, J.A., Sette, L.D., Sato, H.H., Optimization of medium composition for transglutaminase production by a Brazilian soil Streptomyces sp. *Electron. J. Biotechnol.*, 10, 4, 618–626, 2007.
39. Kumar, A., Gupta, R., Shrivastava, B., Khasa, Y.P., Kuhad, R.C., Xylanase production from an alkalophilic actinomycete isolate Streptomyces sp. RCK-2010, its characterization and application in saccharification of second generation biomass. *J. Mol. Catal. B: Enzym.*, 74, 3-4, 170–177, 2012.
40. Eldeen, K., II, Ibrahim, H.M., Elkhidir, E.E., Elamin, H.B., Optimization of culture conditions to enhance Nattokinase production using RSM. *Am. J. Microbiol. Res.*, *3*, 5, 165–170, 2015.
41. Chauhan, M., Dutt, S., Manjul, A.S.,., Singh, B., Garlapati, V.K., A sustainable approach of turning potato waste towards bioethanol production using indigenous microbes of Himachal Pradesh, India. *Chemosphere*, 299, 134429, 2022, https://doi.org/10.1016/j.chemosphere.2022.134429.
42. Elumalai, R.S., Ramanujam, P.K., Tawfik, M.A., Ravichandran, P., Gurunathan, B., Optimization and kinetics modeling for enhancing the bioethanol production from banana peduncle using Trichoderma reesei and

Kluyveromyces marxianus by co-pretreatment methods. *Sustain. Energy Technol. Assess.*, 56, 103129, 2023.
43. Nalini, S., Inbakandan, D., Dhas, T.S., Sathiyamurthi, S., Optimization of biosurfactant production by marine Streptomyces youssoufiensis SNSAA03: A comparative study of RSM and ANN approach. *Results Chem.*, 3, 100223, 2021.
44. Tangahu, B.V., Abdullah, S.R.S., Basri, H., Idris, M., Anuar, N., Mukhlisin, M., Lead (Pb) removal from contaminated water using constructed wetland planted with Scirpus grossus: Optimization using response surface methodology (RSM) and assessment of rhizobacterial addition. *Chemosphere*, 291, 132952, 2022.
45. Murugesan, M.P., Kannan, K., Selvaganapathy, T., Bioleaching recovery of copper from printed circuit boards and optimization of various parameters using response surface methodology (RSM). *Mater. Today: Proc.*, 26, 2720–2728, 2020.

6

Evaluating Mass-Spring-Damper Systems and Models for Reinforcing Engineering Designs: A Qualitative and Quantitative Approach

Prashant Lahre, Kundan Meshram*, Shailendra Kumar and Rajendra Kumar Choubey

Department of Civil Engineering, School of Studies of Engineering and Technology, Guru Ghasidas Vishwavidyalaya Bilaspur, Chhattisgarh, India

Abstract

In the present chapter, a mass-spring-damper (MSD) system is evaluated for reinforcing the capabilities of engineering designs based on qualitative and quantitative information. It is important to model the dynamic characteristics of different types of systems including structures, bridges, vehicles, and aeroplanes where MSD systems are commonly applied. MSD systems are utilized extensively in a variety of engineering disciplines, including aeronautical, mechanical, and civil engineering. It is prominent to find the appropriate values of system parameters in MSD like mass, stiffness, damping coefficient, and excitation frequency that minimize a given objective function like displacement, acceleration, or energy, which is a necessary step in the optimization of MSD systems. The same will help in finding the optimal values of the system parameters for attaining maximum outputs. In this chapter, MSD models are evaluated on the basis of brief review of optimization models for mass damper systems. Moreover, the benefits of optimizing mass damper systems and role of qualitative optimization along with quantitative information are presented. To effectively analyze the characteristics of the MSD model and to reinforce the abilities of mass damper systems, the importance of qualitative cum quantitative information in a MSD system for societal utility is presented. The MSD model parameters like spring rigidity and attenuation coefficient are presented, and algorithmic optimization technique like particle swarm

**Corresponding author*: kundan.transpo@gmail.com

Atul Kumar Sahu, Rakesh D. Raut, Rohit Raja, Anoop Kumar Sahu and Nitin Kumar Sahu (eds.) Industrial and Manufacturing Designs: Quantitative and Qualitative Analysis, (173–200) © 2024 Scrivener Publishing LLC

optimization is presented. In the present chapter, many criteria related to MSD are presented by the authors that can help in optimized manufacturing of spring dampers, which will induce safety, cost decrements, long life, sustainability, material savings, etc. In the present chapter, an extensive review of various existing optimization models for mass damper systems is presented with the intention to define crucial facts.

***Keywords*:** Mass-spring-damper (MSD), particle swarm optimization (PSO), liquid-circular rotational accelerometers (LCAAs), frequency, acceleration

6.1 Introduction

The mass-spring-damper (MSD) systems are extremely prevalent in the world of engineering and are used to sustainable replicate the dynamic behavior of a wide variety of structures and equipment. The MSD is usually referred as MSD, which consists of distinct mass nodes spread all through an object and connected via arrangement of springs and dampers. MSD sustainable mechanism consists of a mass that is put under the influence of an external force and is coupled to a spring as well as a damper. Both of these components are connected to one another. The numbers of the system characteristics such as its mass, elasticity, attenuation coefficient, and stimulation frequency determine the dynamic behavior of that system, i.e., the mass of a system determines that is springy. The values of these parameters are needed to be optimized and combined to produce the greatest reduction in a particular objective function such as the amount of displacement, acceleration or energy. Accordingly, it is required to build decision-making models for the process of optimizing MSD systems, which can help in achieving aforesaid functions and assist in locating the appropriate values that produce this reduction.

MSD system is required to be evaluated to attain long-term benefits and maximization of outputs in various applications, i.e., structure building, vibration control, and energy gathering. The elements in MSD system should also be structurally constructed to minimize the displacement and acceleration by distributing the load more evenly. In MSD, optimization among elements and input parameters is needed to reduce the overall vibration amplitude in vibration management. The same is used to maximize power generation by calibrating the device properties to the stimulation frequency. The optimization and development of MSD systems is a process that is plagued with difficulty because of the nonlinearity and complexity of the dynamic behavior of the system. Manufacturing process allied with MSD elements and the right structuring of said elements in a closed loop system is extremely challenging as the wrong structuring will lead in poor characteristics and deficient outputs. The functional behavior of the MSD system is crucially influenced by several factors, i.e., mass distribution, attenuation ratio, stiffness, and stimulation frequency in comparison to many others. As a consequence, an efficient optimization procedure is necessary in order to determine the values of the system parameters that produce the best results. There is a need to build algorithm that can have the capacity to handle the nonlinearity and complexity of the underlying system.

In recent years, a number of optimization strategies that are tailored particularly to manage MSD systems have been found. These strategies include simulated annealing, genetic algorithms, particle swarm optimization (PSO), and gradient-based approaches. The aforesaid are the different types of algorithms that can be utilize to determine the values of the system parameters that produce the best results. Each algorithm has several benefits and drawbacks, and the choice of selection of proper algorithm to use is dependent on the kind of problem that needs to be solved as well as the objective function that needs to be met. The optimization of MSD systems is an important subject that has a wide range of applications in a variety of engineering subfields. Accordingly, it is always needed to develop efficient optimization techniques and the implementation of a variety of optimization strategies in conjunction with one another for different scenarios to yield appropriate output with implemented MSD.

6.2 Extensive Review of Existing Optimization Models for Mass Damper Systems

Today, it is required to reduce vibration effects from the system to maintain the functional characteristics for efficient working and to impart sustainable life span to it. It is possible to reduce the number of vibrations that are experienced in buildings, different types of structures and diverse system by utilizing devices that are classified as mass dampers. Implementation of mass dampers in the working characteristics of the system is an effective technique to reduce vibration. It is challenging and indeed requirement to optimize the variables associate with MSD for attaining efficiency and effectiveness. The optimization of these systems can be a challenging endeavor due to the complexity of the underlying structure as well as the large number of variables that are involved in the process.

Optimization in MSD is required to be done by utilizing optimization algorithms for the purpose to achieve a higher level of performance. The combination of different kinds of optimization algorithms would be feasible choice to optimize MSD in comparison to any one algorithm. It can be significant to develop hybrid optimization algorithms that incorporate several different optimization algorithms to attain maximum efficiency, but the same is challenging and difficult in attainment. This is required to be done to achieve a higher level of performance. Researchers have utilized composite optimization strategies to optimize the mass, attenuation, and flexibility of the mass damper system. The aforesaid three aspects of the system have been considered by the researchers to determine their optimal levels. The same will assists in attaining elevated structure's response toward handling dynamic pressures successfully with reduced vibration by the application of these algorithmic models. The performance of the mass damper systems can be successfully improved by the application of algorithmic models. The objective of reduced vibration can be successfully completed by MSD for handling respective situations.

The MSD model can be prepared on the basis of qualitative characteristics, and long-term benefit can be acquired by the built qualitative model. The same can be used to plan the behavior of mechanical systems specifically those that exhibit oscillating motion. An MSD model that exhibits the oscillatory motion of a mass, which stands for the thing that is being examined, is connected with a spring and an absorber that stand for the pressures being exerted on the thing. Together, these components can be used to make up the model, where it is required to collect qualitative or quantitative information for said components. The MSD model has seen significant application across a variety of domains, including mechanical engineering, physics, and control systems. After understanding the course of this literature analysis, it is found that the research studies that concentrate on the investigation of a selection of the MSD model's applications and developments are indeed requirement for vibration sustainability.

The MSD paradigm can be utilized in a wide variety of contexts and settings. The model can be applied in the field of mechanical engineering to investigate the behavior of mechanical systems like automobiles, aeroplanes, and other types of industrial machinery manufacturing. The behavior of a suspension system in a vehicle for instance can be analyzed with the help of MSD model, where the mass component of the model represents the body of the vehicle, the spring component represents the suspension springs, and the damper component represents the shock absorbers. The MSD model can also be applied in the field of physics to investigate the behavior of oscillating systems like pendulums, simple harmonic oscillators, and wave transmission. In the discipline of biomechanics, the model can be use to investigate how the human body reacts to its surroundings and how it interacts with the world around it.

The sustainable MSD system can be utilized in the field of control systems engineering for the purpose of designing control systems for a wide variety of applications including automation, aerospace, and long-lasting automobile systems. The model can be used to create feedback management systems that stabilize the system and minimize fluctuations produced by perturbations. These systems are needed to be optimized for reducing cost based on algorithmic design based on variables under the said model. The MSD model variant can make the advancements both in terms of its precision and its adaptability to a variety of different kinds of systems. The sustainable system characteristics and life can be improved in a number of ways by developing a right model. The incorporation of nonlinearities exists into the MSD model, which is one of the examples that guarantee the need to develop MSD model. Here, nonlinearities can be caused by a number of different variables including friction, the nonlinearity of the substance, and the nonlinearity of the geometry. Incorporating these nonlinearities toward developing the MSD model allows for a more accurate representation of the behavior related to real-world systems.

In addition to this, the MSD models are also needed to be developed by incorporating the utilization of fractional derivatives. The concept of derivatives is expanded by fractional derivatives, which include values that are not integers. This allows a more accurate representation of the behavior of systems that have long-term memory or complicated dynamics. The MSD model is a flexible mathematical model that is utilized in a broad variety of contexts and has many different implementations in a variety of areas. Because of its high applicability in use under different areas, a high degree of precision is an appealing characteristic needed for the study of the behavior of mechanical and oscillating systems. Incorporating nonlinearities and fractional derivatives into the MSD model is one way in which the applicability of the model can be further enhanced for different used cases.

It is found that the conventional dampening control techniques in semi-active suspension systems possess limited characteristics. Accordingly, a new strategy based on enhanced PSO is proposed by Li *et al.* [1] to improve efficiency and precision in the solution. The strategy uses a dynamic non-linear inertia coefficient and maximizes non-linear damping curves for various types of expression. The algorithm is then tested in a computational model and is found to achieve high vibration isolation at a high frequency and powerful suppressive resonance ability at a low frequency while maintaining the movement of the spring mass. The study suggests that this approach can improve the comfort of operating a vehicle while traveling [1].

Chen *et al.* [2] examined and addressed the reduction of oscillation in a dual-mass flywheel (DMF) rotational damper under various operating circumstances. Their study constructs a 6–degree-of-freedom (DOF) non-linear transmission system model that includes the engine, DMF, gearbox gear pair, and clutch friction model to design a multi-condition simulation strategy to analyze DMF's vibration reduction performance. Chen *et al.* [2] validated the accuracy of the transmission system model through simulations tests and also investigated the impact of DMF performance characteristics on a rotational oscillation in different operating conditions, where they primarily found that DMF performs best in reducing disturbance during drive situations and additionally found that the open angle has a minimal effect on DMF's dampening effectiveness, whereas other characteristics have a more significant impact [2]. Kandil *et al.* [3] focused on the use of a servo-controlled linear actuator to control the dynamics of a mass-spring-damper system. The position of the mass is detected using a linear variable differential transformer (LVDT), which generates an output signal proportional to the mass position. The output signal is processed by a positive position feedback controller, which generates a control signal that is sent to the actuator, which then regulates the mass position. To obtain the system's dynamical equations for the amplitudes and phases, Kandil *et al.* [3] used the Krylov–Bogoliubov averaging perturbation technique. Their study included 2D and 3D representations to aid in visualizing the system's behavior before and after control. Their study presented a comprehensive approach toward controlling the dynamics of a mass-damper-spring system using a servo-controlled linear actuator and LVDT feedback. The Krylov–Bogoliubov averaging perturbation technique is found as a valuable tool for obtaining the system's dynamical equations and understanding its behavior [3].

Takayama *et al.* [4] have introduced the two types of eddy current dampers named as perpendicular-motion–type and parallel-motion–type, where they proposed a method (named as A method) for easily obtaining damping coefficients using magnetic vector potentials in a stationary-conductor coordinate system. The article also compares their A method with the previously suggested coil method and 3D-FEM (3D finite element method) and demonstrated good agreement between the results obtained through the A method and the 3D-FEM. Finally, their study suggested the application of an equivalent mass of the disc to further improve the accuracy of the method in calculating damping ratios [4]. Xia *et al.* [5] suggested the use of auto-parametric internal resonance systems for mass detection in scenarios, where the excitation amplitude varies due to factors such as temperature drift. Their systems used a slope-based working mechanism that transitions into a flat roof, when modal damping increases. Their system is independent of excitation amplitude and frequency. The slope of the flat roof depends on mass perturbations, where the frequency range expands or contracts based on excitation amplitude. Consequently, a method for excitation-insensitive mass detection based on the swaying of flat roofs is presented, which is mutually validated by theoretical and empirical findings. The proposed sensing system is insensitive to slow changes in excitation amplitude and has potential in engineering scenarios with varying environmental temperatures [5].

Hao *et al.* [6] proposed an electrical control system using an eccentric mass block–type actuator to reduce the intermittent resonance force generated by the primary spiral blade of a non-fixed-wing aircraft. Their study provided a detailed process of constructing a prototype for the proposed system and validated its efficiency and dependability using sensitivity control theory. Their results demonstrated that the prototype meets the vibration control requirements of the primary spiral blade of the aircraft and suggested that the proposed system is an innovative solution to a challenging problem in aviation [6]. Fang *et al.* [7] proposed a new theory to address the issue of cross-axis sensitivity in micro-electro-mechanical-systems (MEMS) devices that are based on parallel plate structures. This problem is caused by errors in material and fabrication processes leading to a decline in device performance. The theory proposed by Fang *et al.* [7] focused on the interaction between cross-axis interference and compression film air damping and developed an associated theory of damping-induced suppression, which is named as a model of a tilting plate with small amplitude and derives analytical expressions to characterize the phenomenon. The analysis demonstrated that the damping-induced enhancement effect exists, which is particularly noteworthy in situations of large mass and small air layer thickness. Here, the cross-axis sensitivity is found reduced by 80%, when measured in terms of air damping. The proposed theory can be used to elaborate the interplay between squeeze film air damping and cross-axis interference in non-vacuum MEMS devices, which is important for achieving high performance and stability [7].

After, identifying the above said literatures [1–7], it is found that many spring damping variables, like movement of the spring mass, DOF, rate of oscillation, drive situations, mass position, damping ratios, slope of the flat roof, frequency range, resonance force generated, cross-axis sensitivity, and material and fabrication processes, are considered by the researchers, which can be qualitative and quantitative for developing MSD model.

Afterward, additional literatures are exploited to understand more critical variables, parameters, and factors for designing MSD, where it is found that De Oliveira and Travaglia [8] optimized the stiffness of circular spring-mass systems with nearest-neighbor (NN) and next-nearest-neighbor (NNN) springs, which are subjected to dampening and intermittent exterior pressures. Their goal is to minimize the average ratio of the square norm of on-site internal forces (reaction) to the square norm of on-site exterior forces, which are averaged over time and across all possible configurations of exterior factors. Their primary finding suggested that, as the number of masses in the system increases, the optimal stiffness matrix converges to a discontinuous and periodic bi-Laplacian operator. This optimal stiffness matrix has a negative stiffness for NNN springs and the relation between the optimal NNN spring constant and the optimum NN spring constant unbalances. This means that installing springs with negative elasticity in certain locations can lower the ratio of response to input. Numerical charts are provided to illustrate the primary acceptable outcome for each of the various situations [8].

Rajaprasanna *et al.* [9] defined a methodology for evaluating system characterization of lightly damped processes using the Kautz model and state space representation. The Kautz model is a two-parameter orthogonal basis function representation that can identify processes with underdamped characteristics. The methodology utilizes subspace techniques and sequential execution of the Kautz model stages to derive process parameters, which are then converted into a convex optimization problem. The eigenvalues (poles) of the process are approximated as constrained within user-defined regions in the convex plane, which improved the optimization efficiency. Their study suggested to use a state observer architecture as an effective control strategy for attaining optimum states. Their methodology was tested using a Coriolis mass flow meter in a no-flow situation, and the results were displayed using MATLAB, where crucial improvements and control are identified [9]. Feng *et al.* [10] suggested to utilize high-rise buildings, which are a type of large flexible structure, where dumping issue is required to take into consideration pertaining to wind resistance and anti-seismic protection in order to ensure the safety of their occupants and their properties particularly in coastal areas. Their study presented a model with an unlimited number of dimensions as well as an adaptive boundary control rule for an active mass damper (AMD). They developed a dynamic model of the high-rise structure, which is a combination of several stories, each of which possesses malleable walls and stiff floors and is subjected to several different physical circumstances. The adaptive boundary controller is applied to an AMD that is installed on the upper level to dampen the disturbance felt on each floor and ensure the inhabitants' sense of well-being. In addition, calculations and experiments are performed on an adaptable structure with two floors to demonstrate the efficacy of the suggested management approach [10].

Tawaki *et al.* [11] presented a MSD model to categorized elderly individuals, who experienced falling based on their center of pressure (COP) trajectories during a silent standing test. They approximated six structural parameters of the MSD model using data from 212 individuals, who completed the standing assessments under various conditions. Their study found that, in the MSD parameter methods, classification precision was highest in the foam condition for categorizing senior individuals with a positive probability ratio of approximately 8.0. The MSD parameters were found able to detect fluctuations in COP behavior caused by aging, altering the floor surface, and changes in viewing circumstances in their study [11]. Accordingly, they stressed on identifying more allied parameters in the form of qualitative or quantitative under their model for attaining more robust solutions.

Zhao *et al.* [12] discussed about the use of impedance detection techniques for measuring mass in resonant mass sensors. They found traditional frequency scanning technique not suitable for identifying incremental mass. They found that the geometrical parameters affect the impedance sensitivity while considering the effects of electromechanical coupling. Zhao *et al.* [12] proposed design rules to enhance sensitivity, which included increasing the cantilever length and the piezoelectric thickness. Their design rules were validated by fabricating three macro-sized sensors with cantilever lengths of 21.0 mm, 26.0 mm, and 30.0 mm, respectively. The measured impedance sensitivities were found to increase with the cantilever length, whereas the frequency sensitivities decreased. Furthermore, the impedance sensitivity was found significantly higher than the frequency sensitivity; accordingly, a general framework for optimizing the sensor structure is presented by them [12].

Adiputra *et al.* [13] investigated the use of an ankle velocity reference to improve the passive controllable ankle foot orthosis (PICAFO) for post-stroke patients. The ankle velocity reference is defined as the average ankle velocity of a healthy individual walking at a particular speed. Their study derived the walking motion data of 16 healthy individuals with different body mass indexes to build an ankle velocity with appropriate damping. The effect of controlled stiffness (based on the ankle velocity reference) on ankle kinematics and muscle activity is presented by comparing it with various settings, which included walking barefoot. They suggested to maintain constant damping stiffness levels toward controlling the ankle stiffness based on the ankle velocity reference to improve ankle biomechanics and to reduce muscle activity during walking, leading to better walking stability. Their study suggests that the ankle velocity reference can be used to determine an appropriate PICAFO prescription for post-stroke patients without relying on the trial-and-error method [13]. Li *et al.* [14] suggested to implement energy-driven damper (EDD) to maintain appropriate damping. They defined EDD controller as a type of semiactive suspension controller that uses the kinetic energy of the sprung mass to power the dampers and improve comfort. They used CarSim-MATLAB co-simulation under a nonlinear magnetorheological damper model to compare the effectiveness of the EDD controller to other popular semi-active suspension controllers such as Skyhook (SH), acceleration-driven damper (ADD), and mixed SHADD. The results demonstrated that the EDD controller is superior in terms of both comfort and energy efficiency. The design mechanism of the EDD controller is explained in terms of energy and a linear quadratic regulator to optimize the energy input and output of the system [14].

Das *et al.* [15] proposed a new method to enhance vibration prediction in permanent magnet synchronous machines (PMSMs) by applying frequency-dependent damping. Their approach involved the use of Rayleigh damping coefficients that depend on the mass and stiffness of the system instead of assuming a constant value. To obtain these coefficients, the authors conducted impact hammer tests on two 12-slot 10-pole PMSM prototypes and measured their frequency response functions (FRFs). The obtained coefficients were then used to predict the vibration behavior of a third prototype, and the results were experimentally validated [15]. To facilitate future construction, their paper also proposes a damping coefficient prediction strategy that considers the changes in mass, natural frequency, and stiffness between different PMSM designs. By using their method, it can be possible to rapidly predict the Rayleigh damping coefficients for new builds and to reduce the reliance on prototype availability.

Bagchi [16] presented a traditional MSD model based on the differential equation of the second order that describes the motion of a single mass that is connected to a spring and an absorber. This model is known as the MSD equation. Here, the author considered the motion of a mass as one of the parameter to determine the effectiveness of the MSD model [16]. Their second-order expression can be identify from Equation 6.1 , where m is the mass, c is the dampening coefficient, k is the spring constant, x is the displacement of the mass from its equilibrium location, F is the exterior force that is applied to the mass, and t is the passage of time. This equation is used to characterize a broad variety of physical systems, such as electrical systems like resistor (R), an inductor (L), and a capacitor (C) (RLC) circuits and mechanical systems like automotive suspensions among others.

$$\left(\frac{md^2x}{dt}\right)^2 + \frac{c*dx}{dt} + kx = F \tag{6.1}$$

Liu *et al.* [17] stressed on MEMS-based electrochemical angular accelerometer, which is based on the electrode structures toward showing the feasibility and promise of angular accelerometer applications in seismology. Here, they suggested to translate angular acceleration into an electrical signal to provide new opportunities in disciplines like navigation, robotics, and motion sensing. They stressed on the collaborative efforts of mathematical evaluation, modeling, and manufacturing to demonstrate the interdisciplinary approach required for creating and furthering cutting-edge sensor technology. They stressed on the identification of essential elements of liquid sensors to describe the inner workings and complexities of sensors' design and performance. Their study provided essential information to researchers, engineers, and experts in the area by providing a complete evaluation of fluid-based technology in the context of inertial sensors, thereby allowing them to make educated judgments about sensor selection and implementation in controlling in seismology [17]. Elaswad *et al.* [18] provided a feasible and viable option for the attainment of outstanding performance with accelerometers and gyroscopes. Their study provided breakthrough ramifications for areas such as inertial navigation and rotating seismology, where accurate angular acceleration measurement is critical. Their research greatly advances the state of the art in radial measurements measuring technology by presenting a revolutionary technique that integrates fiber-optic technology with interferometric concepts [18]. They considered signal-to-noise ratio as a parameter for evaluation.

Xiangdong *et al.* [19] demonstrated that that proportional–integral–derivative (PID) controllers can enhance system behavior through PSO. Their study proposed a complete technique for designing and optimizing PID controllers in complex systems by PSO, heuristic optimization, and analytical insights [19]. Additionally, a comparison of the impacts based on classical fixed-point theory and the intelligent PSO algorithm is made by Shukla and Sinha [20]. Their results showed that combining classical theory with an intelligent algorithm can yield improved results for active vibration control. This combination, in particular, can aid in achieving almost identical resonance peaks on amplitude–frequency response curves, which is a difficult achievement under fixed-point theory [20].

The above-said literatures presented the wide applicability of mass dumping system in various application scenarios and additionally presented various indexes, which are determined and used by the researchers based on accumulation of qualitative and quantitative data information onto the model for evaluating the MSD. After understanding the extensive review of the existing optimization models for mass damper systems, it is found that, for designing a more efficient MSD model, it is crucial to explore maximum number of qualitative and quantitative factors, parameters, variables,

Table 6.1 Directory of qualitative and quantitative parameters for MSD modeling.

S. no.	Qualitative or quantitative parameters	S. no.	Qualitative or quantitative parameters
1.	Movement of the spring mass	13.	Internal forces
2.	Degree of freedom (DOF)	14.	Ratio of response
3.	Rate of oscillation	15.	Derivation of process parameters
4.	Drive situations	16.	Number of dimensions
5.	Mass position	17.	Rate of oscillation amplitudes
6.	Damping ratios	18.	Number of linkages
7.	Slope of the flat roof	19.	Aging
8.	Frequency range	20.	Heat by friction generation ratio
9.	Resonance force	21.	Stressing of material
10.	Cross-axis sensitivity	22.	Designing of geometrical parameters
11.	Material and fabrication processes	23.	Cantilever length
12.	Stiffness	24.	Piezoelectric thickness

etc., whose information collection is very easy for building a more robust model. The quality of the build model depends on the number of variables, which is precisely needed to be explored and identified more in the context of development of the robust MSD model.

In the present chapter, after identifying the critical literatures, many spring dumping variables like movement of the spring mass, DOF, rate of oscillation, drive situations, mass position, damping ratios, slope of the flat roof, frequency range, resonance force generated, cross-axis sensitivity, and material and fabrication processes are found crucial for building MSD model, which can be qualitative and quantitative. Additionally, stiffness, internal forces, ratio of response, derivation of process parameters, number of dimensions, rate of oscillation amplitudes, number of linkages, aging, heat via friction generation ratio, stressing of material, designing of geometrical parameters, cantilever length, and piezoelectric thickness are also found critical parameters for the dampers to prevent bouncing up and down in various application fields. The directory of qualitative and quantitative parameters that are used for the MSD modeling and their evaluations are shown in Table 6.1.

6.3 Use of Mass Damper Systems: Active and Passive

A mass damper system is a component that can diminish the amplitude of vibrations generated by external factors such as mechanical shocks, wind, or earthquakes. These vibrations can occur when an object is subjected to conditions related to the movement of object from one extreme position to another extreme and back again. In majority of the time, it is needed to install mass damper devices in buildings with multiple stories and mechanical machineries and in other system, where movement among components is required. The primary mass of the system is intended to move in one direction, whereas the secondary mass of the system moves in the opposite direction of the primary mass. Because of this, the overall amplitude of the fluctuation can be reduced when the second mass is added to the system. The operation of the mass damper device adheres to the tailor-made mass damper principle. In order to apply this principle, it is necessary to adjust the mass of the system as well as its flexibility, so that it is consistent with the structure's natural frequency of oscillation.

There are many uses of active and quiescent mass damper devices and each has its own unique set of applications. Active system requires energy for working, whereas passive systems are those that can continue to function even in the absence of any external source of energy by having their frequencies tuned to a specific point in order to reduce fluctuations. They are composed of a mass that is linked to a spring and an absorber that is attached to the framework in some fashion. They are also attached to the framework in some manner. The capacity of the spring to absorb the energy of the structure in conjunction with the capacity of the damper to dissipate that energy results in a diminution in the overall magnitude of the perturbation.

Active systems require an input of external energy and make use of sensors and control systems to make real-time modifications to the movement of the mass. The sensors can detect any movement that the structure might be making and the management system will then adjust the movement of the mass in order to cancel out any oscillations that the structure might be causing. Active systems are more expensive than passive systems, but they offer greater control over the vibration of the building and can be adjusted to compensate for changes in the frequency of oscillation of the building. Although active systems cost more, they also offer greater control over the building's vibration. Active systems are more expensive as they offer greater control over the vibrations that are transmitted through the structure mass damper systems. These systems are an effective technique that can be utilized to reduce the negative effects that are caused by external variables on towering structures, bridges, and industrial machinery. Wind and earthquakes are two examples, where active system is required to control variables. The system functions according to the principle of the customized mass damper, which involves adjusting the mass and flexibility of the system so that it is consistent with the frequency of oscillation of the structure. This allows the system to effectively dampen the oscillations of the structure. There are many uses for active and quiescent mass damper devices, and each has its own unique set of applications. Active systems possess the caliber to make real-time modifications to the movement of the mass. Passive systems are those that can continue to function even in the absence of any external source of energy by having their frequencies tuned to a specific point to reduce fluctuations. Accordingly, one can say that active and passive, both the system structures, are nowadays used by the practitioners for successfully damping undesired vibration, which depend on the multiple qualitative or quantitative means. It is now a common practice in a wide range of applications including the building of tall structures, bridges, and even some types of industrial machinery to make use of mass damper device sets.

6.4 Brief Review of Optimization Models for Mass Damper Systems

Mass damper devices are utilized in the construction of large structures such as skyscrapers, bridges, and wind generators with the goal of reducing the detrimental effects of fluctuations brought on by the structures' moving components. With the assistance of an optimization model, one can able to determine the optimal values for the design parameters of the mass damper system, but the same depends on the integration and consideration of numbers of qualitative or quantitative characteristics for designing robust model. These characteristics can include things like the bulk of the damper and where it sits on the framework. To effectually utilize the mass damper devices, a variety of different optimization techniques as shown in Sections 6.4.1 to 6.4.4 are developed that can be used for attaining a robust MSD model.

6.4.1 Modal Analysis–Based Optimization

The first method is called modal analysis–based optimization, where the goal of reducing the detrimental effects of fluctuations can be achieved by taking into consideration the natural frequencies and oscillation patterns that are already present in the structure. It is possible to easily determine the mass and position of the damper that will be most effective way of analyzing the various modes of vibration and determining which mode is most important to the performance of the structure. The determination of critical mode related to the performance of the structure will assist in making strategic decisions.

6.4.2 Optimization in the Frequency Domain

This model discovers the optimal design parameters by analyzing the FRFs of both the framework and the damper. The model considers the natural frequencies of the structure to build model related to the optimization in the frequency domain. The purpose of the optimization model function is to determine the greatest amount of acceleration or displacement that the structure is subjected to majority of the time to bring it down to an acceptable level.

6.4.3 Time-Domain Optimization

This model replicates the behavior of the structure and suspension system over time by employing computational techniques. This optimization considers vibration data as a function of time. In this optimization, vibration signals are collected in the form of digital values to understand about the proximity, velocity, or acceleration in the time varying plot. Here, the optimization function needs to be built and integrated to minimize or condense the greatest amount of displacement or acceleration subjected by the structure.

6.4.4 Multi-Objective Optimization

This method takes into account a number of different objectives such as lowering the peak displacement while simultaneously lowering the cost of the suspension system. This is an optimization strategy that considers a number of different objectives. Here, the optimization was determined to discover the Pareto front of optimal solutions, which illustrates the equilibrium that needs to be achieved between the different objectives.

In general, optimization models are built on the basis of qualitative and quantitative variables for mass damper systems to assist the designers in optimizing the design parameters and enhancing the performance of structures that are prone to oscillations. This can be done by analyzing the frequency response received by the structure and comparing it to the predicted models to optimize frequency response. The specific requirements allied with qualitative and quantitative characteristics in relation to the application as well as the system architecture will determine which model should be chosen for the various application cases.

6.5 Algorithm of Particle Swarm Optimization (PSO)

The present chapter advocates the utility of PSO-based MSD model for attaining sustainability in reducing the detrimental effects of fluctuations and vibrations. Particle swarm optimization (PSO) is a popular optimization technique inspired by the social behavior of birds and fish, particularly their flocking and swarming behavior. PSO is used to find optimal solutions to optimization problems, particularly in search spaces with multiple dimensions [21]. In PSO, a group of particles represents potential solutions that can be achieved, and, thus, optimization is needed to understand which position is the best. In PSO, each particle adjusts its position in the search space based on its own experience and the experiences of its neighbors. The algorithm tries to guide the particles toward the best solution by updating their velocities and positions iteratively. The PSO considers the below-mentioned steps for the process optimization with the help of formulas:

Step 1: Initialization:
Initialize a population of particles with random positions and velocities within the search space. Each particle's position represents a potential solution to the optimization problem. Initialize the personal best position (pBest) of each particle as its initial position.

Step 2: Evaluation:
Evaluate the fitness of each particle based on its current position. Update the personal best position and fitness for each particle.

Step 3: Global Best Update:
Identify the particle with the best fitness in the entire population.

Step 4: Velocity and Position Update:
An algorithm to determine the best velocity and positioning vector using n particles is required to fix.

Step 5: Termination:
Repeat the velocity and position updates for a certain number of iterations or until a convergence criterion is met.

The core PSO algorithm parameters like inertia, cognitive coefficient, and social coefficient are tuned to influence the exploration and exploitation behavior of the algorithm. There are various variations and modifications of the PSO algorithm, where each are suited for different types of optimization problems or to enhance performance and representation and to provide the PSO algorithm's core steps. The actual implementation of PSO might involve additional considerations of parameter tuning and enhancements to improve its performance on various optimization problems. In PSO, individuals (particles) within the swarm adjust their positions according to their own best-known position and the best-known position of the entire swarm. This movement allows the swarm to explore the search space and converge toward the optimal solution. In MSD, PSO algorithm can impart best positioning of multiple qualitative and quantitative that can assist in reduction of forced vibrations.

6.6 Benefits of Optimizing Mass Damper Systems

Mass damper systems can be very effective in reducing the vibrations that occur in structures, making them more stable and safer. By adjusting the damping force of the mass damper, the intensity of the vibrations can be reduced, which can lead to a number of benefits including increased comfort for inhabitants, lower maintenance costs, and improved safety. The device is made up of a weight that is attached to the structure by a spring and a closure and its frequency can be modified to match the natural oscillation of the structure, which thereby helps to reduce the intensity of the shaking. Overall, mass damper systems can be an excellent investment for anyone who wants to ensure the safety and stability of their building or structure.

When mass damper systems are optimized to their full potential, one of the primary benefits that can be obtained is an improvement in structural safety. Vibrations in multiple structures and systems can be caused by the weather, earthquakes, and other external variables, which thereby can cause damage to a structure, and, in extreme cases, it can also lead toward failures and less life and can cause the structure's complete disintegration. Installation of a mass damper device, which functions to reduce the amplitude of the structure's natural oscillations, can alter the natural oscillatory frequency of the structure. This can be accomplished by implementing the device, which allied on certain qualitative and quantitative parameters and thereby will help in attain several benefits. This not only helps to protect the building from damage but also has the potential to save lives.

In addition, mass dampers systems will help in reducing maintenance costs and improved safety. Optimizing mass damper systems can also lead toward a higher level of comfort for occupants. Vibrations caused by external factors can be disruptive and uncomfortable for people who are allied with the system like inside a building with no dampers, which can lead to health issues in extreme failures. A mass dampening system can reduce the severity of these vibrations, and occupants are more likely to feel comfortable and at ease. This can lead to higher levels of satisfaction and overall well-being for those, who work or live in the building. Therefore, the optimization of mass damper systems has numerous benefits that can improve the performance, safety, and comfort of buildings and structures.

Indeed, research toward maximizing the use of mass damper systems is required and that will have a significant impact on both the safety and sustainability of buildings, machineries, and structures. By reducing the risk of structural damage, these systems can potentially reduce the amount of waste generated by the construction and repair processes, thereby contributing to a more sustainable built environment. Additionally, by improving the comfort and satisfaction of building occupants, mass damper systems may also contribute to increased productivity and well-being. As such, incorporating these systems into building design and construction should be seriously considered as a means of promoting safety, sustainability, and overall occupant satisfaction. The crucial benefits of optimizing mass damper systems can be understand from Sections 6.6.1 to 6.6.4, where the authors have summarized some of the numerous reasons to explicate why it is essential to maximize the mass damper systems for any particular piece of work:

6.6.1 Vibration Reduction

Mass dampers are designed to reduce vibrations in buildings and other structures, which can help to prevent damage and improve safety. By optimizing the mass damper system, researchers can improve performance and effectiveness of the system, which can help to reduce the amplitude of vibrations caused by external factors and protect the structure from potential damage or even catastrophic failure.

6.6.2 Maintenance and Repair Costs

By reducing the wear and tear on buildings and structures through the use of proper mass damper systems, maintenance and repair costs can be reduced. Additionally, by extending the life span of the structure, the need for costly replacements can be delayed or avoided altogether. All of these factors can lead to significant cost savings for building owners and operators.

6.6.3 Health and Well-Being

People who are confined within a building or structure for the sake of living or work may experience discomfort as a result of vibrations, which may have a negative impact not only on their level of productivity but also on their quality of life. By optimizing the systems based on the regulation of mass dampers, researchers can develop the ability to make living spaces cozier for locals, which, in turn, can have a positive impact on the residents' overall health and well-being.

6.6.4 Repercussions for the Natural World

Optimizing the performance of mass damping systems can indeed have a positive impact on the natural world. By reducing the need for frequent maintenance and repairs, the amount of waste generated by the construction industry can be reduced. This, in turn, can lower the carbon footprint of the building or structure over its life span. Moreover, it can contribute to the conservation of resources and energy that would have been required for repairs and replacements. Overall, the optimization of mass damping systems aligns with the goals of sustainability and can help to create a more resilient, efficient, and eco-friendly built environment.

6.7 Role of Qualitative Optimization and Discussions

Qualitative structure is usually required to describe the qualities or characteristics holding by the individual, system, or property [22]. Qualitative analysis assists in relating the nature or standard of something in vocal language rather than to its quantity. Qualitative analysis can facilitate in collecting a wide range of complex data, which can assist in understanding the system certain behaviors [23]. There are various reasons that infer the utility of qualitative information in MSD as discussed in Sections 6.7.1 to 6.7.11.

6.7.1 Language of the Developer

Qualitative optimization plays a crucial role in the engineering design process, particularly when dealing with complex systems like the MSD system. The same assists in collecting parameters information for building the MSD model. This approach involves refining and enhancing the design using insights, analysis, and discussions based on the language of the designer rather than relying solely on quantitative measures. Thus, it is necessary to consider both qualitative and quantitative parameters simultaneous for attaining robust modeling [24].

6.7.2 Conceptual Understanding

Qualitative discussions help engineers to gain a deeper understanding of the system's behavior, interactions, and limitations. This understanding is crucial before diving into quantitative optimization techniques.

6.7.3 Trade-Off Analysis

Qualitative discussions allow engineers to explore critical trade-offs between different designs choices such as spring stiffness, damper settings, and mass properties. These discussions can lead to better-informed decisions by considering the pros and cons of various options.

6.7.4 Identifying Critical Factors

Through qualitative analysis, engineers can identify critical factors that might not be immediately obvious through quantitative analysis [25]. This includes considering how changes in one parameter might affect the overall system behavior.

6.7.5 Non-Linear Effects

The behavior of systems like MSD models can sometimes exhibit non-linear effects that are challenging to capture solely through quantitative optimization, where qualitative measures can help to uncover these non-linear behaviors and their implications.

6.7.6 Sensitivity to Assumptions

Engineering designs often involve making assumptions and simplifications. Qualitative optimization helps to identify the sensitivity of the design to these assumptions, prompting engineers toward questions and validation of their choices.

6.7.7 Incorporating Practical Constraints

Real-world engineering designs must often account for practical constraints such as manufacturing limitations, cost considerations, and safety requirements. Qualitative discussions help to ensure that these constraints are taken into account [26].

6.7.8 Iteration and Iterative Learning

Qualitative discussions encourage iterative design improvements. Engineers can learn from previous iterations, adapt the design, and refine it based on qualitative discussions, which are insights based and easy to capture.

6.7.9 Interdisciplinary Collaboration

A complex system often requires input from multiple disciplines. Qualitative discussions facilitate interdisciplinary collaboration, where experts from different fields can provide valuable insights to refine the design.

6.7.10 Communication with Stakeholders

Communicating complex engineering concepts and design decisions to designers, manufactures and stakeholders with technical background is a critical aspect of engineering projects. Qualitative discussions help convey these ideas in a more accessible manner [27].

6.7.11 Risk Assessment and Mitigation

Qualitative discussions can lead toward the identification and assessment of potential risks associated with the design. By discussing these risks, engineers can develop strategies to mitigate them effectively [28].

Qualitative optimization complements competent analysis for evaluating the efficiencies of the engineering design process. They provide a holistic view of the system, enable engineers to explore trade-offs, uncover non-linear behaviors, and incorporate practical considerations [29, 30]. Through collaborative discussions, iterative learning and qualitative measures, the MSD system model can be better designed, refined, and optimized to meet the desired engineering objectives [31].

6.8 Conclusion

In the present chapter, the directory of qualitative and quantitative parameters that can be used for the MSD modeling and evaluation is presented. In the present chapter, many spring dumping variables like movement of the spring mass, DOF, rate of oscillation, drive situations, mass position, damping ratios, slope of the flat roof, frequency range, resonance force, cross-axis sensitivity, and material and fabrication processes are presented as crucial for building MSD model, which can be qualitative and quantitative. Additionally, stiffness, internal forces, ratio of response, derivation of process parameters, number of dimensions, rate of oscillation amplitudes, number of linkages, aging, heat via friction generation ratio, stressing of material, designing of geometrical parameters, cantilever length, and piezoelectric thickness are also presented as critical parameters for the dampers to prevent bouncing up and down in various application fields. The chapter presented the utility of mass dumping system in various application scenarios and additionally presented various indexes, which are determined and used by the researchers based on accumulation of qualitative and quantitative information onto the model for evaluating the MSD. The same will assist is preparing various optimization models, which can include evolutionary algorithms, PSO, fuzzy logic, artificial neural networks, and combinatorial optimization algorithms for preparing robust mass damper systems. The chapter affirmed that the adaptability of these sustainability models will lead to successfully optimize the mass, stiffness, and dampening of the mass damper system, which thereby can improve the performance of the structure, when it was subjected to dynamic stresses. The same, in turn, will result in improvement in the overall performance of the structure. The chapter portrayed a picture of understanding about diverse MSD framework as well as the optimization challenges, which can assist to determine the best MSD optimization model. Additionally, the chapter presented a clear picture about the optimization parameters in mass damper systems and affirmed that, by utilizing these parameters, significant enhancements to the safety, comfort, and cost-effectiveness of buildings and other structures can be attained by MSD models. These enhancements can be achieved by reducing the amount of energy that is wasted for different scenarios. The evaluation of the MSD system based on the MSD model for reinforcing engineering designs through qualitative and quantitative optimization yields valuable insights and benefits that complement traditional quantitative approaches. The MSD model based on qualitative optimization approach embraces a holistic understanding about the system's behavior, interactions, and constraints, which thereby will lead toward enhanced engineering designs. By leveraging qualitative optimization techniques in MSD models, a deeper conceptual understanding of the MSD system can be understand, which will assist in uncovering intricate behaviors that might not be readily apparent in quantitative analysis alone. The presented chapter advocated toward considering both qualitative and quantitative parameters simultaneous for attaining robust modeling, which will allows us to explore trade-offs, identify critical factors, and consider non-linear effects that contribute to a more refined design. The iterative nature of qualitative optimization facilitates a continuous learning process in MSD models, where each iteration builds upon the knowledge gained from the previous one. This chapter ensures that the design evolves through qualitative and quantitative measures optimization can meet or exceed engineering objectives while addressing unforeseen challenges that arise during the process. The chapter advocated that the evaluation of the MSD system through qualitative and quantitative optimization represents a robust methodology that not only refines engineering designs but also encourages innovation, fosters interdisciplinary collaboration, and enhances the overall quality and reliability of the final product. As engineering endeavors continue to evolve, embracing qualitative optimization in conjunction with quantitative analysis will undoubtedly contribute to the advancement of engineering designs more holistically and effectively. The chapter limitedly found few parameters in MSD modeling and admires that, for designing a more efficient MSD model, it is crucial to explore maximum number of qualitative and quantitative factors, parameters, variables, etc., to reciprocate toward quality and efficiencies. The quality of the build model depends on the number of variables, which is precisely needed to be explored and identified more in the context of development of the robust MSD model.

References

1. Li, D., Liu, F., Deng, J., Tang, Z., Wang, Y., Nonlinear Damping Curve Control of Semi-Active Suspension Based on Improved Particle Swarm Optimization. *IEEE Access*, 10, 90958–90970, 2022, doi: 10.1109/ACCESS.2022.3201867.
2. Chen, L., Shi, W., Chen, Z., Research on Damping Performance of Dual Mass Flywheel Based on Vehicle Transmission System Modeling and Multi-Condition Simulation. *IEEE Access*, 8, 28064–28077, 2020, doi: 10.1109/ACCESS.2019.2951618.
3. Kandil, A., Hamed, Y.S., Awrejcewicz, J., Harmonic Balance Method to Analyze the Steady-State Response of a Controlled Mass-Damper-Spring Model. *Symmetry*, 14, 6, 1247, 2022.
4. Takayama, Y., Kijimoto, S., Ishikawa, S., Eddy Current Damper Model of Ring Magnet and Coaxially Moving Conducting Disk. *IEEE Trans. Magn.*, 57, 9, 1–13, Art no. 8107513, 2021, doi: 10.1109/TMAG.2021.3058739.
5. Xia, C., Wang, D.F., Ono, T., Itoh, T., Maeda, R., Autoparametric Internal Resonance in Coupled Oscillator: An Excitation Amplitude Insensitive Mass Sensing Scheme With a Roof Tilting. *IEEE Sens. J.*, 22, 3, 1998–2005, 2022, doi: 10.1109/JSEN.2021.3134464.
6. Hao, Z., Wang, T., Cao, X., Zhang, Q., Design of Eccentric Mass-Type Vibration-Damping Electric Actuator Control System for Non-Fixed-Wing Aircraft. *IEEE Access*, 8, 219415–219429, 2020, doi: 10.1109/ACCESS.2020.3042854.

7. Fang, W., Cai, J., Bai, J., Chen, J., Xv, X., Wang, C., Lu, Q., Correlated Theory of Cross-Axis Interference of MEMS Devices Based on Parallel Plate and Squeeze Film Air Damping Induced Suppression. *IEEE Sens. J.*, 22, 7, 6698–6705, 2022, doi: 10.1109/JSEN.2022.3149979.
8. De Oliveira, L.L.A. and Travaglia, M.V., Optimizing the spring constants of forced, damped and circular spring-mass systems—characterization of the discrete and periodic bi-Laplacian operator. *IMA J. Appl. Math.*, 86, 1, 785–807, 2021, doi: 10.1093/imamat/hxab021.
9. Rajaprasanna, R., Begum, K.G., Balasubramanian, G., System Identification of Lightly Damped Systems Using Recursive Kautz Functions-Case Study With Coriolis Mass Flow Meter. *IEEE Access*, 9, 70924–70934, 2021, doi: 10.1109/ACCESS.2021.3077608.
10. Feng, J., Liu, Z., He, X., Fu, Q., Li, G., Adaptive Vibration Control for an Active Mass Damper of a High-Rise Building. *IEEE Trans. Syst. Man Cybern. Syst.*, 52, 3, 1970–1983, 2022, doi: 10.1109/TSMC.2020.3036891.
11. Tawaki, Y., Nishimura, T., Murakami, T., Classification of Older and Fall-Experienced Subjects by Postural Sway Data Using Mass Spring Damper Model. *IEEE Trans. Neural Syst. Rehabil. Eng.*, 30, 40–49, 2022, doi: 10.1109/TNSRE.2021.3139966.
12. Zhao, J., Huang, Y., Wen, X., Zhang, S., Liu, P., Identification of Geometric Parameters Influencing the Impedance Sensitivity of Piezoelectric Mass Sensors. *IEEE Sens. J.*, 20, 24, 14740–14746, 2020, doi: 10.1109/JSEN.2020.3011878.
13. Adiputra, D., Rahman, M.A.A., Ubaidillah, Mazlan, S.A., Improving Passive Ankle Foot Orthosis System Using Estimated Ankle Velocity Reference. *IEEE Access*, 8, 194780–194794, 2020, doi: 10.1109/ACCESS.2020.3033852.
14. Li, Z., Sun, W., Gao, H., Energy-Driven-Damper (EDD): Comfort-Oriented Semiactive Suspensions Optimized From an Energy Perspective. *IEEE Trans. Control Syst. Technol.*, 28, 5, 2069–2076, 2020, doi: 10.1109/TCST.2019.2954793.
15. Das, S. *et al.*, Modeling of Frequency-Dependent Damping for Fast Vibration Prediction in Permanent Magnet Synchronous Machines. *IEEE Trans. Transp. Electrif.*, 9, 1, 561–574, 2023, doi: 10.1109/TTE.2022.3180289.
16. Bagchi, S.N., IA-15: Vibration Isolation Systems For Optical And Laser Instruments And Experiments Related To Surface Engineering And Metrology. *Proc. Incosurf*, pp. 355–358, 2004.
17. Liu, B. *et al.*, A MEMS-based Electrochemical Angular Accelerometer based on Planar Electrodes with Micro-Pillar Array Structure. *IEEE Sens. J.*, 23, 16, 18032–18041, 2023, doi: DOI: 10.1109/JSEN.2023.3293107.

18. Elaswad, R., El-sheimy, N., Abdulmajeed, M., Basic and Advanced Inertial Navigation Fluid-Based Technology. *ECS Sens. Plus*, 2, 2, 2023, DOI 10.1149/2754-2726/acd0c0.
19. Xiangdong, M. *et al.*, Angular accelerometer based on a dual-polarization fiber-optic Sagnac interferometer. *Opt. Lett.*, 48, 4, 859–862, 2023.
20. Shukla, S.R. and Sinha, A.K., Study of PID controller gain for active vibration control using FEM based particle swarm optimization in COMSOL multiphysics. *J. Micromanuf.*, 6, 1, 40–49, 2023.
21. Gao, T. *et al.*, H∞ Optimization of Three-Element-Type Dynamic Vibration Absorber with Inerter and Negative Stiffness Based on the Particle Swarm Algorithm. *Entropy*, 25, 7, 1048, 2023, https://doi.org/ 10.3390/e25071048.
22. Li, N., Prabhu, M., Sahu, A.K., Investigating economic replacement policy under uncertainty for managerial application based on grey-reliability approach using QCC. *Grey Syst.: Theory Appl.*, 13, 2, 297–321, 2023, https://doi.org/10.1108/GS-07-2022-0075.
23. Wang, W., Huang, L., Zhu, Y., Jiang, L., Sahu, A.K., Sahu, A.K., Sahu, N.K., Decision support system toward evaluation of resilient supplier: A novel fuzzy gain-loss computational approach. *Kybernetes*, 49, 6, 1741–1765, 2019.
24. He, Z., Ma, X., Luo, J., Sahu, A.K., Sahu, A.K., Sahu, N.K., Exploitation of the advanced manufacturing machine tool evaluation model under objective-grey information: a knowledge-based cluster with the grey relational analysis approach. *Grey Syst.: Theory Appl.*, 11, 3, 394–417, 2021.
25. Sahu, A.K., Sahu, N.K., Sahu, A.K., Laminating STRATH block chain technology- SWOT architectures to endure business strategy between digital transformation, firms and supply chains capabilities for sustainability. *J. Clean. Prod.*, 383, 1–20, 2023, https://doi.org/10.1016/j.jclepro.2022.135531.
26. Bag, S., Viktorovich, D.A., Sahu, A.K., Sahu, A.K., Barriers to adoption of blockchain technology in green supply chain management. *J. Glob. Oper. Strateg. Sourc.*, 14, 1, 104–133, 2021.
27. Kang, D., Prabhu, M., Ahmed, R.R., Zhang, Z., Sahu, A.K., Digital-IIoTs spheres approach toward public development: an exploiting fuzzy-grey mathematical modeling of IIoTs spheres. *Grey Syst.: Theory Appl.*, 12, 2, 389–416, 2022.
28. Bag, S., Sahu, A.K., Kilbourn, P., Pisa, N., Dhamija, P., Sahu, A.K., Modeling barriers of digital manufacturing in a circular economy for enhancing sustainability. *Int. J. Product. Perform. Manage.*, 71, 3, 833–869, 2021, https://doi.org/10.1108/IJPPM-12-2020-0637.
29. Koch, S., Duvigneau, F., Orszulik, R., Gabbert, U., Woschke, E., Partial filling of a honeycomb structure by granular materials for vibration and noise reduction. *J. Sound Vib.*, 393, 30–40, 2017.

30. Wang, S.J., Lee, B.H., Chuang, W.C., Chiu, I.C., Chang, K.C., Building mass damper design based on optimum dynamic response control approach. *Eng. Struct.*, 187, 85–100, 2019, https://doi.org/10.1016/j.engstruct.2019.02.053.
31. Manzoni, S., Argentino, A., Lucà, F., Berardengo, M., Vanali, M., SMA-based adaptive tuned mass dampers: Analysis and comparison. *Mech. Syst. Signal Process.*, 186, 1–21, 2023, https://doi.org/10.1016/j.ymssp.2022.109883.

7

A Fuzzy Decision Optimization of Wire-EDM Process for Reinforcing Manufacturing Design Under Quantitative Data

Neha Verma[1]*, Vinay Sharma[2] and Rityuj Singh Parihar[1]

[1]Department of Mechanical Engineering, Shri Shankaracharya Institute of Professional Management and Technology, Raipur, Chhattisgarh, India
[2]Department of Production Engineering, Birla Institute of Technology, Mesra, Ranchi, Jharkhand, India

Abstract
In the present chapter, quantitative data are utilized in Wire-EDM process for evaluation based on the fuzzy decision optimization technique. The same will assist in attaining robust manufacturing design, which will lead in reinforcing capabilities of the manufacturers by reducing failures. In the present chapter, a fuzzy design optimization approach is presented using bond-graph approach for operating dielectric circuit maintenance for wire electrical discharge machining (WEDM) operations. Here, experiments are performed to obtain quantitative information under different parameters, i.e., electrical, electrode, dielectric, and workpiece for evaluation and optimization. In WEDM processes, it is always appropriate to choose the better machining condition for acquiring desired properties to the machined parts and to induce increased efficiency. Here, four main variables (parameters), which possess the capability to impact the Wire-EDM method, are presented. Under these aforesaid parameters, crucial features like peak current, gap voltage, servo feed, spark gap, wire material, feed rate, tension, material, height, flow rate, and conductivity of dielectric are considered for conducting experiment and recording quantitative data. These parameters are used to evaluate the machining process in both qualitative and quantitative terms. A fuzzy logic (FL) controller is employed for receiving fuzzy guidelines to calculate

**Corresponding author*: n.verma@ssipmt.com

Atul Kumar Sahu, Rakesh D. Raut, Rohit Raja, Anoop Kumar Sahu and Nitin Kumar Sahu (eds.) Industrial and Manufacturing Designs: Quantitative and Qualitative Analysis, (201–232) © 2024 Scrivener Publishing LLC

the lessened amount of sparking power by using the physical characteristics of the WEDM. The model can be used to create the fuzzy rules that can be used by an FL controller to determine the lowered percentage of sparking force. The model is presented in this chapter with the primary application to organize maintenance jobs into reasonable packages that may be completed within predetermined time frames as well as to enhance the accessibility of machine components during longer maintenance periods. The presented model is easy to understand, is simple to maintain, and saves time in implementation with learning new things. With the help of the presented model, automatic recommendations can be made on the occasions that require servicing of machines. During the condition of receiving fault, the suggestions will be received to eliminate it based on the priority rules and realization. The chapter presented the experimental validation to determine the viability and effectiveness of the suggested approach.

Keywords: Electrical discharge machining (EDM), computerized numerical control (CNC), optimization, Wire-EDM (WEDM), fuzzy logic

7.1 Introduction

Wire electrical discharge machining (Wire-EDM) has evolved over the last three decades and is now a common procedure. Today, Wire-EDM machines are getting better and better with three of the most important machining characteristics, i.e., speed, accuracy, and surface finish [1]. Nowadays, reductions in total process cycle durations is needed to be achieved thoroughly for sustainability while maintaining dimensional precision and repeatable tasks by these machines utilizing for satisfying diverse manufacturing requirements from the side of manufacturers, customers, and assemblers. WEDM is a modern technology wonders that can automatically thread wires with a diameter of less than 0.001 inches through holes, which is equal to the size of a human hair. Moreover, they can make mirror-like surface finishes with micro-machined parts with precise specifications, which are very small and can only be seen with a microscope. In order to be able to produce reliably and consistently performance at this level, one should follow the manufacturer's recommended maintenance routine for these modern technology wonders [2].

Wire-EDM is not a new comer under the scene of manufacturing precise thresholds. It was developed in the late 1960s and has completely changed the dimensions of manufacturing, metal work, and mould-based industrial sectors. The machine has many benefits for manufacturing and is perhaps the most interesting and versatile machine source created for receiving effectual industry-based manufacturing shapes and designs in the previous 50 years. Any electric conductive workpieces can be machined with the said machine tool, which are, irrespective of any hardness values, possessed by the work material. Any hard material can be easily machined with WEDM. This includes materials as hard as polycrystalline diamond compresses and conductive ceramics, as well as more exotic space-age alloys like hastelloy, waspaloy, carbide, and tool steel.

An EDM technique is said to be named as Wire-EDM, which uses a moving wire electrode to create the form of the workpiece's surface. WEDM is a relatively recent concept in the area of EDM. The machining is completed automatically with Wire-EDM, and, thus, the designers and managers have additional freedom for creating dies [3]. The machine does not require several skill levels or numerous pieces of equipment to make parts with complex geometry and liberty. Once the workpiece starts processing during manufacturing, the machine can be left unattended and significant gains in productivity can be made by allowing operators to focus on other tasks. A "lights-out" setting is the norm for these machines, which is required as they run overnight. Shorter jobs are planned during the day, whereas longer ones takes overnight or over the weekend to be processed, cut, or machined due to less metal removal rate in comparison to conventional machine tools.

Producing manufacturing projects with intricate shapes, through holes, straight taper, and other features demonstrates the usefulness of the WEDM machining process [4]. Additionally, producing complex shapes with high accuracy and preciseness also reveals the utility and critical characteristics of WEDM process. WEDM has been identified to be the most promising non-traditional electro-thermal machining method due to its capacity to suit the needs of today's product production industries [5]. Furthermore, it is always required to optimize multiple characteristics of the said machine for receiving elevated outputs. Thus, optimization of elements is needed in the said domain. Fuzzy logic (FL) concept can be used for optimization, which is originally developed by Lofti Zadeh, a Berkeley professor to represent human intuition and experience. Traditional set theory, according to Zadeh, was unable to adequately estimate the inherent vagueness in the real world, and Zadeh found that there are no middle-ground sets that exist, which can be called as the standard set theory, and, thus, he proposed the concept of FL to mould-inherent vagueness and uncertainty with middle datasets. According to him, the degree to which a component is a member of a set is a question of degree in fuzzy set (FS) theory [6]. In some cases, an element can be a member of many sets. In many cases, IF-THEN logical principles are frequently used as a technique to describe human laws of thumb, when using FSs. Only the components that are being established are fuzzy in an FL device, where the laws of logic are logically established [7]. In the primary phase, for the diagnosis of computation hydraulic power system failures, an FL topology has been devised. Analytical and FL approaches are combined in the suggested strategy to diagnose the said problem, where, to assess the remainders generated through the nonlinear observer, FL is applied. The membership function (MF) and rule bases of the FL system influence the severity of its inherent flaws. Fuzzy logic can be used as an addition to an analytical system in order to improve its overall performance [8]. In the present chapter, FL is applied in Wire-EDM. The main parts of Wire-EDM can be identify from Figure 7.1. A Wire-EDM device is usually categorized into four primary parts:

(1) **Computerized Numerical Control:** Think of this as "The Brains".
(2) **Power Supply:** This provides power to the spark. Think of this as "The Muscle".
(3) **Mechanical Section:** This refers to the worktable, taper device, and wire drive structure with workstand. This is the cutting tool itself. Consider this to be "The Body".
(4) **Dielectric Device:** This refers to the dielectric fluids, its circulating system liquid reservoir that offers and sustains filtration, temperature, or quality (resistivity/conductivity). Consider it "The Nourishment" [2, 3].

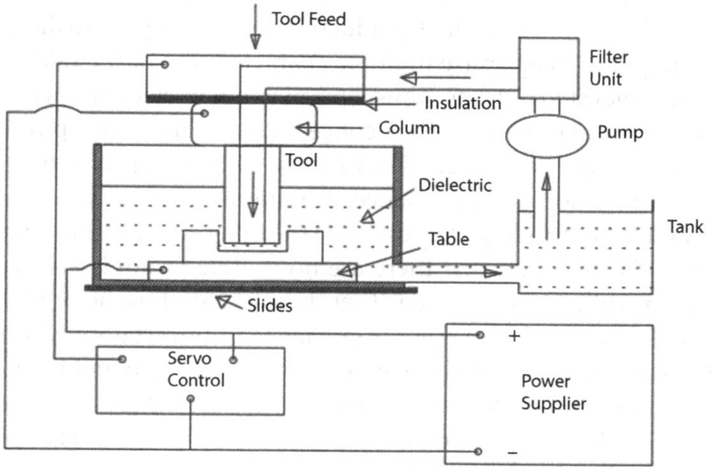

Figure 7.1 Block diagram of the Wire-EDM machine [9].

The main principle of Wire-EDM relies on inducing the utility of electrically charged thin wire, which is submerged into dielectric fluid to cut a metal piece into special profile. The spark idea for Wire-EDM and vertical EDM processes are essentially equivalent. In Wire-EDM, the workpiece and an accurately placed moving wire (the electrode) create a series of electrical sparks that mill the conductive materials. Using an insulated dielectric fluid, high-frequency pulses of AC current are discharged from the wire to the workpiece [4]. Instantly, many sparks could be seen. This is because of the fact that concrete discharges happens more than 100,000 times per second with discharge sparks lasting only 1/1,000,000 of a second or less. The amount of metal destroyed during this brief spark discharge is based on the required surface polish and applied cutting speed. The heat of each electrical spark is estimated to be around 15,000°F to 21,000°F, which erodes away a small portion of the material from the workpiece that vaporized and melts [5]. These small particles are swept away from the cut using a stream of deionized liquid by the top and bottom flushing nozzles (some of the wire material is also eroded away). Additionally, the liquid aids in accumulation of chips and stops the workpiece from becoming too warm, which thereby acts as cooling agent. Without the utility of cooling agent, the part's thermal expansion would have an impact on size and positioning accuracy, and, thus, it is desired to utilized cooling agent in WEDM.

7.2 Review of Literature

The wide literature has been studied under the shadow of WEDM, machining, manufacturing, maintenance, optimization, and FL, wherein Lautre and Manna [10] provided a binary relational assessment and expert device base modules for computerized numerical control (CNC)–based Wire-EDM maintenance and fault diagnostics problem. An integrated maintenance and fault diagnosis device is proposed in their module for diagnosis the fault. In accordance with their study, issues in WEDM are anticipated and identified with the help of binary coded vector approach. The forward and backward propagation algorithms were used in their study to identify eight possible causes from a total of 15 inputs. The input and output matrices were viewed as square matrices in their study [10]. The identification of defects is examined by them on the spot using forward and backward propagation of matrix transformation to describe fault diagnosis and to comprehend the need of maintenance through advice-based structure. Their characteristics offer simultaneous backup that can be focused on when the parameters for the input and output do not match. It is a time-saving, information-gathering, and low-maintenance technology that can also teach itself, but the accuracy of the build module is difficult to predict. Makis *et al.* [11] suggested a condition-based upkeep approach to take maintenance decision for retaining system manufacturing efficiency. There are multiple advantages of using condition-based maintenance (CBM), such as being able to build decisions depends on the current state of the device, where monitoring approaches are required to develop to gather data related to CBM. Additionally, there are still very few mathematical designs that can make effective use of the gather required data for maintenance decisions. In the study by Makis *et al.* [11], a CBM design is presented for handling occasional fault, where only partial data are accessible by controlling a signal process that does not necessarily exhibit monotone behavior. Random factors and modest maintenance actions among inspections influence the growth of the signal process. In the study presented by Makis *et al.* [11], a design related to effective replacement policy is presented to yield the highest total predicted profit over the life span of the network.

Huang and Liao [12] created a system for scheduling maintenance and diagnosing problems in Wire-EDM that incorporates an artificial neural network (ANN) and an expert system (ES). However, Wire-EDM still has several problems such as wire breaking and inadequate precision, which can be caused by conduction of wrong operation, utility of inaccurate execution, or incorrect equipment maintenance policy. Accordingly, Huang and Liao [12] attempt to develop a fault-diagnosis device for preparing maintenance schedule by blending ANN with ES. The same saves time in the acquisition of knowledge and can learn on its own, but it is difficult to implement. The same, at the time to perform routine maintenance on the machine, goes off automatically, and, at the time of occurrence of problem, solutions are offered by the developed device in a logical order based on the priority rules that have been assigned to it. Accordingly, it is needed to design a model, which is easy to understand, is simple to maintain, and saves time in implementation with learning new things.

An ES built by Liu [13] attempt toward utilizing fuzzy technique to evaluate information on different elements of bearing monitoring, including diagnostic procedures, fault frequencies, feature selection, and bearing classification. As a result of their research, a fuzzy ES was created to manage roller bearing defects, where the utility of FL has provided remarkable outputs. Their experts system has used a PC to operate. It has a simple menu structure and is easy to use. Their ES possess the caliber to handle a wide range of bearing controlling components, such as methods for diagnosis, fault frequency ranges, and fuzzy categorization. The user can get the most precise information by answering a series of questions in the most efficient manner possible in their ES, where experiments are presented to illustrate the efficiency of bearing ES, where bearing expert and a computer program have been used to identify the best features for avoiding defects. Using a combination of FL and Sugeno-style inference engine, Tony and Liang [14] suggested a simple, efficient, and strong fusion solution. Here, dependable equipment under four condition monitoring indicators is fused into a single comprehensive fuzzy-fused index (FFI), which is created for the detection of transitory and progressive anomalies. Both milling operations and bearing condition evaluation have been successfully tested and validated using this method. Using the same fuzzy rule (FR) framework, the FFI makes a clear distinction between normal and abnormal circumstances, where FFI adaptability and robustness clearly demonstrated the efficacy and capabilities of fuzzy means for handling a problem [15]. It is found easier to set thresholds FFI value for different tools or equipment components due to the generation of FFI value among zero and one. As a consequence of this investigation, it is determined that fuzzy and fuzzy index can be utilized for exploiting the characteristics of the system. Thus, in the present chapter, the FL is utilized to prepare a design for handling uncertainty and fault detection in Wire-EDM. There are four main variables found, as shown in Table 7.1, which can impact the WEDM machining process. Table 7.1 shows the variables, which are needed to be evaluated in WEDM. Additionally, to understand critical ranking of the parameters in terms of research inclination, Figure 7.2 is projected by the authors, which shows the WEDM research inclination.

Table 7.1 Main variables considered in WEDM.

1. Electrical	**2. Electrode**	**3. Dielectric**	**4. Workpiece**
Peak current	Wire material	Flow rate of dielectric	Material
Gap voltage	Wire size	Conductivity of dielectric	Height
Servo feed	Wire tension		
Spark gap set voltage	Wire feed rate		
Pulse-on time			
Pulse-off time			

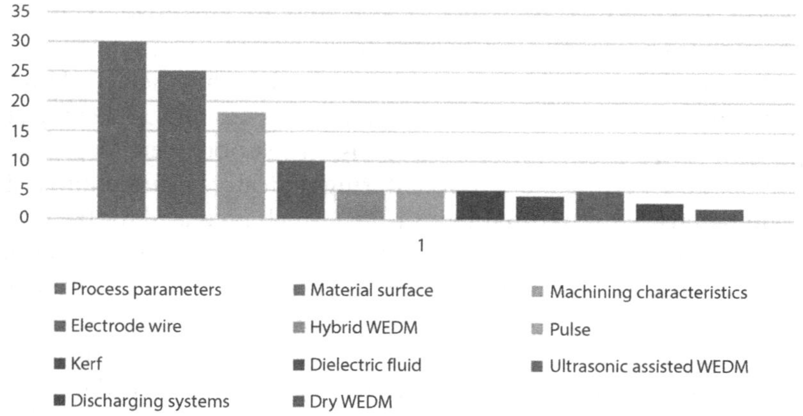

Figure 7.2 WEDM research inclination.

Figure 7.2 demonstrates the outline of the literature review, where it can be seen that the majority of researchers focused on the impact of process variables, machined surface to be attained, modeling of machining features, and assessment of machining characteristics for ensuring effectiveness. Here, one can identify that, for receiving accurate surface finishing, accurate machining, close tolerances, and economic machining of completed parts, the WEDM machining process can be used. The best optimum process parameters should be required to be achieved by inducing research dimensions based on optimization, synchronization, or fuzzy modeling by collecting quantitative values for attaining maximum utility by the machine.

7.3 The Significant Facts Related to Design, Implementation, and Importance of Total Productive Maintenance Programs in Manufacturing Operations

i. It is found that significant attainment can be receive by preparing maintenance item allocation model (MIAM) for wire feed system of Wire-EDM machine. MIAM for wire feed system is based on an FR-based model for minimizing wire wastage and improving product quality. Thus, FL can play a significant role in modeling the characteristics and elements of Wire-EDM.

ii. It is found that bond-graph (BG) approach can be prominent for handling dielectric circuit maintenance.
iii. It is found that Wire-EDM machine requires two consumables:

a) WIRE (acts as an electrode), which is used for cutting purpose, and
b) dielectric fluid, which enables the wire to cut through metal by the use of heat from electrical sparks.

Wire-EDM is a potential electro-thermal method in the category of non-traditional machining methods. Wire-EDM has the immense capacity toward meeting the requirements of the present day from the side of customers and for satisfying the needs of the product manufacturing industry. Even though Wire-EDM is a good alternative for industrial application, still there are some problems related with the best utility and evaluation of associated elements with the machine. It is quiet complicated and expensive machine, and only few units of EDM are found in an industry. It is a time-consuming process, and some faults such as wire breaking and losing capabilities of dielectric exaggerated the total processing time and decreased efficiency. The productivity decreases as a result due to the machine breakdown, which is generally caused by high consumption of wires, wire breaks, and dielectric as follows:

i. Significant wire wastage as a result of the requisite unavailability of federate.
ii. Number of times the wire breaks and/or improper cutting (having a bow like structure rather than a straight one).
iii. Inaccurate working of dielectric device because of unavailability of crucial elements and lose strength of dielectric as follows:

a. Backflush required for cleaning the filter. Various problems and required steps associated with the filter are as follows:

- time wastage (2–3 h) due to unwanted backwash cycle,
- chocking of filter, and
- time to change the filter.

b) Maintaining the dielectric strength of liquid.

In order to maintain the dielectric strength of deionized liquid, typically involves regular monitoring, testing, and if necessary, replacing or purifying the liquid to ensure it remains free of contaminants that could compromise its dielectric properties. In order to overcome such problem, an FR-based model and BG model are developed in this chapter to avoid wire breakage and to diagnose the fault in dielectric system, respectively, which helps in quick detection of fault and hence improves maintenance efficiency.

7.4 Primary Objectives

The primary and secondary objectives of this chapter lie in investigating an appropriate model that can deliver favorable functional properties for the machined parts with increased efficiency. The chapter presented the following objectives:

- Wire-EDM maintenance is researched and investigated for preparing a decision-making model.
- Investigations related with the necessity for WEDM maintenance and the causes of wire breaking are performed.
- Critical features of BG theory are presented in this chapter, and the study integrated with BG and FL is presented.
- The evident capabilities toward optimizing a Wire-EDM process with various performance metrics are demonstrated. The concept of FL is presented by the utility of an input FS and based on the investigation of the MF.

In the present chapter, a fuzzy-based design optimization approach by using BG concept for dielectric circuit maintenance is presented for increasing the efficiency of wire electrical discharge machining operations. Here, experiments are conducted and investigated, where experiments are performed under different parameters domain, i.e., electrical, electrode, dielectric, and workpiece. Under these parameters domain, critical features like peak current, gap voltage, servo feed, spark gap, wire material, feed rate, tension, material, height, flow rate, and conductivity of dielectric are considered. An FL controller is employed for implicating the fuzzy guidelines to calculate the lessened amount of sparking power by using the physical characteristics of the Wire-EDM.

7.5 Research Methodology

This study examined and reviewed many reports on the learning and research relevant to the subject matter. A critical search has been done on web databases to receive a collection of journal articles related to the study. Scrutiny of the articles is done to select and identify the subject-related facts and figures, where it is understood that four main variables named as electrical, electrode, dielectric, and workpiece are significant for modeling Wire-EDM capabilities. The various features under the said variables are needed to be explored to make positive influence on the abilities of the Wire-EDM method. Accordingly, it is needed to focus on the process variables based on integrated modeling of machining features for the assessment of machining characteristics for ensuring effectiveness. The best optimum process parameters is found as research dimensions for preparing driving model based on optimization, synchronization, or fuzzy modeling by collecting quantitative values for attaining maximum utility by the machine. The subjects, methods, and conclusions of scientific research were recognized, as well as the accuracy of the classification techniques are recognized on the basis of the literature review and to understand a systematic overview of the knowledge, research dimensions, and useful themes for model creation and utilizing the results for the development of the machine and total productive time. This research is descriptive rather than prescriptive in nature or relies on primary as well as secondary sources.

7.5.1 Maintenance Method for Wire-EDM Machine

Precision machine tools like Wire-EDM need regular service to ensure a long life span and trouble-free performance. Thus, a machine may be kept from breaking to produce higher production for a long time with adequate upkeep. It is advised that the machine be put in a monitored environment with a room temperature of 20°C in a way to obtain greater machining accuracy. Through properly configuring the temperature controller of the dielectric cooling device, the level of the dielectric fluid must also be kept within a range of 1°C below the heat of the machine origin [16]. Before beginning any work alignment or machining operations, the room, machine tool, or task temperatures need to be stabilized by air conditioning. Additionally, the machine tool power switches and the dielectric cooling device must be turned "ON" at least 30 min after the machining operation begins. It is advised that both the lower and upper flushing valves be kept fully open as well as the guide nozzles of the dielectric flushing should be in close contact with both sides of the work surfaces (0.1 mm to 0.2 mm away from the workpiece) for trouble-free operation as well as for receiving a greatest cutting rate. To attain the necessary flow rate, it is primarily required to set the upper flushing nozzle correctly.

Additionally, it is advisable to keep the dielectric fluid splashing while working particularly at the workspace. It is advised to maintain the lower flushing valve fully open and modify the flow through the top flushing valve, if the upper guide nozzles are unable to fit in close proximity to the work surface. The low-pressure flushing switch can be activated, if neither of the guiding nozzles is in a position to make close contact with the work surface. While using the machine tool, the electrical conductivity of the dielectric fluid needs to be tested, and it must be monitored by testing it occasionally while manufacturing.

7.5.2 Maintenance Method and Remedial Actions for Wire Feed System of Wire-EDM

A flexible and complicated machine named as CNC-WEDM is available in the market that is embedded with number of mechanical and electrical systems and elements, which contents the power device, fluid device, CNC part programming device, voltage subsystem, guide nozzle subsystem, floppy drive subsystem, etc. The WEDM manufacturers have included an alarm element within the machine structure as an integrated part that sends out notifications to indicate the occurrence of fault in advance while machining [17]. Additionally, the manufacturers of the machine are now stepping to integrate a maintenance specific step to be taken during or after the procedure to fix these flaws. However, occasionally, the alarm messages are unable to immediately aid in the detection of problems occurring during machining on the wire electrode of the WEDM, which results in breakdown. There are many explanations for breakdown, which can be understood by Figure 7.3. According to Figure 7.3, one of the most momentous reasons for failure during machining is wire breakage. It could happen as a result of incorrect variable settings and their identification. Accordingly, it is needed to balance multiple factors, including wire deflection, mechanical properties, reliability, spark gap width, and excessive

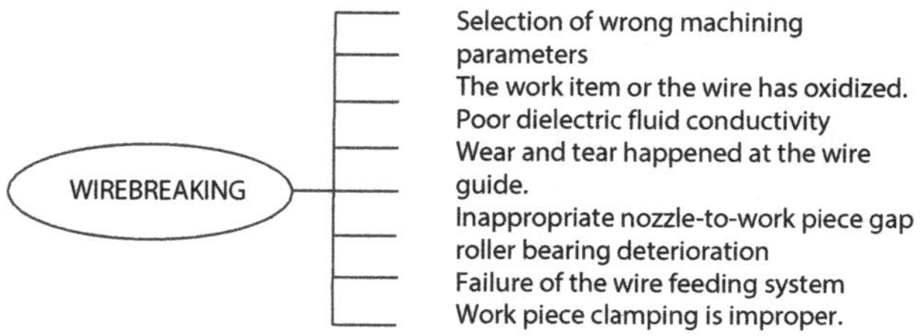

Figure 7.3 Causes of faults for wire breaking.

Table 7.2 Probable reasons of wire breakage in WEDM with remedial solutions.

Components	Probable origins/reasons	Remedial
Checking machining condition	Overly high settings for peak current, pulse-on time, no load voltage, etc., result in increased electrode wear.	Specify appropriate values according to the cutting method.
Lower flushing nozzle	The outer edges of the wire cannot be wrapped with enough dielectric fluid to prevent wire breaking at the current pickup.	Clean the lower flushing nozzle.
Dielectric water	When a wire breaks within a machining gap, the liquid dielectric can evade or mix with the air due to excessive injection pressure, leading to aerial release. If the spark gap's upper and lower flushes are balanced, then air would not get stuck. Wire breakage could be caused by the air pressure in the gap between the sparks.	Dielectric flushing level should be adjusted.
Conductivity	Deionized water has a low conductance and is dielectric. Potential cause of wire breaks.	Decide on an electrical conductivity value of 20 and keep it within 2 units.
Workpiece surface and clearance	Development of stacks on the workpiece's surface.	Make sure not to machine stacked workpieces.
Wire	A wire that has been bent or twisted can generate a discharge concentration that breaks the wire.	Replace twist or bent wire.

free length of wire across the workpiece, to remove breakdowns. Locating the pre-failure signs on the wire in WEDM devices can help in removing the likely reasons of a specific malfunction. The possible reasons for wire breakage in WEDM along with corrective measures that can aid in recognizing issue during machining can be identify in Table 7.2.

7.5.3 Fuzzy Logic

The degree to which a part belongs to a set is an idea in the FS theory [18, 19]. Additionally, one element could belong to multiple sets. Fuzzy sets are frequently used to describe human knowledge or general rules of thumb, where IF-THEN logical assertions are referred to as rules. Only the parts are changed in a system of FL, where the logic principles are clearly stated to define them [20]. Crisp is the idea that contradicts fuzzy, where Yes and No can be defined with the consideration of both 0 and 1 to convey the binary world of crisp. Crisp in FL can be used to handle uncertainty for evaluating numerous uses. Fuzzy logic can be used to define more clear reasoning in areas like expert decision-making, recognition of patterns, product design, and machine tracking [21, 22]. The extent to which data value in the FS belongs is generally represented by MFs, which spans a variety of structures. Membership functions used for decisions are arbitrary and can have any shape. Membership functions cannot be distributed at random. The levels of membership in MF for a particular output would help in assuming a fuzzy variable. Expert systems built on fuzzy reasoning have a natural tendency to be more resilient toward inaccurate modelling, missing of input data, and unexpected shocks. It is ascertained that, to get the right outcomes though input data, knowledge base model are always necessary, where undoubtedly FL is an expert tool for fruitful outputs [23, 24]. The following are the main characteristic of FL systems that improve their efficiency for implementing in particular applications.

i. Fuzzy devices are useful for reasoning that is uncertain, especially when the system has a challenging mathematical model.
ii. Fuzzy logic enables estimation-based decision-making in the presence of imperfect or ambiguous data.
iii. Fuzzy logic also provides instant result despite of giving the results at such conditions where the required result may not be crisp.

7.5.4 Design of a Fuzzy Rule

To design an FR, data collection is the initial step. The collection of the available data is the primary step in the construction of a fuzzy system. The data come in two different forms, i.e., qualitative data based on subjective collection by human experience or quantitative data based on sampled input-output pairings that are objective in nature and are documented by an expert [25, 26]. Here, in FL, human experience is described as a series of IF-THEN rules that specify what must be done and under what circumstances. The sampled pairs provide quantitative data on the basis of inputs and outputs [27]. Although the nature of the various sorts of information differs, they all share the common trait of typically being insufficient on their own. Despite the fact that a human operator is capable of running the system, a part of the knowledge will be lost when it is transferred to an ES structure. However, prior numerical information alone is insufficient to forecast upcoming control circumstances. Thus, it is always needed to design a system through the integration of these two sorts of data in a single decision-making model. In FL, FRs are needed to design for executing a system.

The FL TOOLBOX of MATLAB is used for maintaining this FR in the present chapter for developing the fault detection framework for wire breaking, whose modeling steps are as follows.

The wire breakages depends on various parameters but, in the present chapter, three variables named as wire tension, wear and tear at the wire guide, and interelectrode gap are defined responsible for resulting in wire breakage and for making FR.

These are the input variable used for FR-based model. Afterward, the FSs of each universe of discourse are labeled considering the term and their linguistic MF as shown in the Table 7.3.

Table 7.3 Fuzzy set terms.

Linguistic variable	Membership function
Wire tension	Very low, low, medium, high, very high
Wear and tear at the wire guide	Very low, low, medium, high, very high
Interelectrode gap	Very low, low, medium, high, very high
Wire breakage	Very low, low, medium, high, very high

The universe of discourse for these variables is [0, max], and the range for input variables is taken at range of [1, 10], and, for output variable, the range is taken at scale of [1, 10].

7.5.5 Application of Fuzzy Logic in Wire Breakage

Fuzzy sets are frequently used to describe human knowledge. In the present chapter, FL is used to define general rules of thumb in the form of IF-THEN logical assertions. The same are referred to as rules for managing the wire breakage. Here, the application of FL is used to define the rule based on the three causes for the wire breakage named as follows:

- Wire tension
- Wear and tear at the wire guide
- Interelectrode gap

7.5.6 Membership Functions

It is necessary to develop MFs in order to link crisp inputs into FSs. For the parameters that are input and output, the triangle functions of membership are deployed for computational effectiveness in the present chapter (please refer to Figure 7.4 to identify fuzzy input and output variables). In this chapter, three FSs, i.e., wire tension, wear and tear at the wire guide,

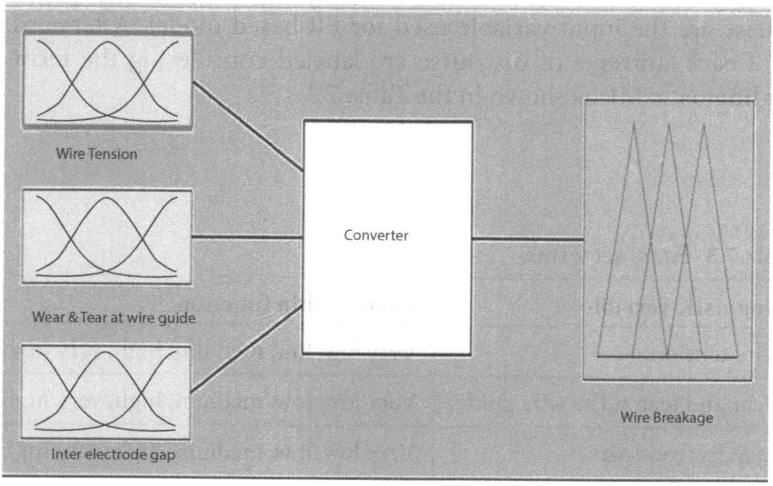

Figure 7.4 Fuzzy input and output variable.

and interelectrode gap, are prepared for the inputs, and one response named as wire breakage is considered for the output. The MFs for input FS are selected in the form of 1 as very low value (VL), 2 as low value (L), 3 as medium value (M), 4 as high value (H), and 5 as very high value

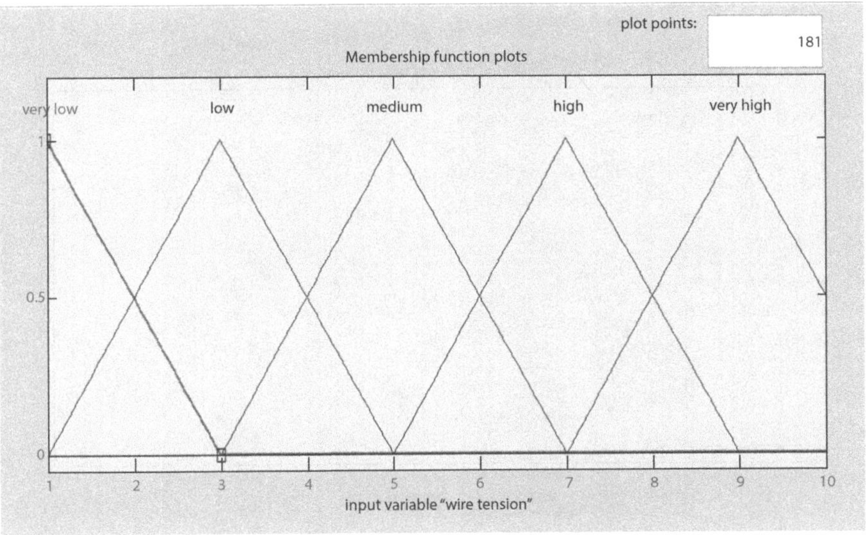

Figure 7.5 MF for input variable (wire tension).

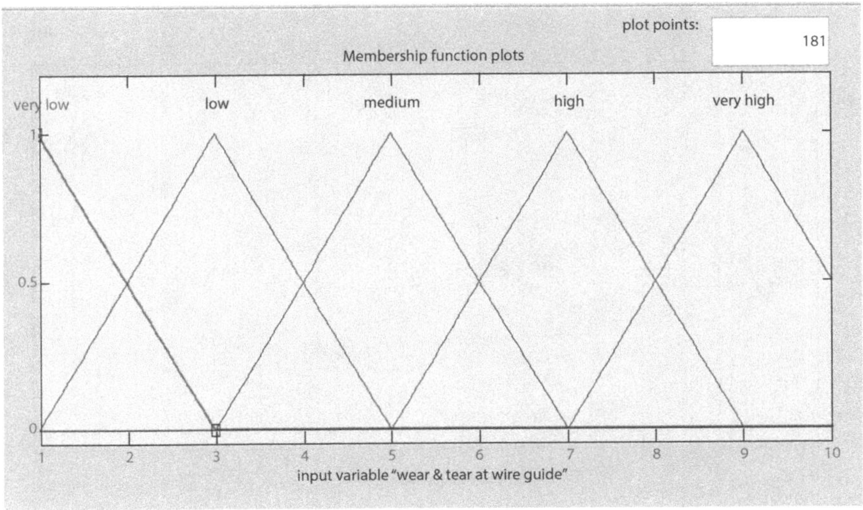

Figure 7.6 MF for input variable (wear and tear at the wire guide).

(VH) (refer to Figures 7.5 to 7.7 to identify MF). Similarly, the output FRs are described in the order 1 as very low value (VL), 2 as low value (L), 3 as medium value (M), 4 as high value (H), and 5 as very high value (VH) (please refer to Figure 7.8 to identify MF for output parameter).

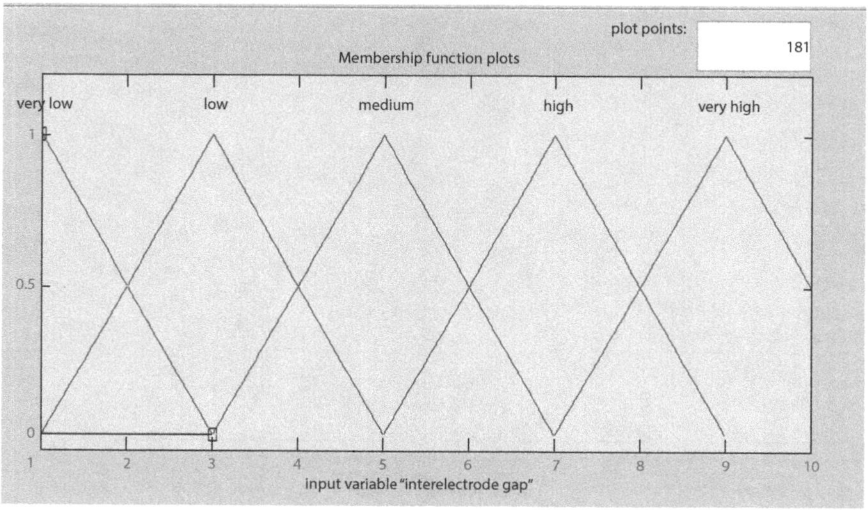

Figure 7.7 MF for input parameter (Interelectrode gap).

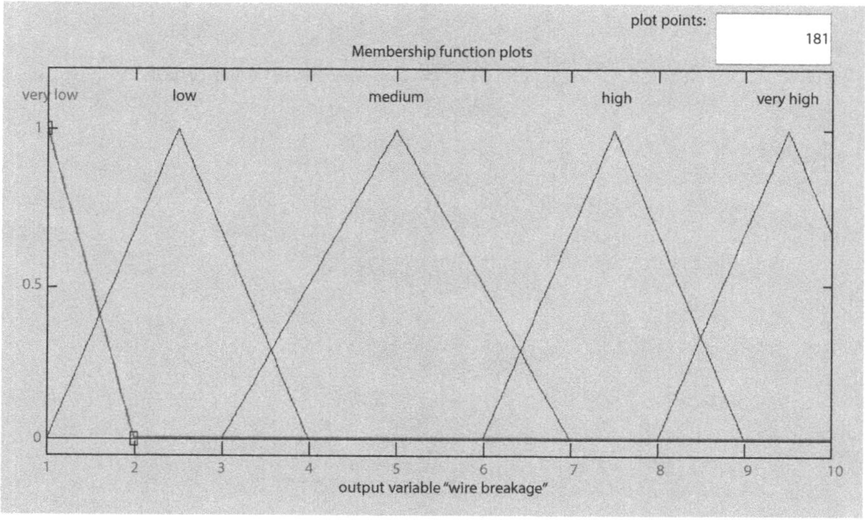

Figure 7.8 MF for output parameter (wire failure).

7.5.7 Detection Rules (DRs)

The DR is one of the most crucial elements in designing fuzzy modules. The FR, which determines the output standard in accordance with the inputs, depends on the knowledge and expertise of domain specialists. Every of these principles is expressed as an IF from input side (range of the three conditions) and THEN from output side (consequence, i.e., the feasibility of wire breaking). There are five condition indicators connected to three input FSs, and, thus, there are 125 regulations produced as a result in this study. The regulations are offered to maintain the case of wire breakage purposes only. A number of FRs are made typically activated for every set of condition indicators. The few guideline rules are shown in Sections 7.5.7.1 to 7.5.7.4.

7.5.7.1 Rule no. 23

Wire breaking is moderate, IF wire tension is extremely low, wire guide wear and tear is substantial, and interelectrode space is moderate.

7.5.7.2 Rule no. 45

IF wire tension is low, wear and tear at the wire guide is high, and interelectrode gap is very high; THEN wire breakage is high.

7.5.7.3 Rule no. 82

Wire breaking is moderate, IF wire pressure is substantial, wire guide wear and tear is inadequate, and the interelectrode gap is small.

7.5.7.4 Rule no. 102

Wire breakage is minimal, IF wire tension is substantial, wire guide wear and tear is very little, and interelectrode gap is small.

7.5.8 Bond-Graph Approach for Dielectric Circuit Maintenance (Mineral Bed Filtration System)

Mineral bed filtration system is shown in Figure 7.9, where one can understand that liquid dielectric with the fine debris from wire EDM machine gets deposited into the dirty tank. The pump in the dirty tank pumps dielectric and metal-chopping dust to the mineral bed filter's input via

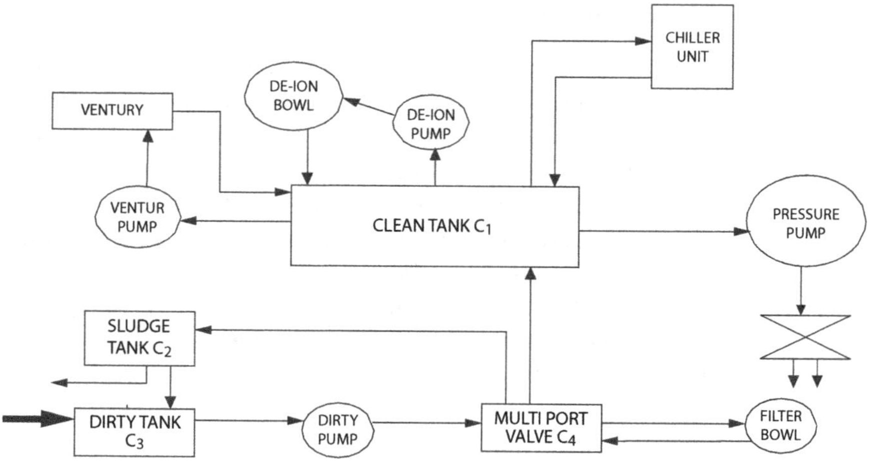

Figure 7.9 Mineral bed filtration system.

five-way MPV (multi-port valve). Filtered liquid again enters to the MPV from outlet of mineral bed filter and is discharged to the clean water tank. This operation continues as the wire EDM machine starts and continues until the mineral filter pressure rises by 0.4 bar (chocking pressure). In the normal running condition, the mineral bed is designed to work for at least 24 h before chocking. However, it entirely depends upon various cycles, units, metal cutting dust, proper backwash, and rinse cycle.

7.5.8.1 Decant Cycle

As soon as the pressure during filtration cycle rises by 0.4 bar (chocked condition), it is time for decant for recovering, where system and machine is required to be stopped for the backwash cycle.

7.5.8.2 Backwash Cycle

When the mineral bed filter is chocked due to cutting dust, which resulted in pressure increase in the system and the liquid flow from dirty water pump reduces. The backwash cycle is the second cycle, which is programmed to clean the filter media of the collected debris, dust, and contaminants.

7.5.8.3 Rinse Cycle

In rinse cycle, the liquid is allowed to enter from top nozzle of filter and comes out again to MPV, which is connected to sludge tank.

7.5.8.4 *Sludge Extraction*

The sludge extraction tank is designed in the system to allow sufficient sedimentation time (24 h) between two filter backwashes. The size of sludge tank and its slope at the bottom are designed to ensure proper sludge settlement.

7.5.8.5 *Conductivity Controls and Deionizer Pump*

The control panel is fitted with conductivity controller for the clean dielectric liquid, and its probe is situated in clean water tank. As conductivity increases beyond the present value (e.g., 20 microsiemens), it will automatically switch on deionizer pump.

7.5.8.6 *Chiller Unit*

The temperature of clean dielectric liquid is to be maintained around 20°C. If this temperature rises, then chiller unit starts automatically and cools the dielectric and dielectric goes back to the clean tank. Once the dielectric temperature is under control, the chiller unit stops automatically.

7.5.9 Fundamental Concepts of Bond Graph

The BG technique usually refers to a graphical way in the direction of modeling. In the said technique, energy ports of the components of the system are connected by bonds, which are specifically responsible for the transfer of energy between these system components. The BG method is simple and logical. In BG, the physical impacts are understood by the way of defining the interaction between the components that are responsible for creating the system, and the same should be taken into consideration in the first place both qualitatively and quantitatively.

System elements are supplied with necessary information at each stage in the modeling process in order to construct and assess a mathematical representation. The BG technique can makes it possible to create a graphical design that is in line with the initial concept of conserving energy without creating or rewriting formulas, when utilized correctly. With BG technique, it is possible to abort the creation of a mathematical framework from a visual representation, which is acceptable from the perspectives of project's goals and a proper software applications that can do this work autonomously. Users of the approach could now concentrate on physical

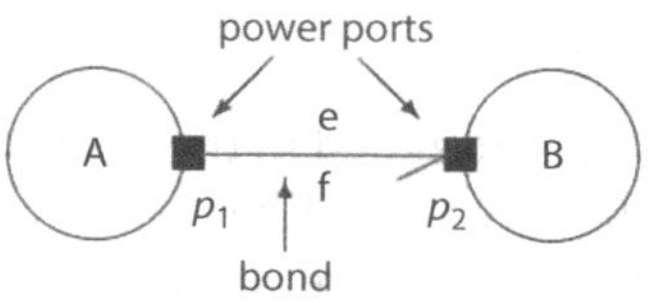

Figure 7.10 Power bond linking two power terminals of two parts designs.

simulation of a device as a result. The basic structure of BG can be understand by Section 7.5.9.1 and 7.5.9.2.

7.5.9.1 Power Variables and Analogies

In BG, a half arrow is attached to every link to represent the starting point of the energy flow as represented by Figure 7.10, because energy can move in both ways across two power terminals of distinct nodes. Here, power can be calculated by Equation 7.1.

$$\text{Power} = \text{Effort} \times \text{Flow} \ldots \quad (7.1)$$

In translational mechanical energy, force or linear velocity is equivalent to endeavor or flow. In electrical engineering, the immediate quantity of electricity generated at a port is calculated as the sum of the voltage drop

Table 7.4 Flow and effort parameters in numerous energy domains.

Bond graph	Translation	Rotation	Electrical	Hydraulic
Effort (e)	Force (F) in N	Torque (τ) in N·m	Voltage (v) in V	Pressure (P) in Pa
Flow (f)	Velocity (v) in m/s	Angular velocity (ω) in rad/s	Current (I) in A	Flow (Q) in m^3
Integrated effort (p)	Momentum (p) in kg·m/s	Angular momentum (h) in kg·m^2/s	Line of flux (λ) in Vs	Momentum per unit region (p) in kg·m/s
Integrated flow (q)	Position (x) in m	Angle (θ) in rad	Charge (Q) in c	Volume (V) in m^3

among two pins of a port as well as the joint current in both pins. The temperature in absolute terms and the rate at which entropy changes are the equivalent parameters in the thermal domain. In every one of the power areas, the elements affecting effort or flow are connected. The flow and effort parameters in numerous energy domains for BG are represented in Table 7.4.

7.5.9.2 Computational Causalities

Each bond connects an array of terminals in different places A and B (Figure 7.10), because each bond has two power variables, one of which may be set by either of the two sections, whereas the other is provided by the other sub-division. In a nutshell, from a calculation standpoint, model A evaluation may be utilized to determine the quantity of effort, but model B evaluation could be used to determine the flow. It could also be the opposite way around. An opposing line connected to the bond represents the method of computing cause and effect, which specifies which sub-model the work is determined in.

7.5.10 Systematic Establishment of BG

One benefit of the BG technique is that the methodical creation of a BG may be guided by the topological connectedness of system elements. For the development of BG for mechanical sub-devices, one of two techniques may be stated, namely:

1. Determine different speeds as well as angular speeds, or depict them with a 1-junction.
2. By using a 0-junction between two 1-junctions, insert C-, R-, TF-, and GY ports, where C stands for capacitive, R stands for resistive, TF stands for transformers, and GY stands for gyrators.
3. Connect the inertia 1-port components to the appropriate 1-junction.
4. Connect the necessary 1-junctions with 1-port sinks and sources.
5. Give each bond an indicative direction (a half-arrow) for the energy flow.

6. Minimize the BG through removing any 1-junctions that indicate a velocity or angular velocity of zero and all incidence bonds.

7.5.11 The BG Design in Wire-EDM Machine

In way to sort out the problems in mineral bed filtration system of wire EDM machine, a BG design is made using software (20-sim). The build BG design of the device using the build rule of BG can be understand by Figure 7.11. In Figure 7.11, MGY is the modulated gyrator and MTF modulated transformer. The volume coefficients of the clean tank, sludge tank, dirty tank, and MPV are C_1, C_2, C_3, and C_4, respectively. The fluid flow resistances in the pipe are as follows:

1. The pipe linking the de-ion bowl and clean water tank is R_1.
2. The pipe connecting the venturi pump and clean water tank is R_2.
3. The pipe connecting the pressure pump and the clean water tank is R_3.

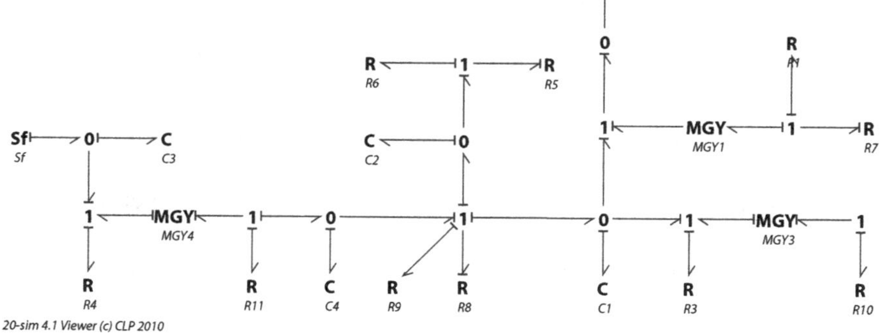

Figure 7.11 BG for mineral bed filtration device.

4. The pipe connecting the dirty tank and dirty water pump is R_4.
5. The pipe connecting the sludge tank and the dirty water tank is R_5.
6. The pipe for drain is R_6.
7. The pipe linking de-ion pump and the clean water tank is R_7.
8. The pipe connecting the clean water tank and the MPV is R_8.
9. The pipe connecting the sludge tank and the MPV is R_9.
10. The pipe connecting the pressure pump to the machine is R_{10}.
11. The pipe connecting the dirty water pump and MPV is R_{11}.
12. The pipe connecting the chiller unit and clean water tank is R_{12}.
13. The pipe connecting to slurry water is R_{13}.

7.6 Result and Discussion

The result for FR-based model for cause of wire breakage indicates that the DR base is one of the fuzzy module's most crucial elements. The success of FRs, which determine the output norm based on the inputs depends upon the knowledge collected through the edification and expertise of domain experts. In the present chapter, these rules are all coded as IF from input side (scales of the three requirements) and THEN from output side (consequence, i.e., cost) statements. Five indicators of condition in the form of very low, low, medium, high, and very high are associated to three input sets (wire tension, wear and tear at the wire guide, and interelectrode gap) of fuzzy data in the present chapter. There are 125 rules produced as a result, and the numbers of FRs are made typically activated for each set of condition signals. An FR-based model for the cause of wire breakage is accordingly developed, which holds the ability of time saving based on ease knowledge acquired and is simple to establish. The model is based on fuzzy inference engine (FIE). The principle of this FIE is to conduct the FRs necessary for the calculation of the fault cause. The model works according to the fuzzy inputs with the activated principles and provides a very quick response. Description for input and output variables along with their MF and range are given in Table 7.5. It is affirmed that, based on the linguistic rules, the output, i.e., possibility of wire breakage, can be measured, which is based on the input variables (Figure 7.12).

Table 7.5 Description for input and output variables along with their MF.

```
[Input1]
Name='wire_tension'
Range=[1 10]
NumMFs=5
MF1='very_low':'trimf',[-1.25   1   3]
MF2='low':'trimf',[1   3   5]
MF3='medium':'trimf',[3   5   7]
MF4='high':'trimf',[5   7   9]
MF5='very_high':'trimf',[7   9   11]
[Input2]
Name='wear_&_tear_at_wire_guide'
Scale=[1 10]
NumMFs=5
MF1='very_low':'trimf',[-1.25   1   3]
MF2='low':'trimf',[1   3   5]
MF3='medium':'trimf',[3   5   7]
MF4='high':'trimf',[5   7   9]
MF5='very_high':'trimf',[7   9   11]
[Input3]
Name='interelectrode_gap'
Scale=[1 10]
NumMFs=5
MF1='very_low':'trimf',[-1.25   1   3]
MF2='low':'trimf',[1   3   5]
MF3='medium':'trimf',[3   5   7]
MF4='high':'trimf',[5   7   9]
MF5='very_high':'trimf',[7   9 11]
[Output1]
Name='wire_breakage'
Scale=[1 10]
NumMFs=5
MF1='very_low':'trimf',[-1.25   1   2]
MF2='low':'trimf',[1   2.5   4]
MF3='medium':'trimf',[3   5   7]
MF4='high':'trimf',[6   7.5   9]
MF5='very_high':'trimf',[8   9.5   11]
```

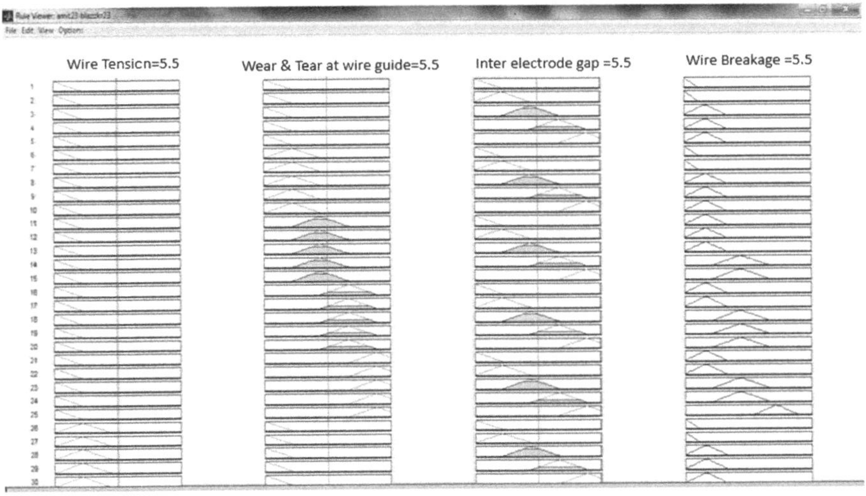

Figure 7.12 Output for fuzzy rule-based model.

7.6.1 Simulation Confirmation for BG

To test the validity and to diagnose the fault in filtration system software (20-sim) is used to generate the BG (refer to Figure 7.13). The simulation of the BG is done on various parameters (R_1, R_2, R_3, R_4, R_5, R_6, R_7, R_8, R_9, R_{10}, R_{11}, R_{12}, and R_{13}), (C_1, C_2, C_3, and C_4) and (S_e and S_f). Here, S_e indicates toward sources of efforts, and S_f indicates toward sources of flows. Various data have been selected from literature survey for simulation and among the various available data. The BG has been simulated for the number of times. Among the number of times, one of the received BG simulation result can be identify from Table 7.6. This confirms the simulation of BG for filtration system of Wire-EDM machine (model number: MAXICUT 734).

7.7 Conclusion

The programming of the fuzzy design technique toward selection of four main variables, i.e., electrical, electrode, dielectric, and workpiece control settings using Wire-EDM, is discussed in this chapter. It is determined that the incorporation of multiple features should be taken into account along

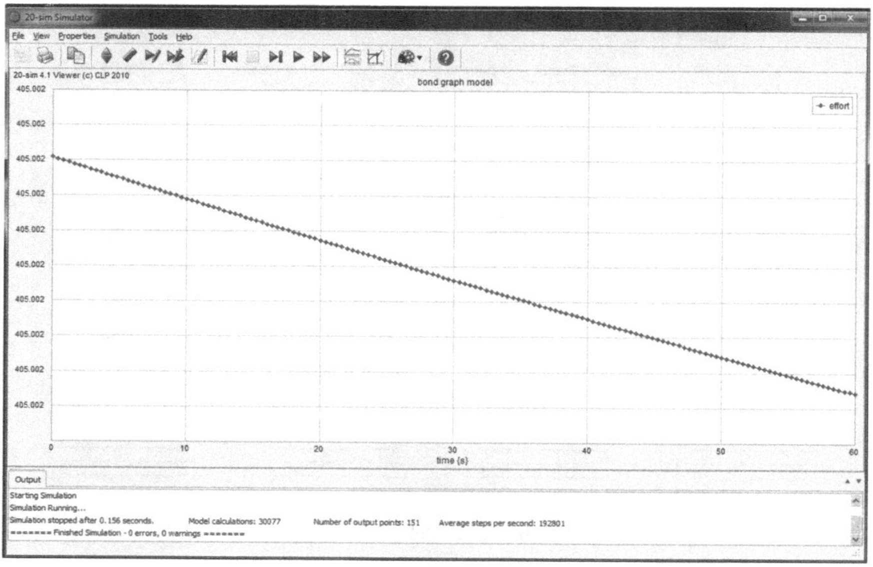

Figure 7.13 Simulation result for bond graph.

with their relative relevance in the fuzzy design process for allowing it to deliver a better and more precise solution toward the options. The strategy discovered with the help of FL and derived rules for regulating filtration system of Wire-EDM is found effective, logical, and reliable when compared to many other MCDM approaches. This approach is found effective to swiftly deliver the best solution, because the computational time required to get the answer is quite minimal. The approach can be used in diverse fields for evaluation and can simultaneously weigh a variety of both quantitative and qualitative choice of features. The presented approach have the capacity to quickly identify problems in Wire-EDM, and, thus, the suggested model can satisfy the needs of the manufacturing sector toward boosting productivity by cutting down lead times. It is found that the wastage of wire can be minimized by reducing the feed rate depending on input parameters. This will help in reducing the down time and increase productivity. The system that was built learns on its own and acquires knowledge while saving time and being simple to operate. Through accurate detection of the Wire-EDM actual defects and the recommendation of appropriate corrective actions against faults during Wire-EDM, functioning can be possible by the adaptability of study presented in this chapter. The designed solution can also assist and direct users or trainees.

Table 7.6 Values of various parameters.

R_1	10 k Pa s/m^3
R_2	130 k Pa s/m^3
R_3	140 k Pa s/m^3
R_4	10 k Pa s/m^3
R_5	12 k Pa s/m^3
R_6	43 k Pa s/m^3
R_7	24 k Pa s/m^3
R_8	13 k Pa s/m^3
R_9	17 k Pa s/m^3
R10	86 k Pa s/m^3
R11	63 k Pa s/m^3
R12	52 k Pa s/m^3
R13	29 k Pa s/m^3
C_1	5.30m^3/k Pa
C_2	6.67 m^3/k Pa
C_3	4 m^3/k Pa
C_4	5.54 m^3/k Pa
S_e	1
S_f	1

The study emphasized on the establishment of the BG for the system, and, once the BG is been established, a fault isolation technique can be developed, which can forecasts the transients brought by faults in the system. The same depends on the qualitative traits of the transients, and the outcomes of the experiment support the approach of BG. This approach requires fewer intricate calculations than theoretical alternatives. This approach is ineffective for multiple faults and only works for abrupt faults. This method is also not appropriate for defects that are only beginning like cavitations, corrosion, and pump leakage. The present chapter affirmed that system predictions can be effectively done using input and output variables in a steady state with the BG design, which can also assist in possible diagnosing the early problems.

References

1. Radhoui, M., Rezg, N., Chelbi, A., Joint quality control and preventive maintenance strategy for imperfect production processes. *J. Intell. Manuf.*, 21, 205–212, 2010, https://doi.org/10.1007/s10845-008-0198-x.
2. Kim, D.H., Kim, S.H., Koh, K.S., A Scheme for an Internet-based Checking Method of Machine-Tools with Variant CNC Architecture. *J. Mech. Sci. Technol.*, 19, 97–105, 2005.
3. Shahanaghi, K., Babaei, H., Bakhsha, A., Nasser, F., A new condition based maintenance model with random improvements on the system after maintenance actions: Optimizing by montecarlo simulation. *Int. J. Simul. Model.*, 4, 230–236, 2008.
4. Sortrakul, N., Nachtmann, H.L., Cassady, C.R., Genetic algorithms for integrated preventive maintenance planning and production scheduling for a single machine. *Comput. Ind.*, 56, 161–168, 2005.
5. Andrew, K.S.J., Daming, L., Dragan, B., A review on machinery diagnostics and prognostics implementing condition-based maintenance. *Mech. Syst. Signal Process.*, 20, 1483–1510, 2006.
6. Mengquan, T., Lingen, C., Jiashan, J., Fengrui, S., Chih, W., Bond-graph-based fault-diagnosis for a marine condensate–booster–feedwater system. *Appl. Energy*, 81, 4, 449–458, 2004.
7. Balino, J.L., Larreteguy, A.E., Gandolfo Raso, E.F., A general bond graph approach for computational fluid dynamics. *Simul. Modell. Pract. Theory*, 14, 884–908, 2006.
8. Sivanaga, M.R.S., Rao, K.V., Reddy, K.H., Parameswara Rao, C.V.S., Prediction and optimization of process parameters in wire cut electric discharge machining for High-speed steel (HSS). *Int. J. Comput. Appl.*, 39, 3, 140–147, 2017.
9. Sethy, R., Experimental investigation and optimisation in EDM process of AISI P20 tool steel (Doctoral dissertation), NIT Rourkela, Orissa, India, 2014.
10. Lautre, N.K. and Manna, A., A study on fault diagnosis and maintenance of CNC-WEDM based on binary relational analysis and expert system. *Int. J. Adv. Manuf. Technol.*, 29, 490–498, 2006.
11. Makis, V., Jiang, X., Jardine, A.K.S., A condition based maintenance model. *IMA J. Math. Appl. Bus. Ind.*, 9, 201–210, 1998.

12. Huang, T. and Liao, Y.S., A wire-EDM maintenance and fault-diagnosis expert system integrated with an artificial neural network. *Int. J. Prod. Res.*, 38, 1071–1082, 2000.
13. Liu., T.I., Detection of roller bearing defects using expert system and fuzzy logic. *Mech. Syst. Signal Process.*, 09, 484–503, 1996.
14. Tony, B. and Liang, M., Mechanical fault detection using fuzzy index fusion. *Int. J. Mach. Tools Manuf.*, 47, 1702- 1714, 2007.
15. Pattison, L. and Hewit, J.R., ACE: A bond graph modelling and development tool for control system design and implementation. *Mechatronics*, 10, 339–351, 2000.
16. Lin, M.Y., Tsao, C.C., Hsu, C.Y., Optimization of micro milling electrical discharge machining of Inconel 718 by Grey-Taguchi method. *Trans. Nonferrous Met. Soc China*, 23, 661–666, 2013.
17. Sharma, N. and Khanna, R., Modeling and multiresponse optimization on WEDM for HSLA by RSM. *Int. J. Adv. Manuf. Technol.*, 67, 2269–2281, 2013.
18. Sahu, A.K., Narang, H.K., Rajput, M.S., Sahu, N.K., A SAW Mechanism for Investigating the Status of Industrial Robots under Comprehensive Sustainable Aspects. *Int. J. Soc. Ecol. Sustain. Dev.*, 10, 3, 69–84, 2019.
19. Wang, W., Huang, L., Zhu, Y., Jiang, L., Sahu, A.K., Sahu, A.K., Sahu, N.K., Decision support system toward evaluation of resilient supplier: A novel fuzzy gain-loss computational approach. *Kybernetes*, 49, 6, 1741–1765, 2019.
20. Kumar, A., Modelling Of Micro Wire Electro Discharge Machining (WEDM) In Aerospace Material, NIT Rourkela, Orissa, India, 2012.
21. Bag, S., Sahu, A.K., Kilbourn, P., Pisa, N., Dhamija, P., Sahu, A.K., Modeling barriers of digital manufacturing in a circular economy for enhancing sustainability. *Int. J. Product. Perform. Manage.*, 71, 3, 833–869, 2021.
22. Sahu, A.K., Sahu, N.K., Sahu, A.K., Rajput, M.S., Narang, H.K., An Investigation Tool for Mounting Sustainable Practice: Modeling Using GIVTFNs in an Indian Context. *Int. J. Decis. Support Syst. Technol.*, 12, 2, 25–49, 2020.
23. Kang, D., Prabhu, M., Ahmed, R.R., Zhang, Z., Sahu, A.K., Digital-IIoTs spheres approach toward public development: an exploiting fuzzy-grey mathematical modeling of IIoTs spheres. *Grey Syst.: Theory Appl.*, 12, 2, 389–416, 2022.
24. Bag, S., Viktorovich, D.A., Sahu, A.K., Sahu, A.K., Barriers to adoption of blockchain technology in green supply chain management. *J. Glob. Oper. Strateg. Sourc.*, 14, 1, 104–133, 2021.
25. Guo, X., Sahu, A.K., Sahu, N.K., Sahu, A.K., A novel integrated computational TRIFMRG approach with grey relational analysis toward parametric evaluation of weld bead geometry of ms-grade: IS 2062. *Grey Syst.: Theory Appl.*, 12, 1, 117–141, 2022.
26. He, Z., Ma, X., Luo, J., Sahu, A.K., Sahu, A.K., Sahu, N.K., Exploitation of the advanced manufacturing machine tool evaluation model under

objective-grey information: a knowledge-based cluster with the grey relational analysis approach. *Grey Syst.: Theory Appl.*, 11, 3, 394–417, 2021.
27. Kapoor, J., Khamba, J.S., Singh, S., Effects of Cryogenic Treated Wire Electrode on the Surface of EN31 Steel Machined by WEDM. *Int. J. Surf. Eng. Mater. Technol.*, 1, 1, 43–47, 2011.

8

The Impact of Corporate Social Responsibility Cost on Financial Performance: A Multi-Level Quantitative Analysis of Corporate Companies for Decision-Making Approach

Kottala Sri Yogi[1]* and Ch. Shankar[2]

[1]*Department of Operations Management, Symbiosis Institute of Business Management, Hyderabad, Symbiosis International University (Deemed University), Pune, India*

[2]*GITAM School of Business, Hyderabad, GITAM (Deemed to be University), Hyderabad, India*

Abstract

Corporate social responsibility (CSR), in general, represents a management impression, where companies are assumed responsible and forced to make integration among social and environmental concerns in their business atmosphere, processes, operations, and interactions. CSR is commonly stated in terms of the course and the way through which a company attains a balance between economic, environmental, and social pillars of sustainability. CSR is the thought that a corporation should perform to deliver a positive responsibility of the company toward community. This chapter delves into the intricate relationship between CSR costs and the financial performance of corporate companies. It employs a multi-level quantitative analysis approach to explore "How CSR expenditures influence various dimensions of financial performance". Here, the quantitative analysis approach is used to help the companies and practitioners to take decision to operate CSR effectively and in an efficient way. The present chapter will assist in thoroughly examining the antecedents, processes, and influence of CSR cost on financial performance based on quantitative analysis. Through a comprehensive

**Corresponding author*: dr.sriyogi@outlook.com

Atul Kumar Sahu, Rakesh D. Raut, Rohit Raja, Anoop Kumar Sahu and Nitin Kumar Sahu (eds.) Industrial and Manufacturing Designs: Quantitative and Qualitative Analysis, (233–264) © 2024 Scrivener Publishing LLC

selection of corporate companies, this chapter aims to provide insights into the potential trade-offs and synergies between CSR initiatives and financial outcomes. The main objective of this chapter is to study the corporate socially responsible activities among selected Indian corporate companies as a financial perspective. The objective of the chapter is to understand the effect of mandatory CSR on the environment using a hypothetical framework. Secondary data were collected and used for the study. To learn more about the effects of CSR initiatives of these selected corporate organizations, CSR reports, financial reports, and report of mandatory CSR expense disclosure in the financial years have been tested using SPSS.

***Keywords*:** Corporate social responsibility, financial performance, emerging economies, environment, content analysis, corporate companies India

8.1 Introduction

Corporate social responsibility (CSR) is a policy undertaken by the corporate companies to grow profits along with the consideration of other dimensions of sustainability like to take an active and positive social role in the country side on behalf of their business activities. The term CSR correlated many time with the term corporate citizenship. The mix of corporate social obligation into business tasks has turned into a fundamental part of contemporary corporate methodology [1]. The allocation of resources to CSR initiatives raises questions about their impact on financial performance as corporate companies continue to acknowledge their responsibilities to society and the environment. In the present quickly developing business scene, the idea of corporate social obligation has acquired a noticeable quality as organizations have started perceiving their more extensive job in the public eye by considering social and ecological development in return of their business activities [2]. Accordingly, decision-makers have started examination toward CSR cost on financial performance. Decision-makers are now expanding their investigation to present a clear understanding about CSR impacts on ecological effect, work practices, and social commitments, where quantitative technique can make a critical contribution to the decision-making process. Accordingly, the present chapter explores the nuanced transaction between CSR costs and monetary execution measurements, utilizing a staggered quantitative investigation. Through a multilevel quantitative analysis, the purpose of this chapter is to investigate the intricate relationship between CSR costs and corporate sector financial performance. The present chapter addresses the following research questions (RQs) to accomplish the research objectives:

RQ1: What is the connection between CSR costs and monetary execution pointers [e.g., return on assets (ROA), return on equity (ROE), and net revenue] in the assembling area?
RQ2: Are there industry-explicit elements that impact the size and heading of the CSR cost–monetary execution relationship?
RQ3: Do organization explicit attributes, for example, size and key direction, moderate the effect of CSR costs on monetary execution results?

CSR in universal represents a practice or impression, which is adopting by the companies for the development of the society and ecology, where they are working. CSR assumed responsible and forced to make integration among social and environmental concerns in the business atmosphere, processes, operations, and activities performed by the companies. CSR reflects company culture and business practices, which includes a hybrid and holistic approach toward development of societal aspects, technical aspect, and human aspects like having ethical labor practices both internally and as part of a larger supply chain. Today, the society's assumptions are for moral and economical strategic policies, which are continuously rising, which thereby producing organizations into constrained boundaries to coordinate CSR drives into their activities [3]. Nonetheless, this reconciliation presents difficulties, as it frequently includes apportioning assets that could, some way or another, be coordinated toward center business capabilities. The crucial question is raised by this: How does a corporate company's financial performance change, "when resources are given to CSR projects?"

8.1.1 The Goals and Connections

CSR reproduces company traditions and their business practices, where it is always required to make investigations on the financial performance of companies, when resources are given to CSR projects. The essential intention of examination related to CSR is to observationally break down the effect of CSR costs on the monetary execution of assembling organizations [4]. It is crucial to quantify the connection between various financial performance indicators and CSR costs. It is needed to investigate the possibility of factors at the industry level and company-specific characteristics moderating CSR relationship. Attainment and presentation of research information is required to be utilized by the corporate companies that can use by them to make informed decisions about CSR investments while considering their financial goals [5]. The chapter presents an examination that will hold a few ramifications for both scholarly world and practice [6]. The chapter focused investigations on the corporate companies to endure critical disclosures to both scholarly world and practitioners, which fall under industrial, distribution, services, and manufacturing sectors. The present chapter contributes to the ongoing academic debate regarding the connection between CSR and financial performance. The chapter has used quantitative information for analysis to enable a more nuanced comprehension of the dynamics involved. The chapter has employing a multi-level quantitative analysis approach to contribute toward the methodology. From a commonsense stance, it is found that the discoveries from examination can direct assembling organization chiefs in settling on very much educated conclusions about their CSR ventures [7]. By understanding about "what various sorts of CSR costs mean for monetary execution measurements", directors can find some kind of harmony between capable strategic policies and financial supportability. In addition, policymakers and administrative bodies can profit from the bits of knowledge acquired to plan arrangements that boost mindful corporate way of behaving. Accordingly, the present chapter delves toward defining the intricate relationship between CSR costs and the financial performance of corporate companies to assist in understanding "how CSR expenditures influence various dimensions of financial performance".

8.2 Literature Review

It is found significant to establish hypothetical points of view on the connection between CSR and monetary execution [8]. Additionally, clashing of discoveries from earlier examination that contributed toward correlations that are positive, negative, or neutral ways, which assist in understanding that CSR contributions could have an impact on financial outcomes, is significant. Nowadays, it is needed to group CSR costs and CSR responsibilities in the areas of philanthropy, ethics, law, and business to reveal the development of CSR consumption and its suggestions [9]. The same is required to offset CSR momentary expenses with long haul benefits. Academic research to contribute into the connection between financial performance and CSR is required to make coverage into a wide range of perspectives and outcomes with CSR. It is found to give an extensive comprehension of the intricate elements hidden in the effect of CSR costs on the monetary execution of assembling organizations [10]. Through a variety of theoretical lenses, the connection between CSR and financial performance is needed to be investigated.

According to survey, proponents of the "positive sum" perspective contend that CSR initiatives have the potential to enhance a company's reputation, increase customer loyalty, and boost employee engagement, all of which can have a positive impact on financial metrics like ROA and ROE [11]. According to this point of view, CSR represents not only a moral obligation but also a strategic opportunity for businesses to generate shared value. Nonetheless, the "unbiased" and "negative aggregate" viewpoints give differentiating perspectives. The impartial view places that CSR drives could straightforwardly affect monetary execution, as they frequently include costs without quick unmistakable returns [12]. The negative total view, then again, contends that extreme CSR uses could diminish an organization's monetary presentation, redirecting assets from center business exercises. The connection between CSR and financial performance has been the subject of numerous theories. The improvement of relationships with stakeholders is one mechanism.

Organizations that put resources into CSR drives pointed toward helping workers, nearby networks, and the climate to construct a more grounded associations with these partners, possibly prompting decreased representative turnover and improved admittance to assets. Identifying and directing decisive variables further entangle the relationship [13]. Industry attributes, for example, can shape the effect of CSR on monetary execution. Businesses with a more prominent cultural like assembling could encounter more articulated impacts of CSR drives on monetary results. Organization explicit qualities like size and vital direction can likewise impact the degree to which CSR costs influence monetary execution. The estimation and revealing of CSR costs will produce advanced understanding toward preparing policies to sustainability that are desirable in terms of the objectives and values of our society, which, in turn, will endure expanded accentuation on truthfulness and responsibility to companies [14]. Organizations should coordinate CSR into their tasks continuously and repeatedly to uncover their CSR consumptions in a normalized and educational way. The ascent of non-monetary detailing structures and supportability revealing rules are needed to be integrated with the correspondence of CSR exercises and expenses for partners. The same are required to be framed by the policymakers and practitioners for coping CSR consumptions massively and actively.

The writing on the connection between CSR costs and monetary execution is set apart by clashing discoveries. While certain examinations report positive connections between CSR endeavors and monetary results, others

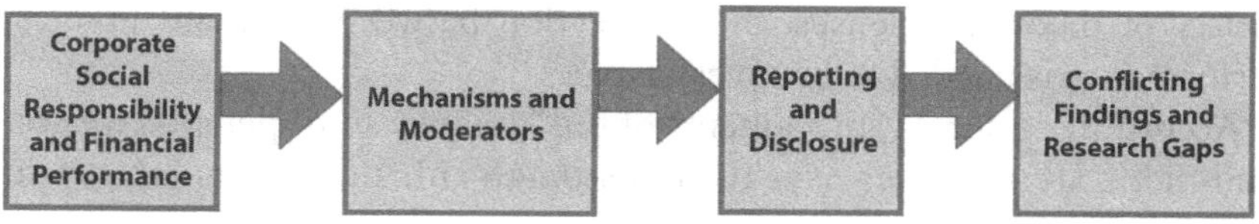

Figure 8.1 CSR and financial performance attainment.

track down no huge relationship or even bad affiliations [15]. Different research methods, contextual factors, and the measures of CSR and financial performance used all contribute to these disparities. In conclusion, the research on CSR and financial performance in the corporate companies sector offers a variety of viewpoints and findings. While certain investigations stressed on the possible cooperative energies between CSR drives and monetary results, others featured the intricacies and compromises included in CSR drives. Many components are still hidden to reveal the CSR relationship with cost elements, to understand the directing impacts of industry and company attributes, and to disclose the advancing scene of CSR to assist in comprehending a social impact beyond profits under exploration region [16]. Accordingly, the chapter objectively presenting a study to deliver quality knowledge about the corporate socially responsible activities among selected Indian corporate companies under a financial perspective. The chapter will assist in understanding the effect of mandatory CSR on the environment using a hypothetical framework. The critical objective of the chapter can be understand from Figure 8.1, where the conflicting findings and research gaps related to CSR and financial performance is suggested to be attended by focusing on CSR mechanism and moderators elements.

8.3 Methodology

Corporate companies should involve in social activities irrespective of diverse industrial activities performed by them, sizes, and their locations. Accordingly, it is required to specifically choose methods for data collection and analysis' timetable for attaining desired outcomes [17]. The same demands the identification and justification for using a multi-level analysis strategy, where quantitative analysis can play a significant role. Monetary execution pointers like return on resources, return on value, and net overall revenue are needed to be calculated in this regard. CSR cost estimation elements like direct expenses, aberrant expenses, and opportunity costs are needed to be asses for making driving equations, mathematical modeling, and analysis. Control factors like size of the business, industry, and economic circumstances are needed for descriptive metrics analysis for financial performance and CSR costs. Multivariate relapse examination to measure the connection between CSR costs and monetary execution is the need of an hour to understand progressive development and indeed demonstration to represent expected cross-level associations [18]. Accordingly, present chapter frames the exploration philosophy to be utilized to research the effect of corporate social obligation costs on the monetary execution of chosen corporate organizations. The research design related with survey using quantitative methods for collecting data based on the selection of variables of interest for literatures to disclose analytical strategies is utilize in this chapter to accomplish the research goals that are all included in the methodology [19].

The configuration of the chapter is organized to give an exhaustive comprehension of the connection between CSR costs and monetary execution in the assembling area. A quantitative examination approach is used to take into consideration the precise investigation of mathematical information and to infer measurable connections [20]. A purposive examining system based on quantitative information is utilized to choose a different arrangement of assembling organizations that address different areas, sizes, and geological areas. The utilize approach will guarantees a decent portrayal of the assembling scene. Monetary information and CSR cost data are gathered from yearly reports, maintainability reports, and other applicable monetary records of the chosen organizations. Monetary execution measurements like profit from resources, return on value, and net overall revenue are done. CSR costs are calculated on the basis of the classification into direct expenses, backhanded expenses, and opportunity costs in view of bookkeeping principles and revelation rehearses [21].

Financial performance metrics such as ROA, ROE, and gross profit margin are the primary variables of interest found for the dependent variables. These metrics show how well business can make money and use resources effectively. The free factor of interest is found as the absolute CSR costs caused by each assembling organization [22]. Philanthropic activities, environmental sustainability initiatives, employee welfare programs, and other socially responsible endeavors are among these costs. To moderate the expected impact of bewildering factors and control factors, organization size, industry area, financial circumstances, and geographic area are thought of. These variables consider variations in the context, which may have an effect on both the financial performance and costs of CSR. Unmistakable insights are processed to give an outline of the circulation of CSR costs and monetary execution measurements across the example of assembling organizations [23]. Multivariate regression analysis is used to quantify the relationship between CSR costs and financial performance. For each financial performance metric (ROA, ROE, and gross profit margin), distinct regression models are created. The models incorporate the absolute CSR costs as the free factor and control factors to represent expected puzzling impacts.

To understand the likely effect of industry-level variables on the CSR cost–monetary execution relationship, progressive displaying procedures are utilized. Industry-level factors are acquainted to investigate that the industry-explicit variables moderated the connection between CSR costs and monetary execution [24]. The examination sticks to moral principles in information assortment and investigation. All information sources are fittingly credited, and the protection and secrecy of the organizations' monetary data are kept up with. It is found significant to recognize specific constraints for the exploration methodology [25]. The study focused on a small number of corporate companies to make its findings applicable to other industries. The cross-sectional nature of the information forestalls the foundation of causal connections, and the presence of unseen factors could present endogeneity in study. The approach embraced for this examination plans to thoroughly analyze the connection between CSR costs and monetary execution in the assembling area [26]. The study presented in chapter aims to shed light on the intricate relationship between financial outcomes and ethical responsibility by employing quantitative methods and considering control and moderating variables. The multi-level quantitative analysis findings, which shed light on the connection between CSR costs and financial performance metrics within the corporate companies

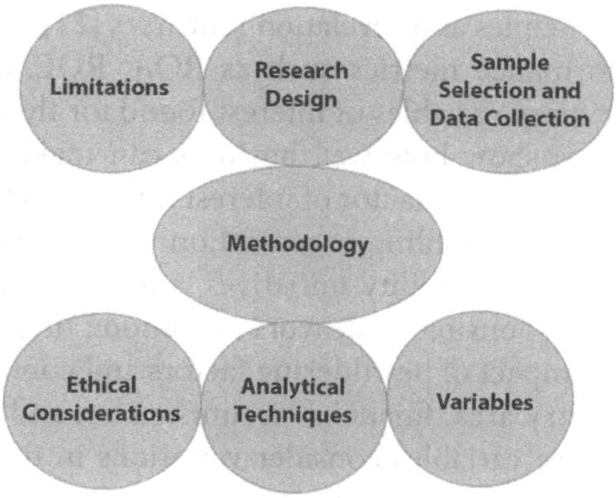

Figure 8.2 Methodology.

under investigation, will be presented by using a hypothetical framework and SPSS. The chapter adopts sample selection and data collection to evaluate CSR research design as shown in Figure 8.2, which consider quantitative analysis to evaluate variable based on ethical practices.

8.4 CSR Initiatives and Discussions

To accomplish the methodological view point (Figure 8.2), a complete target of the review is formed to understand a bunch of speculations with a goal of deciding the impact of obligatory CSR on natural improvement. Utilization of a variety of approaches is done to obtain a clear result to carry out the study scientifically [27]. The efficient procedure, which embraced to lead the examination, is found as exploration philosophy and collection of auxiliary information. Auxiliary information was gathered and utilized for the review. Auxiliary information was gathered from sites, yearly reports, papers, diaries, public CSR gateway, and so on. SPSS software was used to test CSR reports, financial reports, and the report of mandatory CSR expense disclosure in the financial years to learn more about the effects of the selected organizations' CSR initiatives [28]. It was found that numerous modifications to the organizations' CSR policies were brought about by the Companies Act 2013 amendment and implementation [29]. The study attempted to incorporate all of these changes, but it did not evaluate them because they are still in the initial implementation stage.

It was discovered that the existing literature on CSR activities adequately highlighted the issues related to environmental consumer rights and the acts mandating the practice of CSR, when referring to previous research on CSR and its impact on environmental sustainability. Nonetheless, the degree to which the CSR exercises are being attempted by the organizations at risk for it has not been adequately assessed. Furthermore, of the exercises that are being attempted, it is not obvious how much these exercises are pointed toward ecological supportability. The research studies also saw that many researchers have compared the impact of CSR with that of other industries but that environmental sustainability as a subfield of CSR activities has only been investigated to a much lesser extent [30]. As a result, the chapter advocated the new Companies Act's schedule 7(vi) section 135 that covers environmental sustainability, which is the primary focus of this study. To achieve objectives related to sustainable development, businesses must effectively incorporate CSR initiatives [31]. Participating in social responsibility initiatives helps businesses today to boost their commercial and brand value. Companies are obligated to the society in which they operate while they exploit its resources. Companies' sustainable development aspects are influenced by CSR activities of any nature and size [32]. According to schedule 7(vi) of section 135 of the Companies Act of 2013, CSR initiatives are now considered sources of development and growth.

In the year 2014, the Organizations Act became effective with a compulsory CSR arrangement. The demonstration orders the organizations, which fall under specific classifications to burn through 2% of their net benefits in CSR [33]. Notwithstanding, according to the investigation led by [33], it came into the image that not many organizations were seen to have met the 2% standard, as a matter of fact. Just 18% of the organizations concentrated by them were found to have followed the 2% standard. Organizations' approaches to environmental protection mediations in India differ from the prevalent CSR trends.

According to a survey, the fact is accounted that the third largest portion has spent on CSR initiatives, which indicates that environmental sustainability is being taken seriously by businesses today. However, according to survey, about a third of the top 100 companies studied by the Ministry of Corporate Affairs tend to support CSR initiatives other than sustainability and ignore conservation and ecological initiatives [34]. Furthermore, the highest spenders in natural conservation were seen to be fundamentally from the fast moving consumer goods (FMCG), energy, and IT portions, which are as often as possible seen as the most terrible ecological transgressors. Thus, it is additionally accepted that these ventures are just making great the harm done by them. Thus, the restricted conversations on the effect of CSR exercises on natural manageability present differentiating sees. Accordingly, attributable to the absence of lucidity in this regard, the current chapter endeavors to basically assess the CSR execution of Indian organizations on natural maintainability after the execution of the new Organizations Act (2013) [35]. The current chapter is an endeavor to concentrate on the effect of CSR drives of chosen 1352 organizations on natural turn of events.

8.5 Sample Description and Discussions

The present chapter attempts to identify the CSR cost on financial performance and accordingly embraces a range of corporate for attaining fruitful information, where housing development finance corporation (HDFC) Bank is found as a leading financial services provider with over 11,000 branches across 19 countries. HDFC stands for Housing Development Finance Corporation and founded in July 1995; the company is listed on the Bombay Stock Exchange and National Stock Exchange of India [36]. HDFC Bank's core objectives are to provide world-class financial products and services to its customers while maintaining a strong focus on sustainable value creation. The bank has been ranked as one of the top three private banks in India by Financial Times Global 500 in 2016, 2017, and 2018. HDFC Bank is the third largest bank in India by market capitalization and fourth largest lender by assets. The company has a significant presence across 19 countries including Nepal, Sri Lanka, and Bhutan. In 2007, the bank became the first Indian financial institution to issue a US dollar bond overseas. HDFC Bank has over 24 million customers, including 6.6 million digital customers. Its products and services include home loans, personal loans, car loans, and education loans.

HDFC Bank has a network of over 11,000 branches across India. It offers its customers a wide range of services including deposits, loans, and investments in various products like mutual funds, insurance (general), home loan, car loan, and personal loan. The bank also offers internet banking facilities to its customers, which allows them to access their accounts online through the bank's website. HDFC Bank is also ranked among the "Best Banks in Asia", "Best Private Banks in India" by Euromoney, and "World's Best Digital Bank" by World Finance. The bank has received several awards for its innovative service offerings and digital banking innovation. HDFC Bank was awarded the "Best Digital Innovation" award at the Indian Banking Awards 2018 for its flagship product "HDFC UPI". HDFC Bank is the only Indian bank to have been ranked among the top three in Asia by Business Insider. The bank was named India's most valuable brand in 2017 and 2018 by Brand Finance. In May 2018, HDFC Bank became the first Indian private sector bank to be listed on the Forbes Global 2000 list of largest public companies.

8.5.1 Indian Oil

It was founded in 1911 as a public limited company an Indian multinational oil and gas company under the ownership of the Ministry of Petroleum and Natural Gas, Government of India. It is headquartered in New Delhi. In 1932, it became a subsidiary of the Government of India. The company operates in the upstream and downstream segments [37]. Its headquarters are located at Mumbai. The company has been involved in exploration, production, and marketing of crude oil and natural gas since its inception. In addition to these activities, it also engages in refining, marketing, and trading of petroleum products such as gasoline, diesel, kerosene, and aviation turbine fuel.

The company also has a presence in chemicals such as fertilizers and petrochemicals. It has also ventured into other areas such as retailing of cooking gas cylinders and LPG solutions for domestic use by consumers. The company operates through its subsidiaries and associates in India, the Middle East, Africa, and the Americas. It has a presence in more than 50 countries across six continents. The company operates across the globe with subsidiaries in India, Africa, and Asia Pacific. It has a market capitalization of INR 1.34 trillion (US$20 billion) as of April 2018 and a revenue of INR 880.9 billion (US$14 billion) for fiscal year 2017–2018. The company also has debt worth INR 107 billion (US$1.6 billion). The company's products are marketed through its various brands such as Bharat Petroleum, Indian Oil, and Hindustan Petroleum, among others. It also has a presence in over 800 outlets across the country. The company has a production capacity of over 120 million tons per annum (mtpa). It also operates refineries and marketing terminals through joint ventures with foreign oil companies. The company has been the recipient of several awards and recognitions in the past. It was awarded the "Best Publicly Listed Company" by Business Today magazine in 2014, 2015, and 2016. It also received an award for excellence in corporate governance from Governance Now in 2017. The company has been ranked at the sixth position in Fortune India 500 list for 2017. It was also ranked at the 20th place in Forbes India's Super 50 list for 2017.

8.5.2 Infosys and ITC

Infosys is India's second largest software exporter after Tata Consultancy Services (TCS). For the financial year ended 31 March 2019, Infosys reported net income of Rs. 11,902 crores ($1.8 billion) on revenues of Rs. 87,790 crore ($13.4 billion), up 5% from the previous fiscal year. ITC Ltd. is among the top 20 consumer goods companies in India with a turnover of INR 21,700 crore and employee strength of 45,000. ITC is one of the fastest growing consumer goods companies in the world with a presence in over 50 countries across six continents [38]. The company has been ranked among the "Top 200 Doers of Businesses" by Forbes Asia for five years consecutively, since 2012. In FY19, ITC's net sales were INR 154,852 crores. The company's operating profit for the year was INR 37,034 crores, and its net profit was INR 29,845 crores.

ITC Ltd. is a leading Indian company engaged in the production and distribution of tea and coffee. The company has been involved in supply chain management since its inception in 1973. The CSR spend in ITC Ltd. is estimated to be Rs. 12,300 crore. This is a substantial amount of money.

The company's CSR activities include raising awareness about the importance of sustainable growth and the need for an integrated approach to environment protection and conservation in developing countries. Additionally, promoting research, education, and training on environmental issues is also included toward promoting public policies that encourage sustainable development [39]. An attempt to reduce environmental pollution through research and development projects is considered and appraised. In 2009, ITC Ltd. launched a CSR initiative called "Envi Team", which focuses on the environment and sustainability. It includes activities such as tree planting, waste management, waste reduction, water conservation, energy conservation, and recycling. ITC Ltd. spends around Rs. 10 crore every year on these activities, which include tree plantation, water conservation, and recycling initiatives.

8.5.3 NTPC

NTPC Ltd. is one of the largest power companies in India, with a total installed capacity of over 46,000 MW and an annual generation capacity of 205 billion units. The company has been ranked as number 1 in terms of market capitalization on the Bombay Stock Exchange in 2018. NTPC Ltd. has been a part of the S&P 500 index since 2001. In the same year, it was also recognized as one of the "Best 200 Global Growth Companies" by Forbes magazine. In 2017, NTPC was granted the "Global Award for Best Practice in Corporate Governance" by the World Bank Group and International Finance Corporation [40]. The company's corporate sustainability initiatives have been recognized by several organizations including Confederation of Indian Industry, UN Global Compact, and Carbon Disclosure Project. NTPC Ltd. is the largest power company in India. The company has a debt equity ratio of 2.8 and a quick ratio of 0.85. It has generated an ROE of 20% over the past 5 years, with an average profit margin of 7%. The company's current ratio is 1.6, and its cash conversion cycle is 55 days.

8.5.4 ONGC

ONGC is an acronym for Oil and Natural Gas Corporation Ltd. It was established in 1974 by the Government of India. It is a Public Sector Undertaking and is headquartered in Dehradun, Uttarakhand. It has more than 14,000 employees and operates in 27 countries across the world. The company itself has been operating since 1956, when it was known as Oil and Natural Gas Development Corporation Limited. It functions under a business model that includes exploration and production of hydrocarbons like crude oil and natural gas, transportation, refining, marketing and distribution of petroleum products, petrochemicals, and power through its subsidiary companies such as Indian Oil Corporation Limited and Hindustan Petroleum Corporation Limited. It also has stakes in various joint ventures with international oil companies such as BP Plc and Essar Energy Plc [41]. The company's total assets were worth Rs. 156,600 crores. The company has an asset base of more than Rs. 20,000 crore (US$3 billion), which includes over 250 oil and gas blocks with an area of over 10 million hectares in India. The company has total proved reserves as on 31 March 2017 of 4,863 million metric tons of oil and gas, out of which its shareholding in oil and gas blocks held by ONGC Videsh Limited is about 50%.

8.5.5 Power Grid Corporation of India

The company has an installed capacity of more than 90,000 MW, which is expected to increase to 1 lakh MW by 2022–2023. The Power Grid Corporation was incorporated on 15 July 1959, under the Companies Act 1956 as a Public Limited Company and came into existence on 15 August 1959. The company's share capital stands at Rs. 10 crore divided into 10 Lakhs equity shares of Rs. 100 each fully paid up. On 15 August 2019, Power Grid Corporation completed 50 years in the power transmission business. To commemorate this milestone event, several initiatives have been taken by the company over the past few years, which include setting up an endowment fund for social causes and promoting sustainable development through environment conservation projects [42]. The company's performance during the year under review was reflective of its core business activities, with an overall increase of 3% in net sales and 5% growth in EBITDA (Earnings Before Interests, Taxation, Depreciation, and Amortization) over the last year. Furthermore, the consolidated PAT for FY2020 stood at Rs. 864 crores against Rs. 790 crores for FY2019 translating into a PAT CAGR of 7% over FY2013–2019E period. This was achieved due to improved performance across all businesses coupled with prudent cost management measures taken by the management team to ensure continuous improvement in financial performance alongside enhancing shareholder value over time.

8.5.6 Reliance

The company's businesses are organized into two major verticals: energy and petrochemicals, and textiles and retail. In FY2021–2022, Reliance

Industries Limited expects to achieve EBITDA of Rs. 100,000 crores on revenue of Rs. 275,000 crores at an oil price of $60/bbl. In the financial year 2020–2021, the increase in revenues by 9.5% reached Rs. 338,000 crore (US$42 billion). This growth is expected to be driven by a combination of factors including the integration of new businesses and projects, as well as organic growth from existing businesses including retailing and petrochemicals. Growth in sales from their retail business will be driven by the continued rollout of Jio stores across India, whereas the growth in pitcher revenues will be supported by higher refining margins and higher volumes.

8.5.7 TCS

TCS Ltd. is one of the world's largest IT services, consulting, and business solutions companies. TCS believes in providing best-in-class services to its clients through innovation and customer delight [43]. They have committed toward creating a sustainable future for our planet by building energy efficient green buildings, reducing CO_2 emissions in our offices and promoting environmental awareness among their employees. TCS focused on improving its financial performance for the fiscal year 2020–2021. The company planned to achieve this by increasing revenue generation and reducing cost of goods sold. TCS will also focus on corporate sustainability by aiming to reduce greenhouse gas emissions by 10% and water consumption by 20%.

8.5.8 Tata Sons

Tata Sons Private Ltd. is a diversified conglomerate with businesses in telecommunication, software solutions and IT enabled services, engineering, manufacturing of steel, power and other capital goods, consumer products and services, energy, infrastructure development, hotels and real estate, and financial services [44]. The company's revenue is derived from four business segments, i.e., automobiles and components, power and steel, services and consumer products, and investments. The automobile and components segment includes Jaguar Land Rover, a UK-based automotive business that Tata Motors acquired in 2008. The power and steel business includes Tata Steel Europe (formerly Corus Group).

8.5.9 Wipro

Wipro Private Limited (Wipro) is a leading information technology, consulting, and business process services Company, which was incorporated in the year 1995. The company is engaged in providing services in various areas such as information technology, software products and solutions, engineering and construction, financial services, and health care. The company's subsidiaries include Wipro IT Services Limited, Wipro Consumer Care Limited, Wipro Infrastructure Engineering Limited, and Wipro Digital Services Limited. Wipro Private Ltd. has always been a company that prioritizes the needs of its employees and clients. This dedication is reflected in their financial performance, which has been steadily improving over the past few years. The company's revenue for 2020–2021 was $64 billion, up 4% from the previous year. Net profit was $1 billion, which is also up 4% from last year. It is found momentous to derive the linkages among various factors and testimonials of industries based on the range of specifications to satisfy the need of the society and to build decision support frameworks [45, 46]. Moreover, it is found significant to identify and simplify the requirements of the industries, firms, and corporations to provide a reinforced foundation of templates for effectual working [47, 48]. It is identified that the researchers should compare the impact of CSR with that of other industries to reveal the reinforced foundation of templates under CSR activities and societal development and to build decision support frameworks for effectual working.

8.6 Statistical Analysis

8.6.1 T-Test (t-test)

The t-test is a statistical test that is used to determine whether two samples are statistically different from each other. Here, hypothesis are tested, where the null hypothesis states that the two populations being compared have the same mean, whereas the alternative hypothesis states that they do not have the same mean. If you run a t-test and discover that your data show that your population means are not equal, then you can reject the null hypothesis and conclude that there is a significant difference between your populations [49]. One can run a t-test in Excel by using the T-Test tool. This will allow you to enter your data into one or more columns. Select the column, which contains the cell that contains your alternative hypothesis, and then choose between two-tailed and one-tailed tests. The t-test is a parametric test, which means that it assumes that the data you are working with are normally distributed. If your data do not meet this assumption, then you should use a non-parametric test instead.

8.6.2 Multiple Regression

The statistical method known as multiple regression enables us to anticipate the connection that exists between several distinct variables. This can be utilized to figure out what makes one variable related to another or what certain amounts mean for each other. The most widely recognized use of regression is for understanding numerous relapses toward monetary demonstration. For instance, if your model uses age and income to predict how much people will earn, then you might want to factor it in a variable like education level. Then, you would be able to test your model in the way of checking that there are other factors that affect it.

In machine learning, multiple regression aids in the development of algorithms, as opposed to being explicitly programmed in advance. Predicting the value of a dependent variable based on one or more independent variables is typically done with multiple regression. For instance, assuming that you are attempting to anticipate how much cash somebody will make in view of their age and pay level, you would utilize different relapse on the grounds that the two elements are influencing your result. At long last, numerous relapses are utilized in many fields for breaking down information. For instance, if you need to check whether there is a relationship between two factors (like age and pay), then various relapse can be utilized to do this for you. The main thing to comprehend about various relapse is that it is not equivalent to effect output from basic direct relapse. In simple linear regression, you only have one independent variable and one dependent variable. You want to find an equation that accurately predicts "how many of the changes in the dependent variable are caused by changes in the independent variable". Numerous relapses are likewise an effective method for testing the reliability of your information. You can use multiple regression to determine whether multiple variables are related or merely coincidentally correlated, if each variable has its own unique set of values. Different relapse is likewise helpful for information investigation, permitting you to decide "how one variable influence another".

8.6.3 Development of Hypothesis

In the present chapter, the hypotheses are developed and quantitative analytical approach is used to evaluate them to help companies and practitioners to understand decision to operate CSR effectively and in an efficient way. The hypotheses are developed to thoroughly examine the antecedents, processes, and influence of CSR cost on financial performance based on quantitative analysis. This chapter attempts to demonstrate the quality information related to mandatory CSR using a hypothetical framework. Secondary data were collected and used for the study. To learn more about the effects of CSR initiatives of 11 selected corporate organizations, CSR reports, financial reports, and report of mandatory CSR expense disclosure in the financial years of the sample considered in the study have been tested using SPSS. Here, in this chapter, the prediction is performed on the basis of the value of a dependent variable based on one or more independent variables, which is typically done with multiple regression. The chapter developed the hypothesis to understand the disaggregate data on numerous aspects of undertaking CSR initiatives. It is tried to demonstrate that CSR is an important and is adaptable based on various aspects depending on the nature of the firm's business.

In the present chapter, the investigation and translation of "Obligatory corporate social obligation and its effect on natural manageability" are introduced considering an example of 11 corporate organizations chosen from various sectors of Indian industries. The outcomes are evaluated on the basis of different measurable apparatuses in order to understand the behavior of the corporate related to CSR based on review. Tests of statistical significance are carried out at a level of 5%. For background classifications, percentage analysis is used to describe the distribution of respondents. In the present chapter, after due understanding of the CSR and allied initiative, three hypotheses as shown in Table 8.1 are developed by the authors. Accordingly, data are collected and evaluated to test the hypothesis as shown in Tables 8.2 to 8.4.

Table 8.1 Developed hypotheses.

Hypothesis	Description
H_1:	Mandatory CSR spending of 2% net profit of the New Companies Act 2013 is the reason for undertaking CSR initiatives.
H_2:	There is a significant impact between CSR and organizational performance.
H_3:	There is a significant impact between CSR and environmental development.

Table 8.2 T-test output.

One-sample statistics						
	N	**Average**	**SD**	**SEM**		
EE	55	33.10	43.28	6.24		
One-Sample t-test			Test value = 420			
	t	Degree of freedom	Two-tailed test	Difference between averages	95% level of significance	
					Low	High
EE	−61.669	54	0.000	−384.896	−397.41	−372.38

(Source: SPSS Output)

8.6.4 Interpretation of Inference: t-test

SPSS stands for Statistical Package for Social Sciences. It is statistical software that is used in the present chapter to generate table of output for the one-sample t-test. Table 8.2 is generated by the SPSS, which contains all the information that is required to interpret the results of a one-sample t-test. To test the data related to an assumption that there were no significant outliers and that the assumption of variable was approximately normally distributed for each category of the independent variable, then there is a need to interpret this table.

Interpretation of the data is done in chapter using the One-Sample Statistics table, which presents the relevant descriptive statistics. It is more common to present descriptive statistics using the mean and standard deviation rather than the standard error of the mean, although both are acceptable. However, by running a one-sample t-test, the interested evaluator can know whether the sample you have (environmental allocation) comes from a "normal".

The One-Sample Test table reports the result of the one-sample t-test (Table 8.2). The table considers the value of the known or hypothesized mean (419.00) and compares it with sample data. The columns of the One-Sample Test table provide information on whether the sample means statistically significantly different.

The table presented the observed *t*-value ("t" column), the degrees of freedom ("df"), and the statistical significance (*p*-value) of the one-sample t-test. In the table, *p*-value is 0.000, which is satisfying the condition of $p < 0.05$.

Table 8.3 Multiple regression: model summary.

		Variables entered/removed								
Model	Variables Entered		Variable Removed		Method					
	Corporate social responsibility				**Enter**					
a: Dependent Variable: Net Profit b: All requested variables entered										
	R	R Square	Adjusted R Square	Std. Error of the Estimate	Change Statistics					Durbin-Watson
					R Square Change	F Change	df_1	df_2	Sig. F Change	
1	0.754 [a]	0.569	0.561	6565.150	0.569	70.017	1	53	0.000	1.331
a : Predictors Constant (CSR)										
ANOVA		Sum of Square		df	Mean Square		F		Sig.,	
1	**Regression**	3017796145.443		1	3017796145.443		70.017		0.000	
	Residual	2284363333.176		53	43101194.966					
	Total	5302159478.619		54						

(Source: SPSS Output)

Table 8.4 Multiple regression output.

ANOVA	Sum of squares	df	Mean square	F	Sig.
Regression	134,449.882	1	134,449.882	3.325	0.074[b]
Residual	2,143,177.865	53	40,437.318		
Total	2,277,627.747	54			
a. Dependent Variable: CSR b. Predictors: (Constant), EE					

Model	R	R square	Adjusted R square	Std. error of the estimate	Change statistics					Durbin-Watson
					R square change	F change	df_1	df_2	Sig. F change	
1	0.243[a]	0.059	0.041							
				45.321	0.059	3.325	1	53	0.74	0. 728
a: Predictors: (Constant), CSR, Dependent Variable: EE										

(Source: SPSS Output)

Therefore, it can be concluded that the population means are statistically significantly different. If $p > 0.05$, then the difference between the sample-estimated population mean and the comparison population mean would not be statistically significantly different.

8.6.5 Interpretation: Model Summary

The summary of the model is represented in Table 8.3. The model summary box helps to evaluate the model. The R Square value tells the variance of dependent variable. The summary box shows R value as 0.754; it means that the independent variables explain 75.4% of dependent variable. Here, the ANOVA helps to assess the results. To check the statistical significance of the model, the significance value must be < 0.05. The significance value of the ANOVA table, i.e., NaN (NaN values represent undefined or unpresentable results from certain mathematical operations), is found as 0.0; it means the model is significant.

8.6.6 Interpretation: Multiple Regression Output

The output of the multiple regression can be seen in Table 8.4, where the model outline box assists in assessing the model. The R Square worth tells the change of ward variable. The outline box shows R esteem 0.243; it implies that the autonomous factors make sense of 24.3% of ward variable. Here, the ANOVA is useful for evaluating the outcomes. To check the factual meaning of the model, the importance esteem should be < 0.05. In Table 8.4, the ANOVA significance value of NaN is found as 0.74, which indicates that the model is not significant. All the developed hypotheses are tested, and the results of their acceptance or rejection can be identified from Table 8.5.

Table 8.5 Results of hypothesis.

S. no.	Hypothesis	Sig.	Results
1	Mandatory CSR spending of 2% net profit of the New Companies Act 2013 is the reason for undertaking CSR initiatives.	0.000	Accepted
2	There is a significant impact between CSR and organizational performance.	0.000	Accepted
3	There is a significant impact between CSR and environmental development.	0.74	**Rejected**

8.7 Implications and Recommendations

The implications of the study presented in chapter can be inferred from Figure 8.3, where it can be identify that the presented study can be implicated in managerial decision-making for understanding future research directions and for making policy and regulatory frameworks. From Table 8.5, it is inferred that there is a positive connection between CSR expenses and return on resources, which recommends that mindful asset portion can improve monetary execution. Organizations should perceive CSR drives not as simple costs but rather as ventures that can yield long haul benefits through superior effectiveness and partner connections. It is found that CSR investments may have a greater impact on financial performance and reputation in industries with a greater social impact. Accordingly, the areas in which CSR efforts can have the greatest positive impact should be identified by businesses. The present chapter highlights the significance of fitting CSR systems to industry-explicit settings.

It is essential to engage stakeholders like employees, clients, investors, and local communities with CSR drives. Organizations can use CSR drives to construct more grounded connections, which subsequently encourages client steadfastness toward drawing socially cognizant financial backers and making a positive brand picture. The discoveries propose that prompt monetary benefits from CSR speculations could be receive, where mindful strategic approaches are needed to add mark versatility and manageability, which can prompt persevering monetary advantages. Corporate companies that demonstrate significant CSR efforts may be eligible for grants or tax breaks that policymakers might consider implementing. Without compromising financial performance, these incentives can encourage businesses to allocate resources to socially responsible initiatives.

Administrative bodies should implement more tough detailing necessities for CSR uses. Investors and stakeholders are now well-informed about

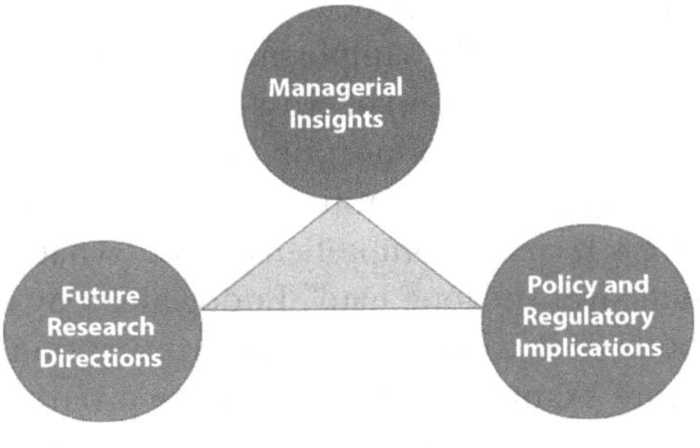

Figure 8.3 Implications of the study.

a company's commitment to social responsibility due to transparent disclosure practices. Comparisons between industries and regions can be made easier by standardizing CSR reporting frameworks. This normalization can direct organizations in benchmarking their CSR endeavors against industry best works and toward upgrading straightforward responsibilities for CSR. More exploration is needed to investigate the development of the CSR cost–monetary execution relationship over the long run. Longitudinal examinations that can reveal insight into "how supported CSR endeavors influence monetary execution over the long haul" are prominent to maintain CSR. Examining the effect of explicit CSR drives, i.e., CSR maintainability measures or CSR local area commitment programs through which an evident knowledge about CSR impacts on monetary results is needed. A multifaceted examination is recommended to inspect "how the CSR cost–monetary execution relationship changes across various social settings and administrative conditions".

The ramifications and suggestions drawn from the examination discoveries presented in the chapter underlines the CSR as a capable strategic approach for assembling organizations. Companies can generate value for themselves, their stakeholders, and society by strategically aligning CSR initiatives with financial objectives and industry dynamics. Future research is needed to uncover the intricate dynamics of CSR relationship in a variety of contexts. It is appraised that many intricate dynamics of CSR relationship in a variety of contexts is needed to aware policymakers about CSR financial attainments as they play a crucial role in incentivizing and regulating CSR efforts. Finally, incorporating CSR into the texture of assembling tasks adds to supportable and moral strategic policies.

8.8 Limitations and Future Research

The sample of corporate companies chosen for analysis limits the study's findings. The results may not be applicable to all corporate companies, *viz.*, sectors or company sizes, and the sample may not fully represent the manufacturing landscape. The utilization of cross-sectional information restricts the capacity to lay out causal connections between CSR costs and monetary execution. Longitudinal examinations following changes over the long run could give a heartier comprehension of "what CSR drives mean for monetary results over the long haul". Endogeneity by which unnoticed factors impacts both CSR costs and monetary execution could influence the outcomes. Future examinations could utilize progressed econometric strategies, like instrumental factors or board information investigation to address endogeneity concerns. Due to variations in reporting standards and accounting practices, accurately measuring CSR costs can be challenging. Future examinations could investigate more extensive strategies for catching both immediate and circuitous CSR expenses to give a more exact portrayal. Even though the study took into account factors at the industry level that affect the relationship between CSR and cost-financial performance, more in-depth analyses of each industry could reveal complex dynamics within each sector.

The present examination did not represent potential outer factors, for example, administrative changes, market patterns, or macroeconomic circumstances that could impact both CSR costs and monetary execution. Counting such factors could upgrade the comprehension of the relationship. The review zeroed in on chosen producing organizations, possibly restricting the generalizability of discoveries to various social and territorial settings. The influence of cultural and regulatory variations on the CSR cost–financial performance relationship could be the subject of future research. Quantifying the connection between CSR costs and financial performance was the primary focus of the study but future exploration could dig further into the components and intervening variables that underlie this relationship, revealing insight into the channels through which CSR influences monetary results. The study did not investigate how stakeholders and CSR efforts interact directly. Stakeholder perceptions and responses to CSR initiatives could be the basis for a deeper comprehension of the observed relationship. Future exploration could investigate elective monetary execution measurements and creative ways to deal with surveying the effect of CSR costs considering aspects past customary monetary markers. This study's constraints highlight the requirement for continuous examination to refine and extend "how we might interpret the connection between CSR costs and monetary execution in the assembling area". Addressing these constraints will add to a more complete and nuanced comprehension of the intricate interchange between moral obligation and monetary results, directing both scholastic talk and viable navigation.

8.9 Importance of Quantitative Analysis in CSR Evaluation

8.9.1 Measuring Impact

Quantitative analysis provides a structured approach to measure the impact of CSR initiatives. It helps organizations to assess whether they are achieving their intended goals and objectives, such as reducing carbon emissions, improving employee satisfaction, or enhancing community well-being.

8.9.2 Comparability

Quantitative data allow for easy comparison over time and across different CSR initiatives. By using standardized metrics and key performance indicators, organizations can assess which initiatives are most effective and allocate resources accordingly.

8.9.3 Accountability

Quantitative data provide a clear basis for accountability. It enables organizations to track progress toward CSR targets and hold themselves accountable to stakeholders, including investors, customers, and the public. Transparency in CSR reporting is increasingly expected and demanded.

8.9.4 Evidence-Based Decision-Making

Quantitative analysis provides concrete evidence for decision-making. By collecting and analyzing data, organizations can make informed choices about the allocation of resources, the selection of CSR projects, and the adjustment of strategies based on performance data.

8.9.5 Risk Assessment

Quantitative analysis can help to identify potential risks and opportunities related to CSR initiatives. For example, by quantifying the environmental and social impacts of a supply chain, a company can better understand its exposure toward risks related to climate change, labor practices, or resource scarcity.

8.9.6 Stakeholder Engagement

Quantitative data can be used to engage stakeholders effectively. By presenting clear metrics and performance indicators, organizations can communicate their CSR efforts to various stakeholders, build trust, and demonstrate their commitment to responsible business practices.

8.9.7 Reporting and Transparency

Many stakeholders, including investors, customers, and regulatory bodies, expect organizations to report quantitative CSR data. Accurate and well-structured quantitative reporting enhances transparency and credibility.

8.9.8 Continuous Improvement

Quantitative analysis supports the process of continuous improvement in CSR initiatives. By analyzing data over time, organizations can identify areas where they can make improvements, set new targets, and refine their strategies.

8.9.9 Resource Allocation

Quantitative analysis helps organizations to allocate resources efficiently. By assessing the cost-effectiveness and impact of different CSR programs, organizations can optimize their investments in CSR activities.

8.9.10 Compliance and Regulation

Quantitative analysis is essential for compliance with CSR regulations and reporting requirements imposed by governments and industry standards. Meeting these requirements can have legal and reputational consequences.

Quantitative analysis is crucial for evaluating CSR efforts because it provides a structured and evidence-based approach to measuring impact, making informed decisions, and demonstrating accountability to stakeholders.

8.10 Conclusions

This chapter presents an exploration part to set out an excursion to investigate the complex connection between corporate social obligation costs and monetary execution inside the assembling area. By utilizing a staggered quantitative examination approach, the chapter looked to uncover the potential compromises and collaborations that emerge while assembling organizations concentrate intently on mindful strategic policies. The discoveries presented in this chapter offer significant bits of knowledge for both scholarly community and practice. The fact that there is a positive correlation between CSR costs and ROA highlights the possibility that better financial outcomes can be achieved through prudent resource allocation. While the relationship with return on value was more nuanced, the chapter highlights the significance of considering numerous monetary exhibition measurements to catch the all-encompassing effect of CSR endeavors. The moderating effects of industry-level factors emphasize the significance of context in determining the relationship between CSR and cost-financial performance. Organizations working in businesses with higher cultural effect could encounter more prominent advantages from CSR speculations, underlining the requirement for customized CSR systems.

The pragmatic ramifications of examination presented in chapter stretch out to administrative independent direction and strategy detailing. Fabricating organizations can decisively assign assets to CSR drives, perceiving that these speculations can prompt superior productivity, partner connections, and long-haul maintainability. Policymakers can consider boosting dependable way of behaving and upholding straightforward CSR announcing practices to direct organizations toward socially capable practices. Even though this chapter adds to our understanding of the relationship between CSR and cost-financial performance, it is important to acknowledge its limitations and the potential for future research to improve on these findings. Longitudinal studies, industry-specific analyses, and investigations into the relationship's mechanisms all have the potential to yield more in-depth insights.

In conclusion, this chapter enhances our understanding of the dynamic relationship between corporate company's financial performance and CSR costs. Corporate companies can navigate the difficulties of ethical responsibility while reaping financial rewards by adopting responsible business practices, thereby contributing to a more sustainable and socially conscious business environment.

References

1. Clarkson, M.B.E., A Stakeholder Framework for Analysing and Evaluating Corporate Social Performance. *Acad. Manage. Rev.*, 20, 1, 92–117, 1995.
2. McWilliams, A. and Siegel, D., Corporate Social Responsibility: A Theory of the Firm Perspective. *Acad. Manage. Rev.*, 26, 1, 117–127, 2001.
3. Orlitzky, M., Schmidt, F.L., Rynes, S.L., Corporate Social and Financial Performance: A Meta-Analysis. *Organ. Stud.*, 24, 3, 403–441, 2003.
4. Porter, M.E. and Kramer, M.R., Strategy & Society: The Link between Competitive Advantage and Corporate Social Responsibility. *Harv. Bus. Rev.*, 84, 12, 78–92, 2006.

5. Waddock, S.A. and Graves, S.B., The Corporate Social Performance-Financial Performance Link. *Strateg. Manage. J.*, 18, 4, 303–319, 1997.
6. Abaeian, V., Khong, K.W., Yeoh, K.K., McCabe, S., Motivations of undertaking CSR initiatives by independent hotels: a holistic approach. *Int. J. Contemp. Hosp. Manage.*, 31, 6, 2468–2487, 2019.
7. Adarkwah, G.K. and Malonæs, T.P., Firm-specific advantages: a comprehensive review with a focus on emerging markets. *Asia Pac. J. Manage.*, 39, 2, 539–585, 2022.
8. Ahlstrom, D., Arregle, J., Hitt, M.A., Qian, G., Ma, X., Faems, D., Managing technological, sociopolitical, and institutional change in the new normal. *J. Manage. Stud.*, 57, 3, 411–437, 2020.
9. Ali, S.S. and Kaur, R., Effectiveness of corporate social responsibility (CSR) in implementation of social sustainability in warehousing of developing countries: A hybrid approach. *J. Clean. Prod.*, 324, 129154, 2021.
10. Aparna, K., Amilan, S., Raj, V.L., Customers' response to mandatory corporate social responsibility in India: an empirical evidence. *Soc. Responsib. J.*, 19, 3, 429–445, 2023. https://doi.org/10.1108/SRJ-04-2021-0174.
11. Arrive, J.T. and Feng, M., Corporate social responsibility disclosure: Evidence from BRICS nations. *Corp. Soc. Responsib. Environ. Manage.*, 25, 5, 920–927, 2018.
12. Awaysheh, A., Heron, R.A., Perry, T., Wilson, J., II, On the relation between corporate social responsibility and financial performance. *Strateg. Manage. J.*, 41, 6, 965–987, 2020.
13. Barauskaite, G. and Streimikiene, D., Corporate social responsibility and financial performance of companies: The puzzle of concepts, definitions and assessment methods. *Corp. Soc. Responsib. Environ. Manage.*, 28, 1, 278–287, 2021.
14. Bergh, D.D., Ketchen, D.J., Orlandi, I., Heugens, P. P. M. A. R., Boyd, B.K., Information asymmetry in management research: past accomplishments and future opportunities. *J. Manage.*, 45, 1, 122–158, 2019.
15. Bhatia, A. and Dhawan, A., A paradigm shift in corporate social responsibility: India's transition from mandatory regime to the COVID-19 era. *Soc. Responsib. J.*, 19, 1, 166–183, 2023. https://doi.org/10.1108/SRJ-01-2021-0020.
16. Cezarino, L.O., Liboni, L.B., Hunter, T., Pacheco, L.M., Martins, F.P., Corporate social responsibility in emerging markets: Opportunities and challenges for sustainability integration. *J. Clean. Prod.*, 362, 132224, 2022.
17. Chaffee, E.C., The origins of corporate social responsibility. *Univ. Cincinnati Law Rev.*, 85, 347–373, 2017.
18. Cherian, J., Umar, M., Thu, P.A., Nguyen-Trang, T., Sial, M.S., Khuong, N.V., Does corporate social responsibility affect the financial performance of the manufacturing sector? Evidence from an emerging economy. *Sustainability*, 11, 4, 1182, 2019.

19. Dana, L.P., Salamzadeh, A., Hadizadeh, M., Heydari, G., Shamsoddin, S., Urban entrepreneurship and sustainable businesses in smart cities: Exploring the role of digital technologies. *Sustain. Technol. Entrep.*, 1, 2, 100016, 2022.
20. Dawar, G. and Singh, S., Barriers to corporate social responsibility implementation in the medium size manufacturing sector: an interpretive structure modelling approach. *J. Entrep. Emerg. Econ.*, 15, 2, 447–479, 2023, https://doi.org/10.1108/JEEE-12-2020-0451.
21. Dharmapala, D. and Khanna, V., The impact of mandated corporate social responsibility: Evidence from India's companies act of 2013. *Int. Rev. Law Econ.*, 56, 92–104, 2018.
22. Djelic, M.-L. and Etchanchu, H., Contextualizing corporate political responsibilities: neoliberal CSR in historical perspective. *J. Bus. Ethics*, 142, 4, 641–661, 2017.
23. Fairhurst, D. and Greene, D.T., Too much of a good thing? Corporate social responsibility and the takeover market. *J. Corp. Fin.*, 73, 102172, 2022.
24. Gatti, L., Vishwanath, B., Seele, P., Cottier, B., Are we moving beyond voluntary CSR? Exploring theoretical and managerial implications of mandatory CSR resulting from the new Indian companies act. *J. Bus. Ethics*, 160, 4, 961–972, 2019.
25. Gaurav, A., Psannis, K., Peraković, D., Security of cloud-based medical internet of things (MIoTs): a survey. *Int. J. Software Sci. Comput. Intell.*, 14, 1, 1–16, 2022.
26. Gomes, M. and Marsat, S., Does CSR impact premiums in M&A transactions? *Fin. Res. Lett.*, 26, 71–80, 2017.
27. Gupta, J. and Das, N., Multidimensional corporate social responsibility disclosure and financial performance: A meta-analytical review. *Corp. Soc. Responsib. Environ. Manage.*, 29, 4, 731–748, 2022.
28. Gupta, S.S. and Wadera, D., Consumer understanding of corporate social responsibility dimensions: an application of CSR pyramid model in Indian context. *Int. J. Indian Culture Bus. Manage.*, 22, 1, 110–136, 2021.
29. Hasan, I., Singh, S., Kashiramka, S., Does corporate social responsibility disclosure impact firm performance? An industry-wise analysis of Indian firms. Environment. *Dev. Sustain.*, 24, 10141–10181, 2022.
30. Hickman, L.E., Iyer, S.R., Jadiyappa, N., The effect of voluntary and mandatory corporate social responsibility on earnings management: Evidence from India and the 2% rule. *Emerg. Mark. Rev.*, 46, 100750, 2021.
31. India-briefing, *Corporate social responsibility in India*, India Briefing, Hongkong, 2020, March 23, Retrieved from https://www.india-briefing.com/news/corporate-social-responsibility-india-5511.html/.
32. India-briefing, *Corporate social responsibility in India: submission of new form CSR-2*, India Briefing, Hongkong, 2022, February 16, Retrieved from https://www.india-briefing.com/news/india-new-csr-reporting-mandate-companies-to-submit-form-csr-2-applicable-fy2021-onwards-24332.html/.

33. Indian Express, *Explained: How CSR expenditure rules have changed for Indian companies*, Indian Express, India, 2021, January 30, Retrieved from https://indianexpress.com/article/explained/corporate-social-responsibility-csr-rules-indian-companies-7163098/.
34. Khojastehpour, M. and Jamali, D., Institutional complexity of host country and corporate social responsibility: Developing vs developed countries. *Soc. Responsib. J.*, 17, 5, 593–612, 2020.
35. Lee, M.T. and Suh, I., Understanding the effects of Environment, Social, and Governance conduct on financial performance: Arguments for a process and integrated modelling approach. *Sustain. Technol. Entrep.*, 1, 1, 100004, 2022.
36. Madhu, S., Padunnavalappil, S., Saajlal, P.P., Vasudevan, V.A., Mathew, J., Powering up an IoT-enabled smart home: A solar powered smart inverter for sustainable development. *Int. J. Software Sci. Comput. Intell.*, 14, 1, 1–21, 2022.
37. Meseguer-Sánchez, V., Gálvez-Sánchez, F.J., López-Martínez, G., Molina-Moreno, V., Corporate social responsibility and sustainability. A bibliometric analysis of their interrelations. *Sustainability*, 13, 4, 1636, 2021.
38. Onyebuchi, A., Matthew, U.O., Kazaure, J.S., Okafor, N.U., Okey, O.D., Okochi, P., II, Matthew, A.O., Business demand for a cloud enterprise data warehouse in electronic healthcare computing: issues and developments in e-healthcare cloud computing. *Int. J. Cloud Appl. Comput.*, 12, 1, 1–22, 2022.
39. Oware, K.M. and Mallikarjunappa, T., CSR expenditure, mandatory CSR reporting and financial performance of listed firms in India: an institutional theory perspective. *Meditari Account. Res.*, 30, 1, 1–21, 2022.
40. Panwar, R., Pandey, V., Suddaby, R., Vidal, N.G., Did India's CSR mandate enhance or diminish firm value? *Bus. Soc.*, 62, 2, 401–433, 2023, https://doi.org/10.1177/00076503221085962.
41. Prasad, K., Kumar, S., Devji, S., Lim, W.M., Prabhu, N., Moodbidri, S., Corporate social responsibility and cost of capital: the moderating role of policy intervention. *Res. Int. Bus. Fin.*, 60, 101620, 2022.
42. Pratihari, S.K. and Uzma, S.H., A survey on bankers' perception of corporate social responsibility in India. *Soc. Responsib. J.*, 16, 2, 225–253, 2019.
43. Roy, P.P., Rao, S., Zhu, M., Mandatory CSR expenditure and stock market liquidity. *J. Corp. Fin.*, 72, 102158, 2022.
44. Saha, R., Cerchione, R., Singh, R., Dahiya, R., Effect of ethical leadership and corporate social responsibility on firm performance: A systematic review. *Corp. Soc. Responsib. Environ. Manage.*, 27, 2, 409–429, 2020.
45. Kang, D., Prabhu, M., Ahmed, R.R., Zhang, Z., Sahu, A.K., Digital-IIoTs spheres approach toward public development: an exploiting fuzzy-grey mathematical modeling of IIoTs spheres. *Grey Syst.: Theory Appl.*, 12, 2, 389–416, 2022.
46. Li, N., Prabhu, M., Sahu, A.K., Investigating economic replacement policy under uncertainty for managerial application based on grey-reliability approach using QCC. *Grey Syst.: Theory Appl.*, 13, 2, 297–321, 2023.

47. Wang, W., Huang, L., Zhu, Y., Jiang, L., Sahu, A.K., Sahu, A.K., Sahu, N.K., Decision support system toward evaluation of resilient supplier: A novel fuzzy gain-loss computational approach. *Kybernetes*, 49, 6, 1741–1765, 2019.
48. Ahmed, K.A., Sahu, A.K., Sahu, A.K., Sahu, N.K., Quantify the Behaviour Intention of Individuals to Control SC Performance by Exploring Cloud Storage Services: An Extended UTAUT2 Approach. *Int. J. Technol. Hum. Interact.*, 18, 7, 1–28, 2022.
49. Sarkar, S., Chatterjee, M., Bhattacharjee, T., Does CSR disclosure enhance corporate brand performance in emerging economy? Evidence from India. *J. Indian Bus. Res.*, 13, 2, 253–269, 2021.

9

Mechanical Characteristics Evaluation of Micro-Alloyed Steel Used in Oil and Gas Industries

Mukesh Kumar Nag

National Institute of Technology Jamshedpur, Jharkhand, India

Abstract

This chapter experimentally presented investigations related to hydrogen-assisted cracking in X80-grade pipeline steels with the aim to comprehend the corrosive impact on the physical and mechanical attributes of aforesaid steel. The chapter catalogs and encompasses a comprehensive analysis of X80-grade steel, which included the base metal, heat-affected zone (HAZ), and weld region through microstructural examinations and an array of tests. During the experiment, several key experimental parameters that are quantitative in nature were employed for the analysis of hydrogen-induced cracking and fracture toughness. These parameters included the control of crack length-to-width ratios, the deliberate pre-cracking and machining of welds, and the exposure of specimens to an extremely low-temperature nitrogen bath to induce specimen embrittlement. The HAZ exhibitions demonstrated finer grain structures than the weld and parent materials, resulting in increased hardness. Moreover, the weld region's fracture toughness is found compromised by the presence of contaminants, including inclusions and porosity. The fracture mechanisms observed within the investigation region encompass river lines, quasi-cleavage fractures, and microstructural cracks instigated by environmental factors, collectively contributing to a decrease in the overall toughness properties. Fractography study is utilized in this chapter to identify fracture causes, types, and potential flaws. The use of chemicals for NDT testings reveals across various crack-initiated locations. Simultaneously, fracture and failure analysis probe is presented in this chapter to demonstrate "how hydrogen presence alters pipeline steel toughness". Findings reveal that a significant reduction in CTOD

Email: nag0859@gmail.com

Atul Kumar Sahu, Rakesh D. Raut, Rohit Raja, Anoop Kumar Sahu and Nitin Kumar Sahu (eds.) Industrial and Manufacturing Designs: Quantitative and Qualitative Analysis, (265–294) © 2024 Scrivener Publishing LLC

and J-R in the weld zone with a 70% decrease in CTOD and a 75% reduction in J–integral crack growth resistance (J_{IC}) as compared to the base metal is found, which attributed to lower weld zone ductility. The weld regions also displayed 35% less ductility as compared to the base metal. This investigation unveils that the fundamental failure mechanism through fractographic analysis of fractured tensile and CTOD specimens from the weld region is elucidating reduced potential for energy absorption, which thus leads to diminished fracture toughness. This experimental study presented in this chapter will help the practitioners in preparing basis for design specification and design optimization, as well as in identifying cost-effective aspects, regulatory compliances, predictive maintenance, long-term durability, research and development traits, and so on. The chapter presents a comprehensive understanding about the micro-alloyed steel used in oil and gas industries.

Keywords: Micro-alloyed steels, coldcracking, safety, cost effective, oil and gas industries

9.1 Introduction

In today's world, petroleum and natural gas stand as the predominant energy resources on a global scale. The vitality of a nation's economy is intricately linked to its energy sector, underlining the critical need for the secure transportation of petroleum and natural gas via pipeline infrastructure [1]. Among the pivotal materials used for this purpose, line pipe steels and micro-alloyed steels emerge as pivotal categories of high-strength alloys. These materials typically comprise alloying elements such as niobium (Nb), vanadium (V), titanium (Ti), and aluminum (Al), where each presents quantities less than 0.1%, either individually or in various combinations [2]. The yield strength (YS) of pipeline steels significantly surpasses that of conventional carbon–manganese (C-Mn) steels and often reaches two to three times the strength levels. Incorporating micro-alloy additives during the thermo-mechanical hot-rolling process facilitates the grain refinement and precipitation mechanisms, resulting in this remarkable enhancement.

Pipeline steel is routinely exposed to corrosive elements like nitrates, chlorides, and carbonates, as well as to potentially deleterious substances like hydrogen, ammonia, and hydrogen sulfide. These environmental challenges give rise to a significant concern, namely, the emergence of residual and cyclic stresses within the pipeline steel. This phenomenon ultimately engenders the potential for leakage or in severe cases breaks and rupture the pipeline, as indicated in previous studies [3, 4]. In addressing the aforementioned challenge, Ossai *et al.* [5] proposed a solution involving the utilization of high-strength low alloy steel pipelines. These pipelines are characterized by their large diameter and operation at high pressures and offer a robust defence against corrosion in sour environments. Additionally, Golahmar *et al.* [6] shed light on the issue of hydrogen embrittlement (HE) and emphasized its adverse effects on their characteristics. Another research subsequently highlighted the significance of fatigue strength in pipeline design and advocated optimization based on the calculation of fatigue life considering varying qualitative or quantitative parameters like stress levels and ferritic and bainitic microstructure [7].

In a positive development, the increasing worldwide need for energy resources has resulted in a significant upsurge in the manufacturing of large-diameter pipelines during recent decades. These pipelines serve as crucial conduits for the efficient transportation of gas and oil across vast distances. To leverage economies of scale, engineers have designed larger diameter pipelines capable of operating at higher pressures, thereby enhancing the carrying capacity of pipeline steels and simultaneously reducing operational costs while extending the service life of these pipelines [8]. Illustrating the significance of this matter, Nag and Kumar [9] pointed out that the United States alone boasts roughly 72,000 operational pipeline networks. As a result, the American Petroleum Institute (API) plays a central role in setting global pipeline standards with API 5L being the primary standard. API 5L includes eight primary grades, spanning from X42 to X70, each defined by distinct criteria for YS and toughness.

The present chapter examines fractures comprehensively with a special focus on crack initiation and propagation, which can have significant consequences including pipeline failure or explosion, especially in environments containing gaseous hydrogen. It is well-documented that gaseous hydrogen has the capacity to significantly diminish the fatigue and fracture resistance of steel. Hence, a comprehensive exploration of the interrelationship between hydrogen pressure and fatigue life mandates a meticulous scrutiny of the characteristics inherent to low-cycle fatigue and the progression of fatigue crack growth, as indicated in prior research [10].

Mohtadi-Bonab *et al.* [11] have contributed significantly to our comprehension of this phenomenon because cold cracking happens mainly in plane carbon steels, characterized by distinct attributes concerning the onset and progression of cracks. Additionally, Hasan *et al.* [12] have illuminated that fractures can initiate at nearly any operational stress threshold due to a manifestation of imperfections and defects originating within the pipeline infrastructure. These defects encompass a spectrum of factors, which includes inherent material imperfections, external forces, excessive corrosion, challenging environmental conditions, and abnormal operating temperatures [13]. In the context of fatigue growth rate tests, it has been discovered that pipeline steels, particularly those belongs to class as X100 and X52, display a much greater rate of fatigue crack development in environments containing sour hydrogen compared to ambient air conditions [14]. Wasim *et al.* [15] have reinforced this finding by demonstrating that HE leads to a reduction in plasticity and a decrease in fracture toughness [15]. The intricate interplay between local hydrogen concentration factors within hydrogen-prone environments and the initiation and propagation of fracture cracks is found prominent and has been scrutinized by Ohaeri *et al.* [16]. Simultaneously, fractures occur in pipeline steel when stress exceeds the tolerance limit, determined by various factors such as operating conditions and material properties (Alamilla *et al.* [17]). This is vital in fracture mechanics [18].

It is widely acknowledged that the presence of macroscopic flaws within a material can significantly influence the probability of fracture initiation. Commonly, non-destructive techniques for defect detection may be employed to identify and pinpoint of these structural concerns. Furthermore, many study stressed to introduce a corrosion-crack correlation model [19] that should be tailored to the context of HE and its impact on the formation of fatigue cracks in carbon steels, which are utilized in pipeline applications and exposed to gaseous hydrogen environments.

Based on an extensive literature survey, many research papers have been written upon that focuses on investigating the impact of a hydrogen-rich environment on the toughness behavior of pipeline steels, where the conducted studies have employed fracture and failure analysis as the primary methodology for evaluation. For the research presented in this chapter, the author has selected X80-grade pipeline steel as the experimental material. The investigations are presented in chapter, which comprises a multifaceted approach, which include many quantitative parameters and analysis like microstructure analysis, hardness testing, tensile testing, as well as CTOD (critical crack tip opening displacement) and J_{IC} (J–integral crack growth resistance) tests. These analyses were conducted across different regions of the X80-grade steel, specifically in the base material, the heat-affected zone (HAZ), and the weld region. The outcomes of these experiments have yielded several noteworthy findings, where microstructural analysis of both the HAZ and the base metal consistently reveals a predominant presence of a combination of ferrite and pearlite microstructures. In contrast, the weld metal exhibits a distinctive microstructural profile characterized by the alignment of ferrite grains alongside the presence of carbides. This variance in microstructure across different regions of the material is analyses to report critical disclosures, which is a critical aspect of the study presented in this chapter. The investigation of fracture toughness, particularly within the weld metal, has revealed a substantial reduction amounting to a significant 68% decrease, when compared to other regions. This decline in fracture toughness can be attributed by evaluating the combined influence of microstructural variations specific to the weld metal and the presence of hydrogen. These findings collectively indicate the importance of evaluation by varying qualitative or quantitative parameters to attain elevated outcomes, where the findings of the chapter enhance our understanding toward the intricate relationship between microstructure, hydrogen exposure, and fracture toughness in pipeline steels, with a specific focus on X80-grade steel. The chapter underscores the practical implications of operating pipelines in hydrogen-rich environments, emphasizing the importance of tailored material selection and maintenance strategies to ensure pipeline integrity and safety. Thus, one can understand that the findings presented in this chapter advocated to determine a critical contribution and connections of the complex interplay between microstructure, hydrogen exposure, and fracture toughness in subjected pipeline steel. The findings of the chapter can be used to devise reinforced pipe designs considering the effects of hydrogen-rich environments on pipeline materials, allowing for more secure and reliable pipeline design and maintenance.

9.2 Novelty of the Work

In this chapter, the impact of hydrogen is evaluated through a comparative analysis of toughness characteristic across the base, weld, and HAZ. Furthermore, the chapter has afforded insights into environment-assisted cracking mechanisms and their influence on the fracture toughness characteristics of pipeline steels. It is plausible that hydrogen may be introduced into the steel during processes such as solidification, welding, or pickling. When HE arises due to external factors, it is referred to as induced HE. Consequently, the chapter deeply demonstrated the connections and complex interplay between microstructure, hydrogen exposure, and fracture toughness in pipeline steel. The chapter offers a comprehensive overview of the investigation into the initiation and progression of both brittle and ductile failure modes within pipeline materials.

9.3 Experimental Setup and Testing Procedure

This study delves into the examination of X80-grade (a low-carbon steel) employed in the production of line pipes. The initial raw materials arrived to produce pipe lines are in the form of tubular structures with the weld junction being meticulously prepared using the shielded metal arc welding process. For the same essential welding, quantitative parameters are required, which include a direct current electrode positive of 30 V, a welding speed of 5 mm/s, and a weld thickness of 14.3 mm. Interpass temperatures were maintained at 220°C with preheating carried out at 100°C. Experimental investigations were conducted to evaluate the influence of the HAZ and the weld zone on various parent materials. To facilitate microstructure analysis and hardness testing, the tubular structure was meticulously sectioned into small rectangular pieces. For the assessment of fracture toughness, specialized CTOD test specimens and J–Integral Crack Growth Resistance (J_{IC}) specimens were meticulously fashioned. The standard specimen of respective tests exhibited the comprehensive evaluations of fracture toughness characteristics.

9.3.1 Chemical Composition Analysis of Base and Weld Metals

The spectroscopy was conducted at the Industrial Metal Testing Laboratory, Kolkata to determine the element present in base and weld material.

9.3.2 Microstructural Analysis

The X80-grade pipeline steel used for metallographic examinations is analyzed by way of involving the preparation of small specimens. These specimens were initially polished using emery paper of varying grit sizes ranging from 50–1000. Following this, an additional polishing step was performed using a diamond disc and satin cloth to achieve a refined surface finish of 6μ. The polishing equipment featured rotating discs enclosed within a soft cloth, infused with diamond particles (6μ and 1μ), and lubricated with an oily medium. A continuous polishing process employing 6μ grade diamond particles effectively eliminated any residual grinding scratches. Following the polishing process, the samples were dried with a drier. Before commencing the microstructural investigation, the samples were etched with an etchant containing 3% nital.

9.3.3 Hardness Test

Testing of the Vickers hardness of pipe line steels is done in three distinct areas, i.e., the base, the HAZ, and the weld zone, where welding was performed. Vickers hardness was tested using a 5-kg load on the sample that had been prepared for metallographic inspection. The sample was penetrated using a pyramidal diamond penetrator, and the force was maintained on the material for roughly 30 s.

9.3.4 Tensile Test

Tensile testing was performed on round bars that were 6 mm in diameter and 45 mm in gauge length as per ASTM8 (Figure 9.1). The base metal, heat-affected zone, and weld metal were all evaluated separately. An extensometer uses elastic bands to securely place a sample. The gauge length is the distance between the notch of the sample. The strain gauge is fitted to the device's deformable head and measures the length's change as a result of the sample's deformation.

9.3.5 Crack Resistance Test

Both the J_{IC} and CTOD tests were conducted using the single-specimen unloading compliance approach as specified in ASTM standard E-1820. Single-edge notch beam specimens from the base, HAZ, and weld were pre-cracked and machined as illustrated in (b) and (a) part of Figure 9.2, respectively. Following the ASTM standard, the crack length-to-width ratio was maintained at (a/W = 0.5), and testing was performed using a 50-kN BiSS servo hydraulic universal testing machine (M/S-SENB) as depicted in Figure 9.3. Subsequently, CTOD experiments were conducted to measure crack-opening displacements (CODs).

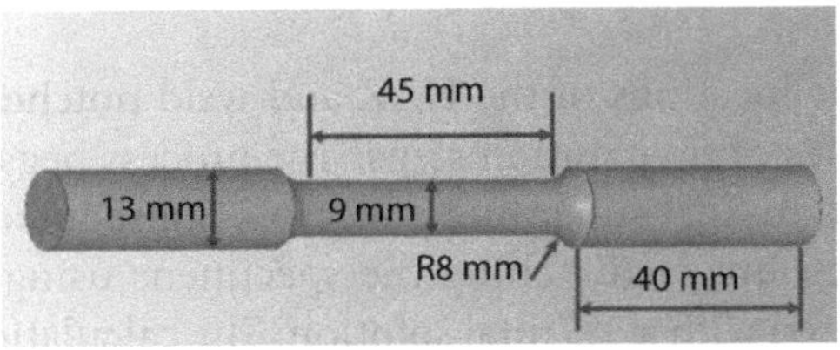

Figure 9.1 Tensile test specimen as per ASTM8.

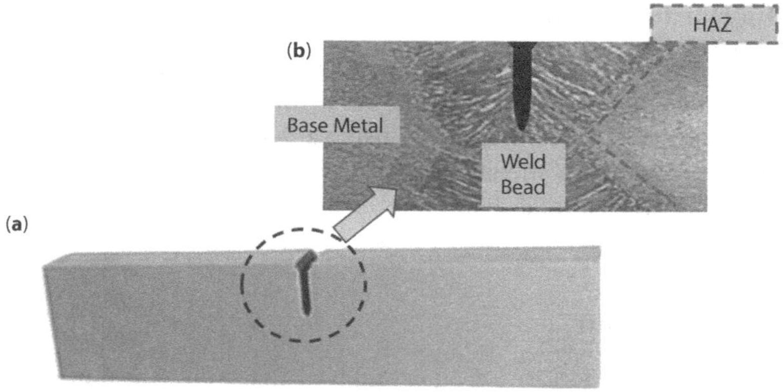

Figure 9.2 Single-edge notched beam (SENB) samples with precision notching.

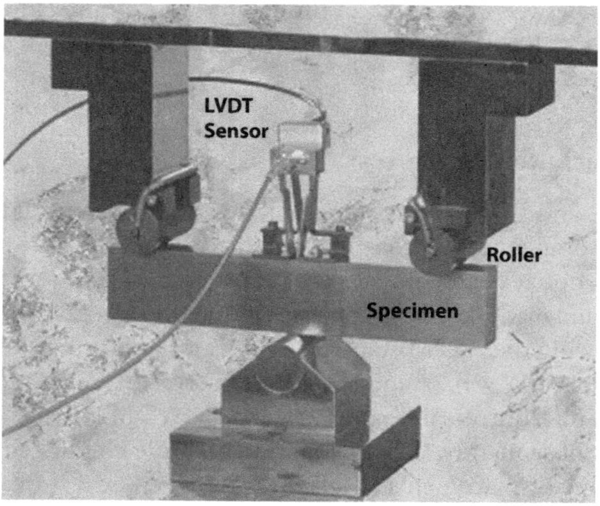

Figure 9.3 Test setup for fracture toughness test.

To identify the locations of the HAZ and weld notches, the specimens underwent a series of preparation steps. The process began with an initial polishing process to final super fine polishing. Subsequently, a final step is performed, which involved etching the specimens using a cotton-tipped applicator saturated with a 3% nital solution. The calculation of J_{IC} involved determining the point at which the blunting line ($J = 2\sigma_f\Delta a$) intersects the J-R curve, where σ_f is flow stress and Δa is crack extension.

9.3.6 Fractography Study

The analysis of fracture surfaces is known as fractography, which was carried out on both the broken surfaces of tensile and CTOD specimens, and the results are presented in (b) and (a) parts of Figure 9.4, respectively. Prior to conducting the fractography analysis, special treatment was applied to the ductile test specimens to render them brittle. This treatment involved subjecting the specimens to extremely low temperatures by immersing them in liquid nitrogen, which causes a transition from ductile to brittle behavior in many materials. This process helps in capturing and examining the characteristic features of brittle fractures. For the fractography analysis itself, sophisticated equipment and techniques were employed. Ultrasonic cleaners, which utilize ultrasound waves, were utilized to gently cleanse the intricate surfaces of the specimens. These cleaners are effective in removing contaminants or debris that might interfere with the analysis. The heart of the fractography analysis lies in the use of high-resolution scanning electron microscopes (SEMs). SEMs offer unparalleled capabilities for fracture analysis, allowing for the examination of surfaces at a much higher level of detail compared to optical microscopes. SEMs use electron beams to create high-resolution images, enabling researchers to study the fine microstructural features and fracture patterns at the microscopic level. This in-depth analysis of fracture surfaces is required for crucially understanding the failure mechanisms and material behavior, which, in turn, aid in the development of safer and more reliable materials and structures.

After obtaining the load-COD plot at the highest load (P_{max}), the parametric values were determined using the relationship represented in Equations 9.1, 9.2, and 9.3 as follows [20]:

$$\delta_t = \delta_e + \delta_p \tag{9.1}$$

$$CTOD = \delta_t = \frac{I_{\max}^2(1-\upsilon^2)}{2E\sigma_{ys}} + \frac{0.4(W-A_0)}{A_0+0.4(W-A_0)}P_c \tag{9.2}$$

Figure 9.4 Fractography of tensile test specimen is shown in (a) and SENB specimens of X80-grade steel in shown in (b).

where; δ_t = Defect assesment in welded structure
δ_e = Elastic component of CTOD
δ_p = Plastic component of CTOD
l = Stress intensity factor
E = Young's modulus
σ_{ye} = Yield strength
W = Specimen width
A_0 = Initial crack length
υ = Poisson's ratio
P_c = Plastic component of crack mouth opening displacement
I_{max} = Maximum stress intensity factor
L_{max} = Maximum length of specimen
S = Span of the sample
B = Specimen width

9.4 Result and Discussion

9.4.1 Chemical Composition Effects on Fracture Toughness

The chapter aimed to assess and demonstrate "how the chemical composition at different sections of X80-grade pipeline steel including the base metal, HAZ, and weld region influences fracture toughness". This investigation is crucial for understanding how variations in chemical composition affect the material's ability to withstand cracking and fractures. Table 9.1 shows the standard chemical composition of X80 micro-alloyed steel.

Table 9.1 Standard chemical composition of X80 micro alloyed steel [21].

% Contribution	Location		
	Base material	**Heat affected zone**	**Weld region**
C	0.04	0.25	0.02
Mn	1.69	1.26	0.53
Si	0.22	0.26	0.31
Cr	0.14	0.04	0.02
Mo	0.17	-	0.27
Ni	0.01	0.03	0.03
S	0.01	0.02	0.02
V	0.01	-	0.01
Al	0.04	-	0.03
P	0.02	0.01	0.01

This table indicated that the compositions should fell within the allowed range, as specified by standards or engineering requirements. This suggests that the material met the expected chemical composition standards.

However, a notable finding was that the weld components exhibited different chemical characteristics compared to the base material. Specifically, the weld had higher concentrations of sulfur and aluminium while having lower amounts of carbon and silicon, when compared to the base material. This divergence in chemical composition is significant because it can have a direct impact on the material's properties including fracture toughness. One key implication of these compositional differences is the cleanliness morphology of the material. In simpler terms, the presence of higher levels of sulfur in the weld can modify the cleanliness of the material. Cleanliness, in this context, refers to the presence of non-metallic inclusions and impurities in the material, which can influence its mechanical properties including fracture toughness. The reference to a greater sulfur concentration in the weld suggests that there may be a higher presence of sulfur-related inclusions or impurities in this region. These inclusions can potentially act as stress concentration points or crack initiation sites, which can reduce the material's resistance to fracture. Additionally, the mention of the non-metallic component primarily comprising rounded oxide particles indicates that these inclusions could be one of the factors affecting fracture toughness. Rounded oxide particles can affect the material's ability to resist crack propagation and make it more susceptible to fracture.

In summary, the chapter demonstrates, "how the chemical composition of different sections of X80-grade pipeline steel" particularly the weld influenced fracture toughness. The differences in composition especially the higher sulfur content in the weld were highlighted as factors that could impact the cleanliness morphology and, consequently, fracture toughness. Understanding these relationships is essential for ensuring the reliability and safety of materials used in critical applications like pipelines.

9.4.2 Microscopy Study Results for X80-Grade Pipeline Material

This section presents an exploration of the results obtained from the microstructure analysis conducted on the X80-grade pipeline steel, encompassing various material zones SEM was employed as the primary tool for scrutinizing the microscopic composition of the material. SEM is a valuable technique utilized for examining the fine-scale structure and characteristics of substances. In this context, images that are presented in Figures 9.5 to 9.7 serve as illustrative examples, which provide photomicrographs that depict the typical microstructures observed within the different zone of the steels. The same are described subsequently. These images are instrumental in offering a visual representation of the diverse microstructural features present in the critical sections of the X80-grade pipeline steel. These images facilitate in a comprehensive understanding of the variations in the steel's microstructure across different material zones, contributing significantly to the assessment of its overall integrity and suitability for its intended application. These images play a pivotal role in our microstructure analysis, providing visual context for the subsequent discussions on the findings and implications of this study.

The micrographic analysis of the base metal is illustrated in Figure 9.5, which unveils the coexistence of ferrite and pearlite structures. A noteworthy observation is the discernible banded configuration within the pearlite structure, a trait widely regarded as undesirable. These discernible bands serve as nucleation sites for the phenomenon known in scientific circles as HIC [22]. To elaborate, the presence of ferrite and pearlite in the microstructure is a common finding in steel materials like X80-grade. Ferrite and pearlite are two distinct phases with different mechanical properties. However, when pearlite is observed in a banded structure, it poses a significant concern. The bands of pearlite create localized areas of mechanical weakness and vulnerability within the material. These areas are more susceptible to the initiation and propagation of cracks, especially when subjected to the presence of hydrogen, which can permeate the steel during its service life. Hydrogen-induced cracking (HIC) is a corrosion-related

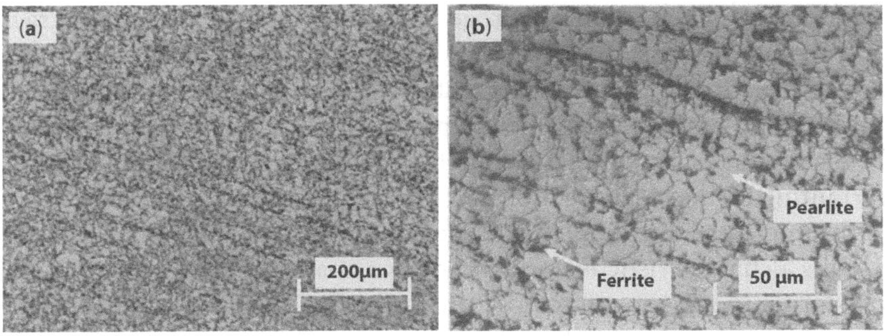

Figure 9.5 Comparative micrographs of base metal at two different magnifications: (a) ×200 and (b) ×500.

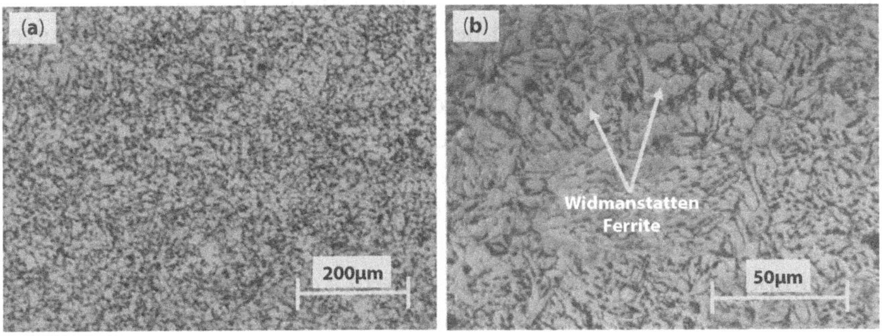

Figure 9.6 Microstructure in heat-affected zone (HAZ): (a) ×200 and (b) ×500.

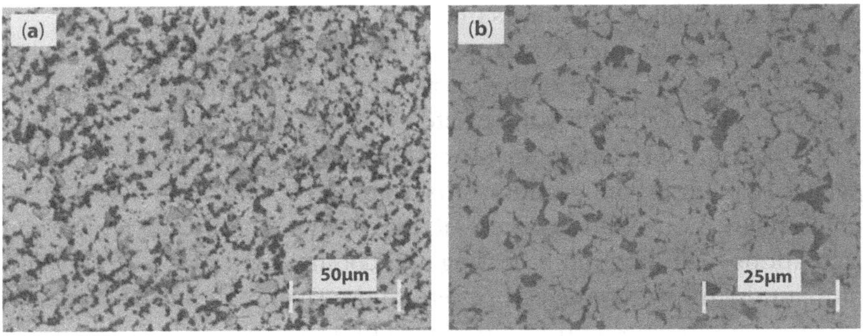

Figure 9.7 Comparative micrographs of weld metal at two different magnifications: (a) ×500 and (b) ×1,000.

phenomenon, where atomic hydrogen infiltrates the steel and causes cracks to form and propagate along these weak points, significantly compromising the structural integrity of the material [23]. Therefore, the presence of banded pearlite in the base metal's microstructure raises concerns about the material's resistance to HIC and its suitability for pipeline applications, warranting further investigation and possibly remediation measures.

The microstructure observed in the HAZ is depicted in Figure 9.6, which exhibits a characteristic feature known as a coarse-grained structure. This transformation in the microstructure under HAZ is significant, as it represents a departure from the original parent metal microstructure, which was essentially an aggregate of ferrite and pearlite and sometimes referred to as Widmanstätten Ferrite. The change in microstructure within the HAZ can be primarily attributed to the rigorous thermal cycling conditions experienced during the welding process. When welding occurs, the base metal in the HAZ is subjected to rapid heating and cooling cycles. These severe temperature fluctuations induced the structural changes in the metal. In the case described, the original microstructure of the parent metal, which consisted of a combination of ferrite and pearlite, undergoes alterations. The transformation leads to the development of a coarse-grained structure in the HAZ. This coarse-grained structure is less desirable and depends on the specific application and requirements of the welded component.

It is important to note that the presence of a coarse-grained microstructure in the HAZ can impact the material's mechanical properties and performance characteristics. Therefore, understanding and evaluating these microstructural changes are crucial steps in assessing the quality and integrity of a welded joint. Further analysis and consideration of the implications of this transformation are typically necessary to ensure the suitability of the welded material for its intended service conditions.

In Figure 9.7, the microstructure of the welded region is unveiled, which thereby is revealing distinctive attributes. Within this microstructural depiction, one can discern the existence of circular and polygon-shaped ferrite phases, complemented by carbides that exhibit alignment. This specific structural configuration arises cold cracking. When the steel is subjected to the welding process, it experiences significant temperature fluctuations including rapid heating and cooling cycles [24]. In this environment, hydrogen atoms can penetrate the steel material. The presence of hydrogen can have a profound impact on the steel's microstructure and mechanical properties.

The circular and polygonal ferrite structures seen in Figure 9.7 are indicative of the crystalline arrangement that results from the cooling of the molten weld metal. These ferrite structures are formed as the steel cools from its molten state to a solid state during welding [25]. The presence of aligned carbides is likely due to the segregation of carbon atoms within the steel, which can occur during the welding process. It is important to note that the influence of hydrogen diffusion on steel microstructure can be detrimental as excessive hydrogen can lead to HIC and reduced material integrity. Therefore, understanding the microstructural changes in welded areas, as depicted in Figure 9.7, is crucial for assessing the quality and potential weaknesses of the welded joint. Further analysis and consideration of the implications of these microstructural features are essential for ensuring the reliability and safety of the welded steel component.

The finding of an SEM analysis conducted on weld metal at a high magnification level of ×500 is depicted in Figure 9.8. The SEM micrograph prominently displays numerous cracks present on the surface of the weld metal. These cracks are attributed to the presence of impurities within the weld material, specifically inclusions and porosity. Inclusions are foreign particles or materials that have become trapped in the weld during the welding process. Porosity, on the other hand, refers to voids or cavities formed due to the entrapment of gases or other contaminants during welding [26]. This observation highlights the importance of quality control in welding processes to ensure the structural integrity and performance of the welds. Further analysis and mitigation measures may be necessary to address these surface defects and enhance the overall weld quality.

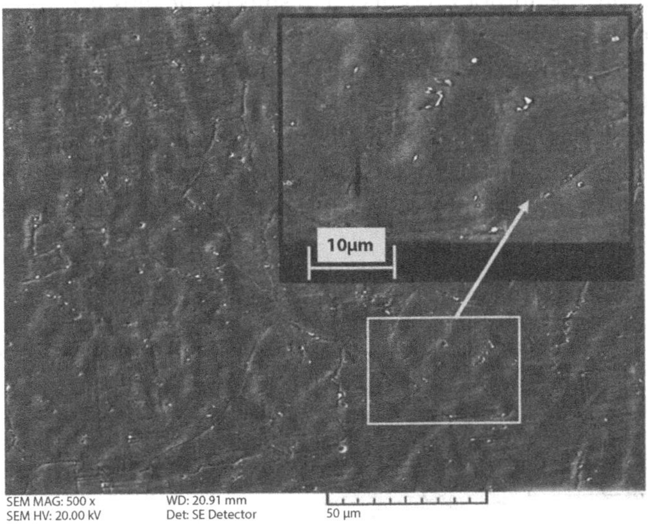

Figure 9.8 Scanning electron microscopy reveals weld metal microstructure at ×500.

9.4.3 Hardness Test Results Analysis

In this section, the author conducted both hardness and tensile tests on X80-grade pipeline steel, focusing on different regions of the material including the base, the HAZ, and the weld area. During these tests, the author considered the Vickers hardness number (VHN) as a key parameter. The VHN was calculated as part of the testing process [27]. The findings from the hardness tests revealed significant variations in VHN values among the different regions of the material. Notably, the weld metal exhibited the highest VHN value, indicating greater hardness, whereas the HAZ metal displayed the lowest VHN value, signifying comparatively lower hardness.

Furthermore, the author concluded that the elevated hardness properties observed in the weld metal were primarily attributed to the microstructure present in the form of ferrite. This microstructure appeared in a carbide form within the weld metal. Finally, the comprehensive overview of the hardness test results is referred in Table 9.2, which provides a detailed summary of the Vickers hardness values obtained for X80-grade pipeline steel in the various regions examined. These findings contribute valuable insights into the material's mechanical properties and its suitability for pipeline applications.

9.4.4 Tensile Testing and Mechanical Properties Assessment

In this section, the author presents the results of the tensile tests conducted on X80-grade pipeline steel on different material regions. The tensile test findings have provided valuable insights into the mechanical properties of the material. The weld metal has a little greater YS and tensile strength (TS) than the base metal as shown by the tensile test results. This indicates that the weld metal possesses a marginally greater ability to resist deformation and external forces. Conversely, the base metal demonstrates

Table 9.2 Vickers hardness test results for X80-grade steel.

	Applied load	**No. of indentation**	**Average hardness value**
Location	**(kg)**	-	**(VHN)**
Base	2	5	156
HAZ	2	5	129
Weld	2	5	180

Table 9.3 X80-grade steel tensile properties at various material regions.

Specimen of regions	UTS (MPa)	Yield strength (MPa)	Elongation (%)
Base	450 ± 16	585 ± 12	35 ± 3
HAZ	420 ± 10	505 ± 15	25 ± 2
Weld	485 ± 15	612 ± 20	16 ± 2

higher ductility, which signifies its capacity to undergo plastic deformation without fracturing. To provide a comprehensive overview of the findings, Table 9.3 is accounted, which presents the detailed results of the tensile tests conducted on X80-grade pipeline steel. These results underscore the subtle differences in strength and ductility properties between the different material regions. Additionally, it is worth noting that the percentage elongation, a crucial measure of ductility, is found to be highest in the base metal achieving a value of 30%. This observation reinforces the superior ductility characteristics of the base metal, which can be a vital factor in assessing the material's suitability for specific applications.

9.4.5 Comparative Fracture Toughness Assessment Across Material Zones

In this section, the author delves into the fracture toughness study conducted on X80-grade pipeline steel, with a particular focus on different material regions, i.e., the base metal, the HAZ, and the weld area. This study is essential for understanding the material's resistance to crack propagation and its fracture behavior under different conditions. To illustrate the findings, the author has provided load versus COD plots for each of these regions as shown in Figure 9.9, labeled as (a), (b), and (c) for the base metal, HAZ, and weld regions, respectively. In Figure 9.9 (a), it is notable that the load (measured in N) exceeds 10,500 N, corresponding to a COD range of 1.5 mm to 6 mm. This suggests the material's ability to withstand significant loads within this specific COD range in the base metal.

However, Figure 9.9 (b) reveals some interesting insights. Here, the shapes of the four R-curves, which depict the relationship between load and COD for different crack depth-to-width ratios, exhibit dissimilar characteristics. This observation indicates variations in fracture behavior based on the crack dimensions. Furthermore, it is worth mentioning that the specimens with a crack depth-to-width ratio (a/W) of 0.1 exhibit the highest scatter, implying greater variability in fracture toughness properties in this particular condition. The fracture toughness study helps us to understand how X80-grade pipeline steel behaves, when subjected to different loading conditions and crack geometries, which is crucial information for assessing its structural integrity and performance in real-world applications.

In Figure 9.9 (c), the author derived J-R curves for four distinct datasets, entailing the evaluation and graphical representation of these curves as indicators of crack resistance or toughness under specific conditions. Through this analysis, it became apparent that the J-R curve corresponding to an a/W value of 0.1 exhibited relatively lower favorable characteristics in contrast to the J-R curves associated with a/W values of 0.6 and 1.3, respectively. In essence, this conclusion indicates that, within the context of the study, the J-R curves corresponding to a/W values of 0.6 and 1.3 exhibit certain advantages or qualities that make them more preferable than the J-R curve linked to an a/W value of 0.1. These findings provide valuable insights into the materials or systems being studied, aiding in our understanding of their performance and behavior under different conditions.

From the generated plots, the author determined the values of CTOD and Δa, which were subsequently illustrated in Figure 9.10, labeled as (a), (b), and (c), respectively. These plots serve to visually represent the variations in CTOD and Δa across different regions of interest.

The J-Δa curves are essential for understanding fracture behavior and toughness attributes, which were derived from the loading-unloading data. Figures 9.11 (a), (b), and (c) showcase these curves individually for the base, HAZ, and weld regions, respectively, shedding light on the distinctive fracture characteristics of each region.

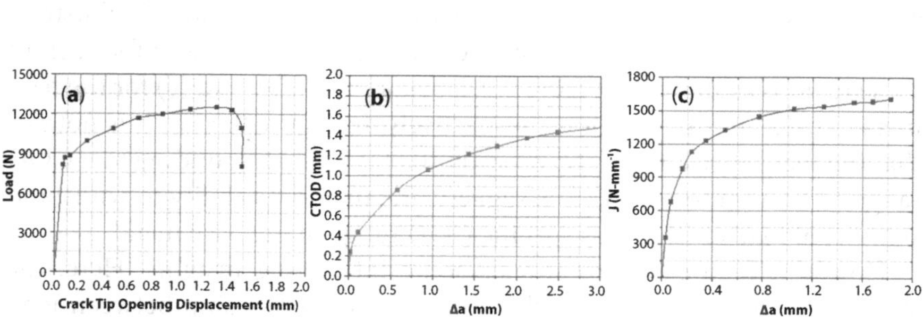

Figure 9.9 Comparative analysis of the base material. (a) CTOD loading, (b) J vs. Δa, and (c) fracture resistance graph.

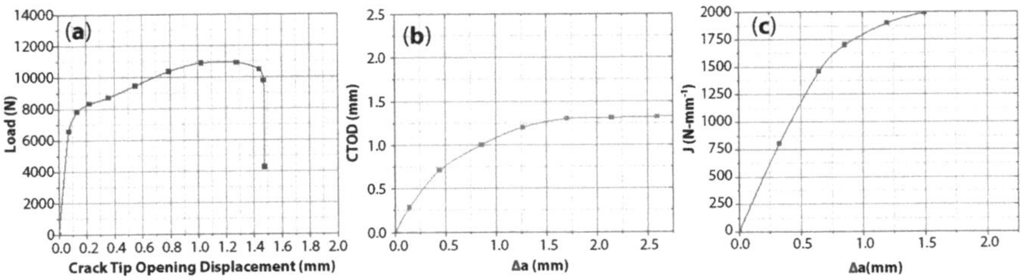

Figure 9.10 Comparative analysis of the HAZ material. (a) CTOD loading, (b) J vs. Δa (mm), and (c) fracture resistance graph.

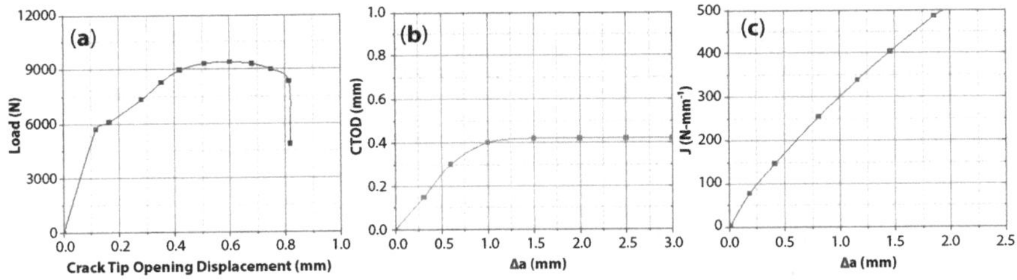

Figure 9.11 Comparative analysis of the weld material. (a) CTOD loading, (b) J vs. Δa (mm), and (c) fracture resistance graph.

From the load versus COD plot, the author calculated the values of K_{max} (maximum stress intensity factor) and CTOD (critical crack tip opening displacement) at the point of maximum load, denoted as Pmax. These calculated values have been organized and presented in Tables 9.4 to 9.6 for the different regions under investigation. Upon examination of the toughness values presented in these tables, it becomes evident that the fracture toughness values for the HAZ and the base metal exhibit only marginal differences [28]. In stark contrast, the fracture toughness of the weld region experiences a significant reduction, ranging from 68% to 74%. Several factors can contribute to this notable difference in fracture toughness. One key factor is the weld region's higher strength, which may result in reduced fracture resistance. Additionally, the weld region's lower ductility can make it more susceptible to brittle fracture. The presence of hydrogen and the specific microstructural characteristics of the weld region may also play significant roles in diminishing its fracture toughness.

In a nutshell, these fracture mechanics tests, including maximum stress intensity factor (K_{max}), CTOD, and J-integral crack resistance (J_{1C}), provide

Table 9.4 Characteristics of base material.

A_0/w	L_{max} (kN)	P_c (mm)	I_{max}	CTOD (mm)	J_{IC} (N/m)
0.5	12.4	4.85	330	1.41	1193
0.5	11.6	4.86	316	1.40	1235
0.5	11.4	4.76	313	1.45	1209

Table 9.5 Characteristics of heat affected zone.

A_0/w	L_{max} (kN)	P_c (mm)	I_{max}	CTOD (mm)	J_{IC} (N/m)
0.5	11.5	4.76	315	1.30	1309
0.5	12.1	4.72	324	1.29	1322
0.5	11.4	4.77	310	1.28	1319

Table 9.6 Characteristics of weld material.

A_0/w	L_{max} (kN)	P_c (mm)	I_{max}	CTOD (mm)	J_{IC} (N/m)
0.5	9.3	4.55	254	0.44	410
0.5	8.0	1.35	226	0.46	390
0.5	7.9	1.42	250	0.41	400

crucial insights into the influence of microstructure, environmental factors, and other variables on conventional mechanical properties. The substantial reduction in fracture toughness observed in the weld region highlights the importance of considering these factors, when assessing the structural integrity and performance of materials in practical applications.

9.4.6 Observations on Fracture Surfaces

In order to gain insights into the micro-mechanisms underlying fracture, the author conducted SEM observations on the failed samples as depicted in Figures 9.12 to 9.14. In both the base and weld materials, a common fracture pattern known as "dimple fracture" was observed. This distinctive dimpled appearance is indicative of ductile fracture behavior. Figure 9.12 (a) illustrates the crack propagation marks observed on the sample and presented voids appears within the material. It is important to highlight that the formation of cracks often initiates from these voids and porosities [29]. Such defects act as stress concentrators, ultimately leading to a reduction in the material's fracture toughness. In Figure 9.2 (b), the author observes a relatively uniform distribution of dimples on the ductile fracture surface. This homogenous pattern indicates a classic ductile fracture behavior, where the material exhibits a high degree of plastic deformation before ultimately failing. Figure 9.12 (c) offers a closer look, revealing dimples clustered around a non-metallic inclusion. This finding suggests that the presence of such inclusions can influence the fracture behavior, potentially acting as nucleation sites for cracks. Lastly, Figure 9.12 (d) presents a fracture surface characterized by "transgranular cleavage". This type of fracture surface exhibits distinct crystallographic facets, indicating a brittle fracture mode. Transgranular cleavage typically occurs, when a material is subjected to high-stress levels, and cracks propagate along the specific crystallographic planes.

In a nutshell, our SEM observations have provided valuable insights into the fracture behavior of the material. The presence of voids, porosity, and non-metallic inclusions and the distinction between ductile dimple fracture and brittle transgranular cleavage fracture are crucial factors to consider in understanding the fracture toughness of the material.

Figure 9.13 (a) reveals a fracture surface devoid of dimples, signifying a distinctly brittle fracture. This lack of dimples suggests minimal plastic deformation before failure, which is a characteristic of brittle materials. In Figure 9.13 (b), a brittle fracture is observed despite the presence of a small number of dimples. This finding indicates that the material exhibited brittle behavior with limited ductility, likely due to factors such as impurities or stress concentration. Figure 9.13 (c) presents a fracture surface marked by brittle cleavage fracture features, which further emphasize a brittle failure mode. Cleavage fracture is indicative of the material breaking along specific crystallographic planes, typical of brittle materials. Lastly, Figure 9.13 (d) illustrates the presence of a small number of voids, with

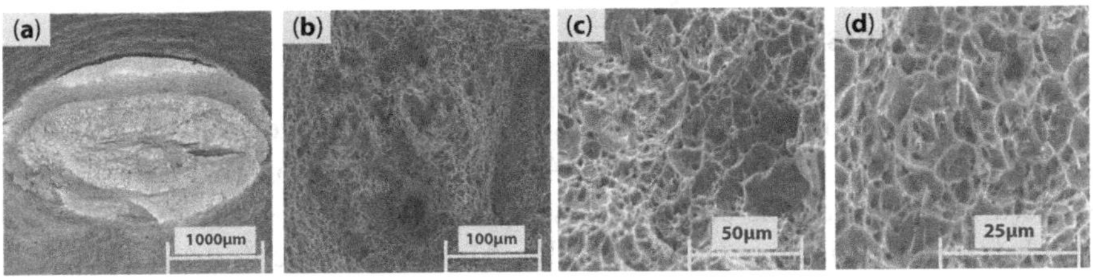

Figure 9.12 Dimpled fracture in tensile failure of the base material sample.

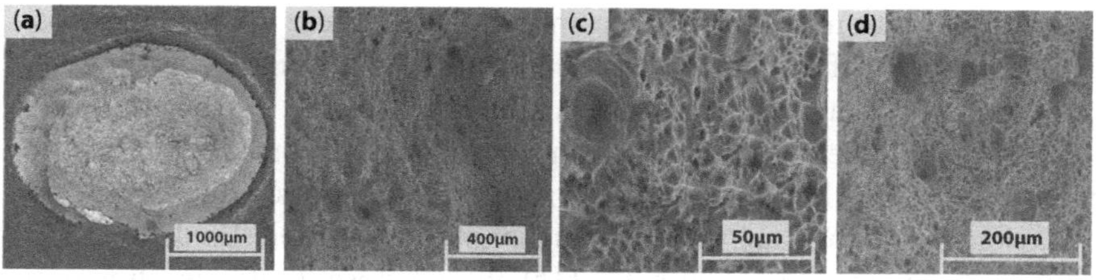

Figure 9.13 Weld sample fracture: small dimples, higher strength, and more brittleness.

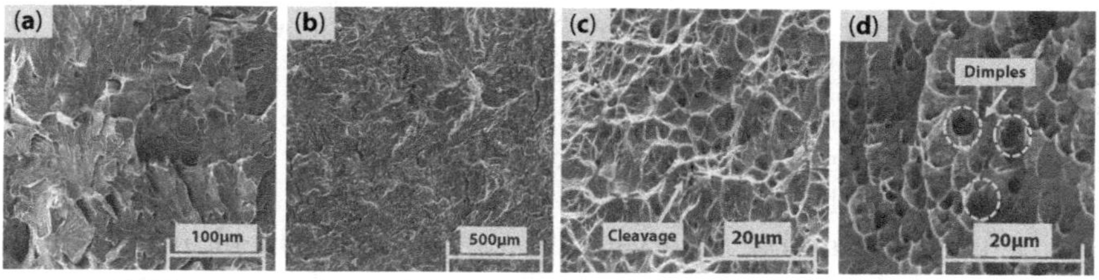

Figure 9.14 Weld region fractographs with reduced toughness (CTOD and J_{1C}) values.

indications of an attempt to create microscopic cups upon fracture. These voids may have played a role in the fracture process, contributing to the overall material failure.

In Figure 9.14 (a), a distinct river-like pattern is observed, which is indicative of the brittle fracture of the specimen. Moving to Figure 9.14 (b), the author observes signs of a river-like pattern, further confirming the brittleness of the joint. In Figures 9.14 (c) and (d), the presence of both cleavage patterns and dimple locations in different regions is evident, suggesting that the fracture occurred via a quasi-cleavage fracture mechanism. These findings collectively provide valuable insights into the fracture behavior of the specimen, shedding light on the mechanisms underlying its brittleness.

9.5 Importance of Qualitative or Quantitative Analysis in Study

The chapter catalogs and encompasses a comprehensive analysis of X80-grade steel under the base metal, HAZ, and weld region through microstructural examinations and an array of tests including hardness, CTOD, and J–fracture resistance (J-R) test value. Here, experiments are conducted considering several key experimental parameters that are quantitative in nature for the analysis of HIC and fracture toughness. The chapter highlighted the consequence of fatigue strength in pipeline design and advocated evaluation of fatigue life based on qualitative or quantitative parameters optimization. Qualitative or quantitative parameters, i.e., critical crack tip opening displacement, load, maximum stress intensity factor, dimple fracture, depth-to-width ratio value, tensile properties, hardness values, and porosity, are presented in this chapter for evaluation and optimization. The same can be used to collect quality facts related with fatigue strength. The importance of qualitative and quantitative analysis in a study of HIC and fracture toughness analysis in X80-grade pipeline steels for the oil and gas sector can be summarized as follows.

9.5.1 Safety and Reliability

The oil and gas sector relies on pipelines to transport critical resources. Understanding the susceptibility of pipeline steels to HIC is crucial for ensuring the safety and reliability of these structures. Qualitative and quantitative analysis helps in identifying vulnerabilities and mitigating potential failures.

9.5.2 Material Selection

Qualitative analysis can provide insights into the nature and mechanisms of HIC in X80-grade steels. This information assists in selecting the most suitable materials for pipeline construction, considering factors like resistance to HE.

9.5.3 Design Optimization

Quantitative analysis, including fracture toughness assessments, helps in optimizing the design and thickness of pipeline components. It ensures that the pipeline can withstand stress, pressure, and potential defects while maintaining its integrity and safety margins.

9.5.4 Environmental Considerations

Qualitative analysis can help in evaluating the environmental conditions and factors that contribute to HIC. This knowledge can inform decisions regarding protective coatings, cathodic protection, and operational practices that reduce the exposure of pipeline steels to hydrogen.

9.5.5 Cost Efficiency

By quantifying the fracture toughness of X80-grade steels, it becomes possible to design pipelines with appropriate safety margins. This can lead to cost savings by preventing over-engineering or unexpected failures that may incur expensive repairs and downtime.

9.5.6 Regulatory Compliance

The oil and gas sector is subject to strict safety and environmental regulations. Qualitative and quantitative analysis helps in complying with these regulations by demonstrating a thorough understanding of materials and potential risks associated with HIC.

9.5.7 Predictive Maintenance

Quantitative analysis can be instrumental in developing predictive maintenance strategies. By monitoring the fracture toughness of pipeline components over time, operators can schedule maintenance and repairs proactively, reducing the risk of unplanned shutdowns.

9.5.8 Long-Term Durability

Qualitative and quantitative analysis contributes to the assessment of the long-term durability of pipeline steels. This is particularly important for pipelines with extended lifespans, and it helps in making informed decisions regarding maintenance and replacement schedules.

9.5.9 Research and Development

These analyses also drive the ongoing research and development efforts to improve the materials used in the oil and gas sector. Understanding the behavior of X80-grade steels under various conditions can lead to advancements in materials science and engineering.

In summary, both qualitative and quantitative analyses are essential for ensuring the safe, efficient, and cost-effective operation of pipelines in the oil and gas sector. They provide the necessary data and insights to make informed decisions about materials, design, and maintenance, reducing the risk of accidents and environmental impact while optimizing operational efficiency.

9.6 Conclusion

The experimental investigation has been presented in this chapter for HIC and fracture toughness in X80-grade pipeline steels for the oil and gas sector under the employed key experimental parameters for the analysis of HIC and fracture toughness. These parameters included the control of crack length-to-width ratios, the deliberate pre-cracking and machining of welds, and the exposure of specimens to an extremely low-temperature nitrogen bath to induce specimen embrittlement. In summary, this study has yielded several significant characteristics as conclusions:

a) Mechanical Characteristics: The examination unveiled notable distinctions in mechanical properties across the studied regions. The weld region displayed the highest levels of both hardness and TS, surpassing the HAZ and the base metal. This underscores the weld region's superior strength attributes.
b) Ductility Disparity: However, an appreciable disparity in ductility was observed, with the weld region exhibiting a marked deficiency. Its ductility was approximately 35% lower than that of the base metal, reflecting a trade-off between strength and ductility inherent to the welding process.
c) Impact on Fracture Toughness: The heightened mechanical properties, encompassing hardness and TS in the weld region, exerted a discernible influence on fracture toughness. Owing to the reduced ductility of the weld region, the CTOD experienced a substantial reduction of 70%, whereas the J–integral crack resistance (J_{IC}) diminished by 75% compared to the base metal. These findings underscore the significance of a holistic consideration that incorporates both strength and ductility when assessing fracture toughness.
d) Fracture Characteristics: Examination of fractographs obtained from failed sample of weld region. This characteristic signifies a diminished capacity for energy absorption within the weld region, ultimately contributing to reduced fracture toughness.

However, this study presented in this chapter provides valuable insights into the mechanical behavior of different regions in the welded structure. The trade-off between increased strength and reduced ductility in the weld region has clear implications for the overall performance and reliability of the weld, particularly in scenarios where fracture toughness is a critical factor. This scientific experiment underscores the paramount importance of both qualitative and quantitative analyses in comprehensively assessing the behavior of X80-grade pipeline steels in the context of HIC and fracture toughness within the oil and gas industry. These analyses play a pivotal role in enhancing safety, reliability, and efficiency in pipeline operations. They inform material selection, design optimization, and long-term durability considerations, ensuring compliance with rigorous regulatory standards and driving advancements in research and development. Ultimately, the application of these analytical methodologies is critical for safeguarding operational pipelines, mitigating risks, and optimizing resource utilization in the oil and gas sector.

References

1. Bouzarovski, S., Bradshaw, M., Wochnik, A., Making territory through infrastructure: The governance of natural gas transit in Europe. *Geoforum*, *64*, 217–228, 2015. https://doi.org/10.1016/j.geoforum.2015.06.022.
2. Konstrukcijska, V.M., High-strength low-alloy (HSLA) steels. *Mater. Tehnol.*, 45, 4, 295–301, 2011.
3. Abohamzeh, E., Salehi, F., Sheikholeslami, M., Abbassi, R., Khan, F., Review of hydrogen safety during storage, transmission, and applications processes. *J. Loss Prev. Process Ind.*, *72*, 104569, 2021. https://doi.org/10.1016/j.jlp.2021.104569.
4. Navaid, H.B., Emadi, H., Watson, M., A comprehensive literature review on the challenges associated with underground hydrogen storage. *Int. J. Hydrog. Energy*, *48*, 28, 10603–10635, 2023. https://doi.org/10.1016/j.ijhydene.2022.11.225.
5. Ossai, C., II, Boswell, B., Davies, I.J., Pipeline failures in corrosive environments–A conceptual analysis of trends and effects. *Eng. Fail. Anal.*, *53*, 36–58, https://doi.org/10.1016/j.engfailanal.2015.03.004, 2015.
6. Golahmar, A., Kristensen, P.K., Niordson, C.F., Martínez-Pañeda, E., A phase field model for hydrogen-assisted fatigue. *Int. J. Fatigue*, *154*, 106521, 2022. https://doi.org/10.1016/j.ijfatigue.2021.106521.

7. Zhao, Y., Seok, M.Y., Choi, I.C., Lee, Y.H., Park, S.J., Ramamurty, U., Suh, J.Y., Jang, J., II, The role of hydrogen in hardening/softening steel: Influence of the charging process. *Scr. Mater.*, *107*, 46–49, 2015. https://doi.org/10.1016/j.scriptamat.2015.05.017.
8. Kurasov, O.A. and Burkov, P.V., Substantiation of methods of improving safety of pipeline gas transportation, in: *E3S Web of Conferences 2021*, vol. *266*, EDP Sciences, p. 01012, 2021, https://doi.org/10.1051/e3sconf/202126601012.
9. He, W., King, M., Luo, X., Dooner, M., Li, D., Wang, J., Technologies and economics of electric energy storages in power systems: Review and perspective. *Adv. Appl. Energy*, *4*, 100060, 2021. https://doi.org/10.1016/j.adapen.2021.100060.
10. Chowdhury, P. and Sehitoglu, H., Mechanisms of fatigue crack growth–a critical digest of theoretical developments. *Fatigue Fract. Eng. Mater. Struct.*, *39*, 6, 652–674, 2016. https://doi.org/10.1111/ffe.12392.
11. Mohtadi-Bonab, M.A., Szpunar, J.A., Basu, R., Eskandari, M., The mechanism of failure by hydrogen induced cracking in an acidic environment for API 5L X70 pipeline steel. *Int. J. Hydrog. Energy*, *40*, 2, 1096–1107, 2015. https://doi.org/10.1016/j.ijhydene.2014.11.057.
12. Hasan, S.M., Khan, F., Kenny, S., Probabilistic Transgranular Stress Corrosion Cracking Analysis for Oil and Gas Pipelines. *ASME J. Pressure Vessel Technol.*, 134, 5, 051701, 2012. https://doi.org/10.1115/1.4006125.
13. Mohammed, A., II, Oyeneyin, B., Atchison, B., Njuguna, J., Casing structural integrity and failure modes in a range of well types—a review. *J. Nat. Gas Sci. Eng.*, *68*, 102898, 2019. https://doi.org/10.1016/j.jngse.2019.05.011.
14. Nakamura, J., Hisamune, N., Fukuba, T., Thodla, R., Scott, C.S., Gui, F., Fatigue crack growth rate and fracture toughness of non-sour service grade X65 and X80 in sour environments, in: *International Conference on Offshore Mechanics and Arctic Engineering*, 2013, https://doi.org/10.1115/OMAE2013-10573.
15. Wasim, M., Djukic, M.B., Ngo, T.D., Influence of hydrogen-enhanced plasticity and decohesion mechanisms of hydrogen embrittlement on the fracture resistance of steel. *Eng. Fail. Anal.*, *123*, 105312, 2021. https://doi.org/10.1016/j.engfailanal.2021.105312.
16. Ohaeri, E., Eduok, U., Szpunar, J., Hydrogen related degradation in pipeline steel: A review. *Int. J. Hydrog. Energy*, *43*, 31, 14584–14617, 2018. https://doi.org/10.1016/j.ijhydene.2018.06.064.
17. Alamilla, J.L., Sosa, E., Sánchez-Magaña, C.A., Andrade-Valencia, R., Contreras, A., Failure analysis and mechanical performance of an oil pipeline. *Mater. Des.*, *50*, 766–773, 2013. https://doi.org/10.1016/j.matdes.2013.03.055.
18. Nag, M.K. and Kumar, P., Optimization of cost and performance of the material used in the mechanism unit of Air Circuit Breaker (ACB) based on

various analysis approach. *Proc. IMechE, Part B: J. Eng. Manuf.*, 238, 549–61, 2024. https://doi.org/10.1177/09544054231168682.
19. Cheng, A. and Chen, N.Z., Fatigue crack growth modeling for pipeline carbon steels under gaseous hydrogen conditions. *Int. J. Fatigue*, *96*, 152–161, 2017. https://doi.org/10.1016/j.ijfatigue.2016.11.029.
20. Ávila, J.A., Ruchert, C.O., Mei, P.R., Marinho, R.R., Paes, M.T., Ramirez, A.J., Fracture toughness assessment at different temperatures and regions within a friction stirred API 5L X80 steel welded plates. *Eng. Fract. Mech.*, *147*, 176–186, 2015. https://doi.org/10.1016/j.engfracmech.2015.08.006.
21. Hashemi, S.H. and Mohammadyani, D., Characterization of weldment hardness, impact energy, and microstructure in API X65 steel. *Int. J. Press Vessels Pip.*, *98*, 8–15, 2012. https://doi.org/10.1016/j.ijpvp.2012.05.011.
22. Elboujdaini, M. and Revie, R.W., Metallurgical factors in stress corrosion cracking (SCC) and hydrogen-induced cracking (HIC). *J. Solid State Electrochem.*, *13*, 7, 1091–1099, 2009. https://doi.org/10.1007/s10008-009-0799-0.
23. Sharma, V.B., Singh, K., Gupta, R., Joshi, A., Dubey, R., Gupta, V., Srivastava, A., Review of structural health monitoring techniques in pipeline and wind turbine industries. *Appl. Syst. Innov.*, *4*, 3, 59, 2021. https://doi.org/10.3390/asi4030059.
24. Spanos, G., Fonda, R.W., Vandermeer, R.A., Matuszeski, A., Microstructural changes in HSLA-100 steel thermally cycled to simulate the heat-affected zone during welding. *Metall. Mater. Trans. A*, *26*, 3277–3293, 1995. https://doi.org/10.1007/BF02669455.
25. David, S.A., Vitek, J.M., Reed, R.W., Hebble, T.L., Effect of rapid solidification on stainless steel weld metal microstructures and its implications on the Schaeffler diagram, in: *Oak Ridge National Lab. (ORNL), Oak Ridge, TN (United States)*, 1987, September 1, https://doi.org/10.2172/5957599.
26. Cao, X., Wallace, W., Immarigeon, J.P., Poon, C., Research and progress in laser welding of wrought aluminum alloys. II. Metallurgical microstructures, defects, and mechanical properties. *Mater. Manuf. Processes*, *18*, 1, 23–49, 2003. https://doi.org/10.1081/AMP-120017587.
27. Zahraei, S.M., Dehmolaei, R., Ashrafi, A., The effect of heat input on microstructure and HAZ expansion in dissimilar joints between API5L X80/DSS 2205 steels using thermal cycles. *Rev. Metal.*, *58*, 2, e222, 2022, July 26. https://doi.org/10.3989/revmetalm.222.

28. Lambert-Perlade, A., Gourgues, A.F., Besson, J., Sturel, T., Pineau, A., Mechanisms and modeling of cleavage fracture in simulated heat-affected zone microstructures of a high-strength low alloy steel. *Metall. Mater. Trans. A*, *35*, 1039–1053, 2004, March. https://doi.org/10.1007/s11661-004-1007-6.
29. Forero, A.B., Núñez, M., Bott, I.S., Analysis of the corrosion scales formed on API 5L X70 and X80 steel pipe in the presence of CO2. *Mater. Res.*, *17*, 461–471, 2014. http://dx.doi.org/10.1016/j.matlet.2003.08.020.

10

Dynamic Response Comparison of Sandwich Panels with Honeycomb and Foam Core Under Blast Loads: A Quantitative Study

Murlidhar Patel* **and Shivdayal Patel**

Department of Mechanical Engineering, PDPM Indian Institute of Information Technology, Design and Manufacturing (IIITDM), Jabalpur, India

Abstract

The present chapter reciprocated in the direction of comparing the dynamic response of sandwich panels with foam and honeycomb cores in blast loads to aid in building efficient industrial designs to avoid blast effects. In many industries, blast is a critical problem, such as chemical plants, aerospace, and armory automotives. In air blast phenomena, a high-pressure gas or steam release occurs due to the failure of equipment such as pipes, valves, or pressure vessels or any explosive explosion. This can result in a sudden and powerful release of energy that can cause damage to nearby structures and equipment, as well as injury or death to people in the vicinity. Hence, the current era needs a superior-quality protective structure. Sandwich panels are commonly used in the construction of protective structures in many manufacturing industries, including those that are designed to withstand blast attacks. The foam and honeycomb cores are mainly preferred for the fabrication of sandwich panels as a protective structure due to their high bending strength with larger porosity. In the context of air blast loadings, the core structure of a sandwich shows a critical role in determining the level of damage that a structure will sustain. Therefore, the present chapter deals with a quantitative numerical comparison of the blast mitigation characteristics of a square honeycomb-core sandwich panel (HCSP) and a crushable foam-core sandwich panel (FCSP) at 1 kg to 3 kg of trinitrotoluene (TNT) explosive blast for a 100-mm standoff distance. The sandwich panels' skins and honeycomb were

**Corresponding author*: murlidharpatel4@gmail.com

Atul Kumar Sahu, Rakesh D. Raut, Rohit Raja, Anoop Kumar Sahu and Nitin Kumar Sahu (eds.) Industrial and Manufacturing Designs: Quantitative and Qualitative Analysis, (295–320) © 2024 Scrivener Publishing LLC

fabricated from high-strength stainless steel. The crushable Al foam was used as a foam core. The effect of honeycomb and foam cores on the blast performance of the sandwich panel was discovered quantitatively. The obtained results revealed that the energy absorption, center deflection as well as the radial deflection of the front skin, core specific energy absorption, and core densification of the HCSP were comparatively higher than that of the FCSP for the same amount of TNT detonation. Additionally, it was discovered that the HCSP's panel mass and back skin peak center deflection were lower than those of the FCSP. As a result, using a honeycomb core instead of a foam core considerably increases the sandwich panel's blast mitigating characteristics. The quantitative findings of the present chapter revealed that the honeycomb core is the most suitable structure for sandwich panel fabrication for various industrial applications.

***Keywords*:** Blast, honeycomb, industries, foam, quantitative numerical analysis, sandwich, TNT

10.1 Introduction

Sandwich panels are playing a crucial role in energy absorption under various impacts and shock-loading conditions. The sandwich panels are the assembly of two stiff skins and a low-density core. The skins are thinner but stiffer than the core structure. These parts of the sandwich panels are assembled by using mechanical fasteners or adhesive. Hence, the sandwich panel acts as an integrated load-bearing unit [1]. Materials including metals (such as titanium and aluminium), composites (such as carbon fiber-reinforced polymer and glass fiber-reinforced polymer), or layered mixtures of various materials are commonly used to make skins [2]. The skins give the panel its main structural strength and defense against external forces, impacts, and blast loads. The tensile and compressive loads are resisted by stiff skins. The core material is positioned between the skins. Usually, it is constructed from low-density, lightweight materials such as balsa wood, aluminium foam, Nomex foam, polyurethane foam, or honeycomb structures. The core structure resists the shear loads. Due to the use of a thicker core in the sandwich panel, its moment of inertia increases. Therefore, the sandwich panel's rigidity, stability, and load-carrying capacity improved. It also has good thermal and acoustic insulation capabilities. These will be achieved by using light-weight cores [3]. Hence, the sandwich panels have an excellent strength-to-weight ratio. This property of the sandwich panels makes them suitable for weight-critical applications like aircraft, spacecraft, and high-strength lightweight components in many manufacturing industries. The stiff skins of the sandwich panel homogeneously distribute the applied loads over a wide area [4]. In terms of material choice, core density, skin thickness, and overall geometry, sandwich panels provide design versatility. Panels can be tailored by engineers to satisfy certain mechanical, thermal, and acoustic needs. The sandwich panels provide better stiffness than the same mass of solid panels [5, 6]. Due to these advantageous characteristics of the sandwich panels, they have lots of industrial applications [7]. To save weight without sacrificing structural integrity, sandwich constructions are utilized in the aerospace industry for fuselage panels, overhead bins, galleys, aircraft wings, and tail sections. Sandwich panels have been exploited more often in the automotive industry to make outer body panels, including the doors, hoods, and roof panels, which help vehicles weigh less and run more efficiently. Boat decks and hulls are made from sandwiches in the maritime industry. Sandwiches are used in the sports industry to make sturdy, lightweight sports equipment, including surfboards, skis, and snowboards. They are employed in the transportation sector for the construction of trailers and containers, as well as interior panels, train flooring, and ceilings. Sandwich panels are used in the renewable energy sector to construct solar panel structural supports and wind turbine blades [8].

In the present era, terrorist blast attacks are a critical problem that affect people's and expensive objects' safety. The challenges and hazards caused by explosive blasts that take place underwater close to or beneath a ship's hull are referred to as "underwater blast issues" on naval ships. Several things, such as depth charges, mines, torpedoes, or underwater improvised explosive devices, might result in these explosion occurrences. Underwater blasts can have serious consequences for the systems, crew safety, and structural integrity of the ship [9, 10]. Strong shock waves, pressure waves, and fragmentation produced by explosive blasts have the potential to seriously harm automobiles and endanger their occupants. Military vehicles are also susceptible to significant structural damage from the explosive blast of a landmine. The vehicle's hull, chassis, and other parts may distort, bend, or shatter due to the shock wave and pressure created by the explosion. Soldiers inside the vehicle are at risk of injuries due to the force of the blast as well as from flying debris and fragments caused by the explosion [11]. Impact injuries can occur as occupants are thrown against the vehicle's interior. In the manufacturing sector, blast-related issues can originate from several causes, such as chemical reaction explosions, gas leaks, dust ignitions, and other dangerous materials. These explosions have the potential to cause serious property damage, safety risks, and manufacturing process interruptions.

Sandwich panels have been employed as protective structures because of their outstanding capacity to absorb energy under impact and blast loadings. The proper selection of the core structure and material is a critical task for an engineer for a particular application. The quantitative characterization of the sandwich panels' dynamic performance under blast loads through experimentation is highly risky, costly, and an environmental pollutant [12]. Hence, the present chapter represents a cost-effective numerical analysis of quantitative blast characterizations of the honeycomb-core sandwich panel (HCSP) and foam core sandwich panel (FCSP). The Conventional Weapon Effects Program (CONWEP) module of Abaqus/Explicit was used for the analysis of blast mitigation performance. The effects of explosive charges and types of cores on blast performance have been discovered. Quantitative analysis is carried out, and the deflections of the skins with regard to time and distance, energy absorption of the panel, specific energy absorption of the core, and core crushing modes of both honeycomb and foam core–used sandwich panels were compared to find out the best core structure.

10.2 Quantitative Numerical Modeling

10.2.1 Geometry Design for Finite Element Analysis

The finite element models of the honeycomb and foam core–used sandwich panels were prepared using Abaqus/CAE software. The sizes of the different parts of the sandwich panels are represented in Figure 10.1. The in-plane sizes of the modeled sandwich panels are 305 mm × 305 mm, which are the dimensions for a quadrant part of the panel. The quarter models of the panels were preferred in this quantitative numerical study due to the regularity of both the construction and loading conditions. The quarter models save computational time and minimize the storage requirement as contrasted to the full model [13]. The full model deformation behavior can also be seen by using the available mirror pattern facility in the software for the quarter model. The core structures' contact surfaces with the skins are joined by using tie constraints. The applied boundary

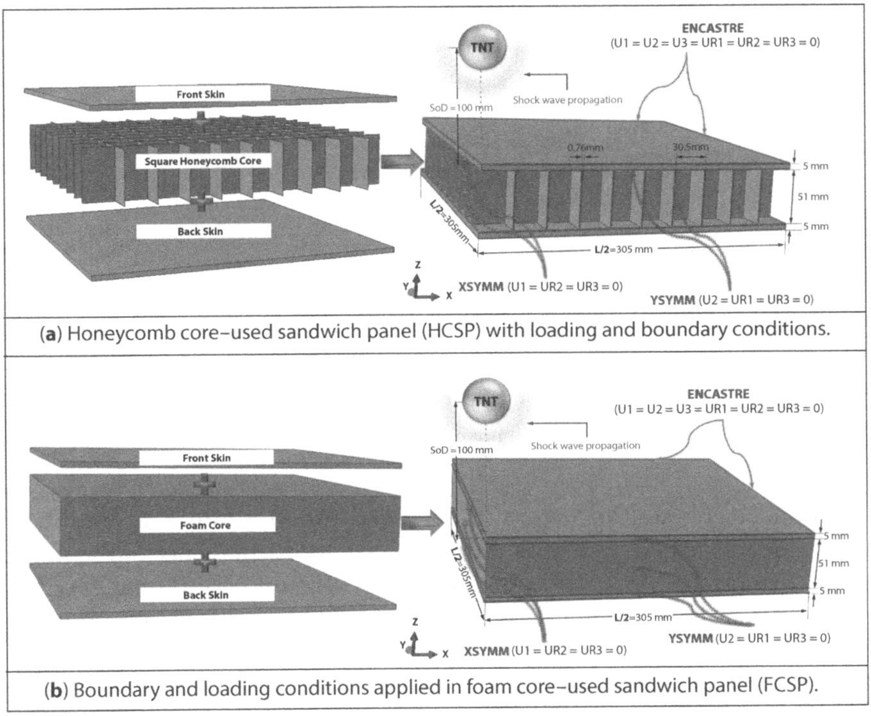

Figure 10.1 Specifications of sandwich panels' geometry, loading conditions, and boundary conditions.

conditions on the different faces of the quarter-model sandwich panels are also represented in Figure 10.1. The constrained degree of freedom is also illustrated in this figure. The CONWEP module of the Abaqus was used to apply the blast loads to the modeled sandwich panels. To apply the blast loads, the topmost surface of the front skin was defined as the target surface, and the center of the blast source was pointed away at 100 mm from the top surface center. This distance between the center of the blast source and the center of the target surface is known as the standoff distance (SoD) and is also shown in Figure 10.1. To apply the blast loads, 1 kg to 3 kg of spherical TNT were used for the explosion.

After a mesh convergence analysis, the skins are discretised with a brick element size of 15 mm × 15 mm × 1 mm C3D8R (continuum three-dimensional eight-noded element with reduced integration) elements.

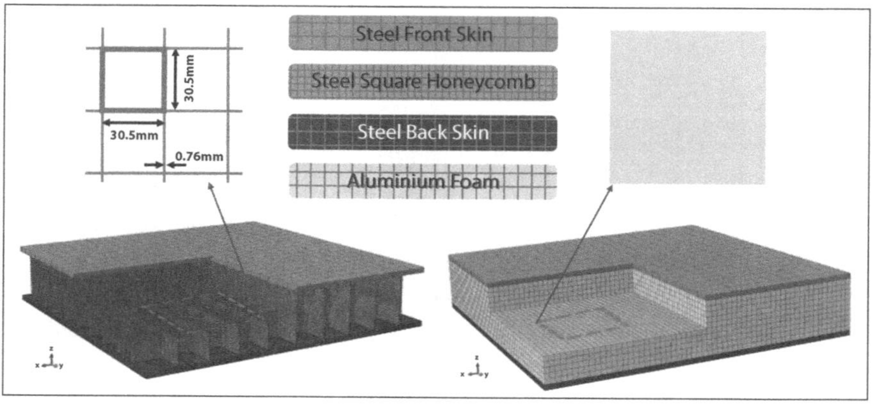

Figure 10.2 Finite element models of honeycomb and foam core–used sandwich panels.

The honeycomb and foam cores are meshed with 1.5 mm × 1.5 mm S4R (four-noded shell element with reduced integration) elements and 5 mm × 5 mm × 5 mm C3D8R elements, respectively. The numerical models of the sandwich panels for the finite element analysis (FEA) are shown in Figure 10.2.

10.2.2 Materials Modeling

A popular material model in computational mechanics, notably in FEA, is the Johnson–Cook (J-C) model. It is used to describe the plastic deformation and failure behavior of materials, when they are subjected to high strain rates and high-temperature conditions that are frequently present in dynamic events like impact, explosions, and metal forming processes [14]. The J-C model is particularly useful for alloy and metal modeling. In the present analysis, the skins and honeycomb core are made from AL6XN steel. Hence, for the prediction of the deformation behavior of these parts, the J-C model is the most suitable model. The J-C model requires material parameters to be calibrated for each specific material. These parameters determine the material's response under varying strain rates and temperatures. The parameters are usually determined through experimental tests, such as tensile tests at various temperatures and strain rates. Based on these parameters, the J-C equation has the following general form (Equation 10.1) [15, 16]:

$$\sigma_y = \left[A + B\left(\varepsilon_p\right)^n\right]\left[1 + C\ln\left(\frac{\dot{\varepsilon}_p}{\dot{\varepsilon}_o}\right)\right]\left[1 - \left(\frac{T - T_{tr}}{T_m - T_{tr}}\right)^m\right] \quad (10.1)$$

Here, σ_y,$\dot{\varepsilon}_p$, T, and ε_p, respectively, are quantitative and stand for effective stress, equivalent plastic strain rate, current temperature, and equivalent plastic strain. The details about the other quantitative parameters are listed in the Table 10.1. Table 10.1 also catalogs the values of various J-C parameters and mechanical properties for the AL6XN steel. It is predicated that the material will behave as an isotropic hardening (B,n) and temperature/strain rate sensitive (C,m) combination, making it suited for modeling dynamic and high-temperature deformation processes.

The metallic foams have a higher crushing strength than polymeric foams. Hence, for the foam core, a lightweight closed-cell Al foam was used for this quantitative numerical study. Under the pressure of the blast wave, the employed Al foam experienced substantial compressive deformation.

Table 10.1 Material properties and details of J-C parameters used for AL6XN steel [17].

Notation	J-C parameter or material property	Value	Unit
E	Young's modulus	161,000	MPa
ρ	Density	7,850	kg/m^3
v	Poisson's ratio	0.35	-
A	Initial yield stress at room temperature and reference strain rate	400	MPa
C	Strain rate sensitivity coefficient	0.045	-
B	Strain hardening stress coefficient	1500	MPa
n	Strain hardening exponent	0.4	-
$\dot{\varepsilon}_o$	Reference strain rate	0.001	s^{-1}
T_m	Melting temperature	1,800	K
m	Temperature sensitivity exponent	1.2	-
T_{tr}	Transition temperature	293	K

For employing Al foam, the crushable foam model, however, was chosen with the volumetric hardening model. The behavior of foam materials, such as foams used in packing, cushioning, and impact protection, is frequently simulated using the crushable foam model. The compression and densification of the foam material as it passes through loading and unloading cycles are considered by the volumetric hardening model within this framework. Tensile hydrostatic stress does not vary throughout the distortion, whereas compression hydrostatic stress does [18]. Equation 10.2 specifies the form of the yield surface (φ) in this model. This yield surface is created by the Mises circle in the deviatoric stress plane.

$$\varphi = \frac{3S}{\sqrt{(3S_T + S)(3 - S)}} \tag{10.2}$$

The yield stress ratios in this case are S_T for hydrostatic yield and S for compression yield, which are calculated by Equations 10.3 and 10.4:

$$S = \frac{\sigma_C^o}{P_C^o} \tag{10.3}$$

$$S_T = \frac{P_T}{P_C^o} \tag{10.4}$$

Valid surface conditions are $S_T \geq 0$ and $0 < S < 3$. The terms P_C^o, P_T, and σ_C^o in the equations above, respectively, stand for hydrostatic compressive yield strength, initial hydrostatic tensile yield strength, and uniaxial compression. Equation 10.5, which connects the pressure stress (p) and Mises stress (q), yields the flow potential (G).

$$G = \sqrt{q^2 + \frac{9}{2}p^2} \tag{10.5}$$

The evolution of the yield surface may be described as in Equation 10.6, when the yield surface size varies adjacent to the hydrostatic stress line [12]:

$$P_C\left(\varepsilon_V^{pl}\right) = \frac{\sigma_C\left(\varepsilon_A^{pl}\right)\left[\sigma_C\left(\varepsilon_A^{pl}\right)\left(\frac{1}{\varphi^2}+\frac{1}{9}\right)+\frac{P_T}{3}\right]}{P_T + \frac{\sigma_C\left(\varepsilon_A^{pl}\right)}{3}} \tag{10.6}$$

where P_C is the compressive strength in this case, ε_A^{pl} is the volumetric compacting plastic strain, and σ_C is the uniaxial compression yield strength's absolute value. In uniaxial compression, the ratio of plastic poisons is zero while taking into account that $\varepsilon_A^{pl} = \varepsilon_V^{pl}$. However, the information needed to identify the strain hardening curve is specified in tabular format. Tables 10.2 and 10.3, correspondingly, display the attributes of the material and hardening curve information for the utilized foam.

10.2.3 Validation Approach for Numerical Model

The most crucial investigation to support the validity of the produced numerical findings is the validation study. Therefore, the experimental findings reported in the literature [6] are compared with the front skin peak deflections of the HCSP, and its deformations at 1 kg to 3 kg of TNT explosions are illustrated in Figures 10.3(a) and 10.3(c), respectively. In the literature [6], an experimental study on the blast performance of the HCSP made of AL6XN has already been performed. Hence, in this chapter, the experimental results of the literature [6] have been used for the validation of the present finite element model of the HCSP. The peak overpressure experienced by the sandwich panels in both the experimental and numerical cases is also compared in Figure 10.3(b). From Figures 10.3(a)

Table 10.2 Material properties of lightweight closed-cell Al foam [19].

E	ν	ρ	S	S_T
5 GPa	0.11	681 kg/m^3	1.53	0.45

Table 10.3 Closed-cell Al foam strain hardening graph data points [19].

Stress	**11 MPa**	**11.9 MPa**	**15.35 MPa**	**16.12 MPa**	**18.39 MPa**	**20.93 MPa**	**23.05 MPa**	**25.18 MPa**	**29.70 MPa**	**69.82 MPa**	**2,100 MPa**
Plastic strain	0	0.03	0.12	0.15	0.25	0.42	0.51	0.60	0.9	1.07	1.8

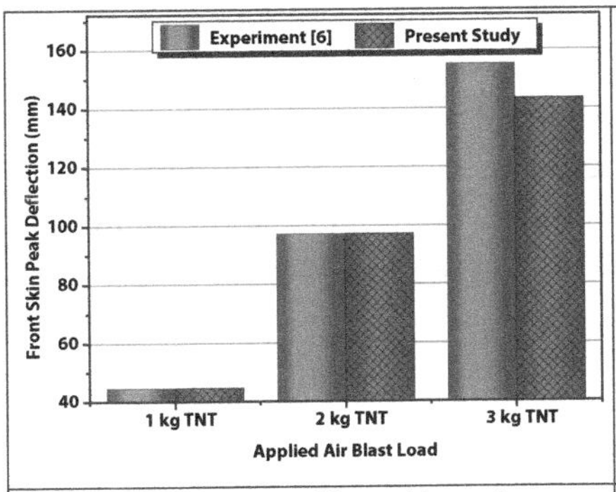

(a) Comparison of front skin peak deflection.

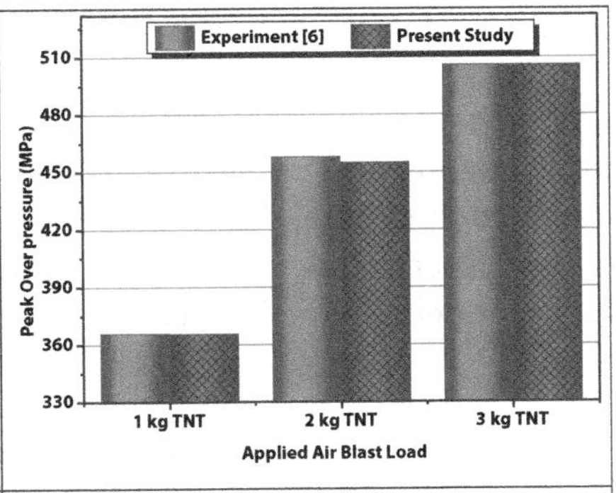

(b) Comparison of peak overpressure on front skin

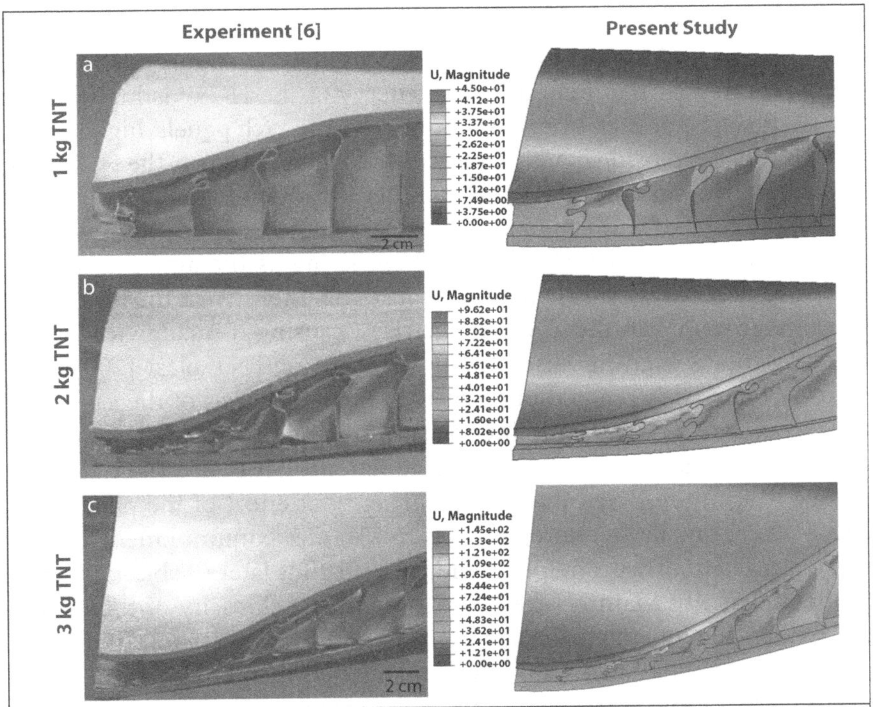

(c) Comparison of deformation of the HCSPs

Figure 10.3 Comparisons of the obtained numerical results with literature [6] experimental results.

and 10.3(b), it is noted that the differences between the experimental and numerical results of front skin peak deflections and peak overpressures are very small. Figure 10.3(c) also shows similar deformation patterns for both experimental and numerical cases. These closed relationships between the obtained numerical results and experimental results showed the validation of the present work and confirmed the pertinency of the present finite element models for further study.

10.3 Results and Discussion

10.3.1 Effect on Kinetic Energy and Pressure Distribution

The variations in the kinetic energies of both the HCSP and FCSP with regard to time at 1 kg to 3 kg TNT blasts are shown in Figure 10.4(a). This graph reveals that the highest kinetic energy of the HCSP is higher than that of the FCSP at each blast condition. The sandwich panels' high kinetic energy under explosion loads denotes its less rigidity. Hence, the HCSP has lower rigidity than the FCSP for similar blast loadings.

From Figure 10.4(a), it can also be observed that the maximum kinetic energies of both the HCSP and FCSP are rising as the amount of TNT explosive used rises. This happens because the intensity of the blast wave pressure grows when the TNT charge rises, causing the sandwich panels to experience a high impulse at a higher mass of TNT blast [20]. From Figure 10.4(a), it is pointed out that the kinetic energies of the sandwich panels become zero at 0.7 ms after blast. At this time, the sandwich panels move in their maximum out-of-plane direction; after that, a rise in kinetic energy is again observed due to the spring-back effect of the panel. This effect is very small; therefore, only up to 8% of maximum kinetic energy was observed in the second peak, and, around 1 ms later the blast, this rise in kinetic energy again reached zero. Due to the elasticity of the panels, their kinetic energy fluctuates many times, and, around 1.5 ms, the kinetic energy at all applied blast conditions finally becomes zero. Hence, at this moment, it is considered that the sandwich panels were deformed permanently, and all the energy and deflection data of the sandwich panels are recorded for this moment of the blast.

The profile of the extreme overpressure faced by the front skin of the sandwich panels throughout the various blast conditions is represented in Figure 10.4(b). From this figure, it reveals that the maximum intensity of pressure is experienced by the center of the sandwich panel, and this intensity of pressure gradually decreases in radial direction from the center of

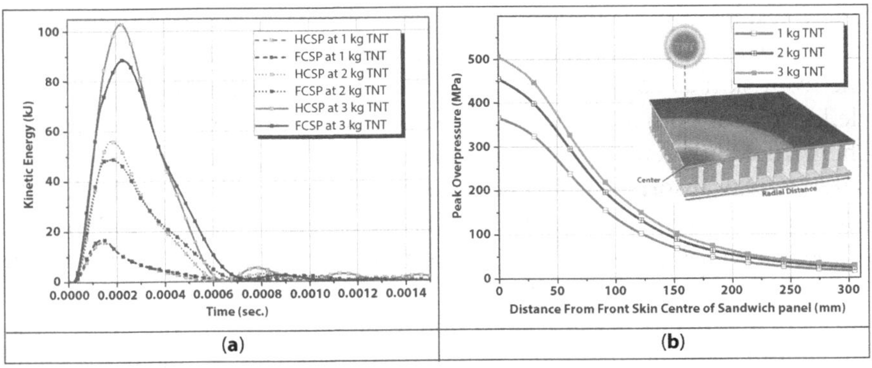

Figure 10.4 Effect of blast loadings on (a) kinetic energy of the sandwich panels and (b) pressure distribution on the sandwich panel's target surface.

the panel. This is due to the spherical blast wave generation during the explosion and its radial propagations. As per the present loading condition, the explosive has been used at the center of the panel and 100 mm away from the front skin. Therefore, the generated blast wave first interacts with the center of the panel, which applies the maximum overpressure to the center of the panel [21]. Due to the increased interaction delay of the blast wave with the radial distance of the panel, the intensity of the overpressure decreases with radial distance. From Figure 10.4(b), it is also found that the concentration of peak overpressure on the front skin surface is directly proportional to the explosive massat any point on the target surface. This could be due to the higher amount of energy released in a high-mass TNT explosion [5].

10.3.2 Effect on Sandwich Panels' Skin Deflection

The effect of blast wave pressure on the sandwich panels' skin deflections is shown in Figures 10.5 and 10.6. The center point deflections according to time for both the front skin and the back skin are illustrated in Figures 10.5(a) and 10.5(b), respectively, at all blast conditions. The observed radial skin deflections for the front skin and back skin of the modeled sandwich panels after 1.5 ms of blast are compared in Figures 10.6(a) and 10.6(b), respectively. It is evident from Figure 10.5(a) that the front skin center point deflection of the HCSP is always extreme than the front skin center point deflection of the FCSP. From Figure 10.6(a), it can also be observed that the radial deflection of the front skin of the HCSP is greater than the FCSP's front skin radial deflection for each blast condition.

These results indicate that the foam core offers stronger base to the front skin and helps to distribute incident overpressure more uniformly on the front skin as compared to the HCSP. The patterns of the center point deflections of both sandwich panels' back skins are more non-linear than their front skin center point deflections. This depends on the core structures' energy absorption and energy transmitting capacity to the back skin. However, this type of non-linear behavior is seen in Figures 10.5(b) and 10.6(b). From Figures 10.5 and 10.6, it is found that both center point and radial deflections of the sandwich panels' back skin are lesser than the front skin. This happens because both honeycomb and foam core absorb the maximum incident energy and reduce the back skin deflection of the sandwich panels with respect to their front skin deflection. It is observed from

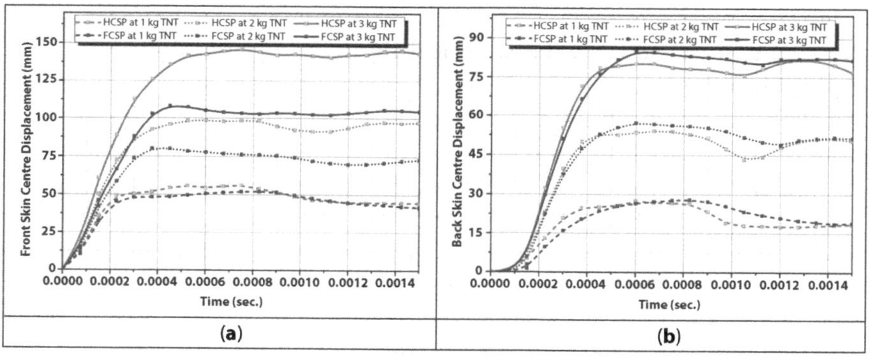

Figure 10.5 Sandwich panels' center point deflection in blast loadings for (a) front and (b) back skin.

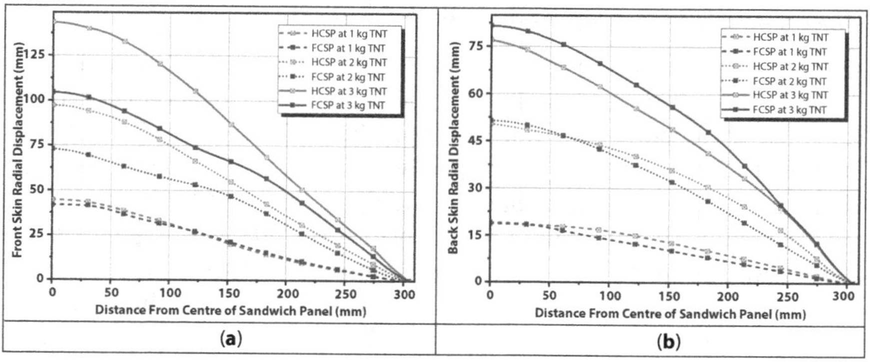

Figure 10.6 Sandwich panels' radial skin deflections in blast loadings for (a) front and (b) back skin.

Figure 10.5(b) that the central deflection of the HCSP's back skin is smaller than the back skin center point deflection of the FCSP.

Figure 10.6(b) shows that the FCSP's central region back skin radial deflection is higher than the HCSP's back skin radial deflection. This higher back skin deflection region of the FCSP is increasing with a rise in explosive charge for blast loadings. This could be due to the use of a foam core that transmits more energy to the central region of the sandwich panels' back skin as compared to the honeycomb core. The back skin deflection profile shown in Figure 10.6(b) represents that the back skin deflections of the HCSP are a little bit more globalized than the FCSP's back skin deflections. This result indicates the honeycomb core is more efficient than the foam core for designing the protective structures. The sandwich panels' skin deflection profiles show that the foam core protects the front skin displacement greater than the honeycomb core, and honeycomb core protects the back skin displacement greater than the foam core. The most crucial layer of safety for blast or impact loads is the back skin of any sandwich panel. Its minimum deflection indicates high blast or impact resistance. Due to the smaller back skin deflections of the HCSP with respect to the FCSP for the same blast condition, the HCSP is the best choice for making protective cabins in many manufacturing industries and power plants to save expensive devices and workers from the unwanted and unexpected blast of any chemical storage tank, boiler, etc.

10.3.3 Effect on Sandwich Panels' Energy Absorption

The various types of energy absorption behavior of both the HCSP and FCSP are illustrated in Figures 10.7 and 10.8. The internal energy of the sandwich panel is the total of its strain energy (energy absorbing during elastic deformation), artificial strain energy (generated as a result of reduced integration use in FEA), and plastic dissipation energy (PDE) (energy absorption during plastic distortion). The variations in the internal energy of both the HCSP and FCSP according to time are displayed in Figure 10.7(a) for all explosion loads. According to this graph, the HCSP has a higher energy absorption capacity as compared to the FCSP. High energy absorption is a good characteristic of sandwich panels for their utilization against blast loads. However, the HCSP has higher blast mitigation characteristics as compared to the FCSP with regard to absorbing energy. From Figure 10.7(a), it is seen that the sandwich panel energy absorption reached a steady-state condition after 0.7 ms of blast. This is due to the sandwich panels deforming mostly within this period under blast loads.

Figure 10.7(b) represents the PDE by each part of the modeled sandwich panels at all applied blast conditions. The total PDE by the whole sandwich panel is the sum of its components' PDE. Hence, every bar of Figure 10.7(b) characterizes the PDE by whole sandwich panel for a particular blast condition. It is evident from the front skin PDE comparison that the HCSP's front skin PDE is higher than the FCSP's front skin PDE for similar blast loading specifications. This is because the overall front skin plastic deformation of the HCSP is higher than that of the FCSP for an explosion with a set weight of TNT. Figures 10.5 and 10.6 can also be used to confirm it. Similarly, given the identical blast loading circumstances, the PDE of the honeycomb core is higher than the PDE of the foam core. Hence, the honeycomb core deforms higher than the foam core for a constant mass of TNT explosion. This high PDE by honeycomb core shows a good blast resistance characteristic. From the comparison of the back skin PDE of both HCSP and FCSP, it is discovered that the rear skin of the FCSP dissipates higher energy than the PDE of the HCSP's rear skin. This happens because the maximum energy is absorbed by both the front skin and the honeycomb core, and a tiny percentage of the energy would be directed toward the back skin. But, in the case of FCSP, the foam core reduced the front skin deflection, but higher energy was transferred to the FCSP's back skin. From Figure 10.7, it is revealed that both the sandwich panels' internal energy and PDE of the as well as their parts are increasing with the increase in explosive mass from 1 kg to 3 kg because the deformation of the panels is directly proportionate to the loaded impulse or blast wave intensity [22].

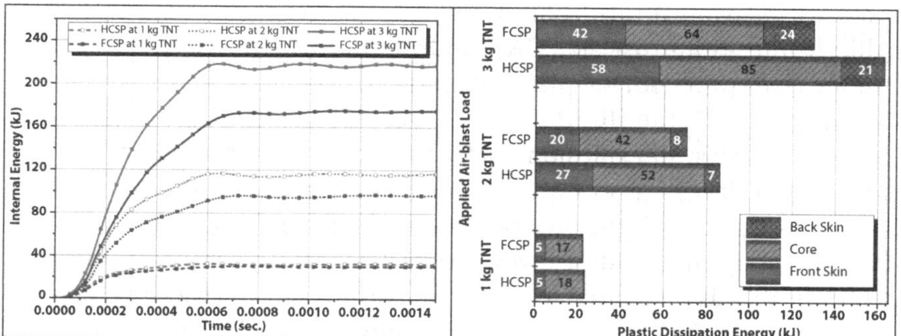

(a) Energy absorption with respect to time. (b) PDE of various parts of sandwich panels.

Figure 10.7 Comparisons of the energy absorption behavior of the modeled sandwich panels.

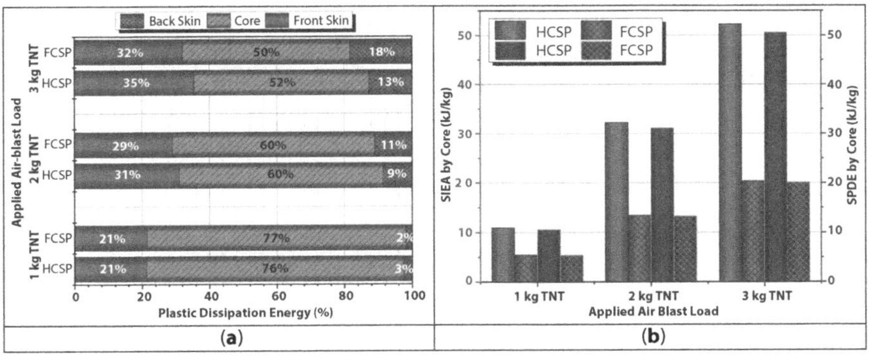

Figure 10.8 (a) Percentage contribution to PDE by sandwich parts. (b) Specific energy absorption by the sandwich cores.

The contributions to PDE by the various parts of the sandwich panels are represented in Figure 10.8(a). This figure indicates that, with regard to absorbing energy from blast loads, the sandwich panel's core makes the most contribution. Hence, the selection of the core structure is a crucial task in designing a good performing sandwich panel. From Figure 10.8(a), it is also noted that the front skin's contribution to energy absorption is higher than that of the back skin. This is due to the fact that the bend of the front skin is higher than the back skin's bend. It is discovered that the proportion impact to PDE by both the front skin and core of the HCSP is higher than that of the FCSP's front skin and core for the same blast load. When back skin contributions are compared in PDE, it is seen that the HCSP's back skin contributes less than the FCSP's back skin. This is because the HCSP's back skin deforms lesser than the FCSP's back skin for similar blast conditions. The higher contribution of the front skin and core of the HCSP and the lower contribution of its back skin represent its better blast resistance characteristics than that of the FCSP. From Figure 10.8(a), it can also be observed that the percentage contribution in PDE by both honeycomb and foam cores decreases and the skins contribution increases with a rise in the amount of explosion loads. This happens because the skin's deflection increases with an increase in the intensity of the blast, but the core is densified only after a certain limit of deformation.

In both the modeled sandwich panels, their skins are made of the same material with the same thickness. Hence, there is no difference in the masses of skin. But the cores were designed with different materials; therefore, there is a difference between the masses of the foam and honeycomb cores. Due to differences in the masses of the core structures, it cannot be confirmed whether honeycomb core has better energy absorption capabilities or foam core has better energy absorption capabilities for the same blast condition. To avoid this effect of core mass, specific energy absorptions of the cores are calculated by using Equations 10.7 and 10.8 [23]:

$$SIEA = \frac{IE_C}{m_C} \tag{10.7}$$

$$SPDE = \frac{PDE_C}{m_C} \tag{10.8}$$

Here, $SIEA$ and $SPDE$ denote the specific internal energy absorption of the core and the specific PDE by the core, respectively. Similarly, PDE_C and IE_C stand for the PDE by the core and internal energy absorption of the core, respectively. m_C is the mass of the sandwich core. The values of m_C are 2.3 kg and 1.67 kg, respectively, for the foam core and honeycomb core of the modeled sandwich panels.

These specific energy absorptions of the sandwich panels' cores are compared in Figure 10.8(b). From this figure, it is noted that both SIEA and SPDE of the honeycomb core are more than two times higher than those of the foam core. These results of specific energy absorptions confirmed that the use of honeycomb core is the most suitable core for making sandwich panels as protective structures in various manufacturing industries. Only a marginal difference between the SPDE and SIEA is seen in Figure 10.8(b). This is due to the fact that the cores of the sandwich panel mainly absorb the energy due to plastic deformation. Hence, the SPDE of the core is a little bit smaller than its SIEA.

10.3.4 Effect on Sandwich Panels' Deformation Behavior

The plastically deformed sandwich panels under the blast loads are illustrated in Figure 10.9. The skins of the sandwich panels were stretched and bent in a transverse direction. The walls of the honeycomb cores of the HCSP are buckled and densified. The foam cores of the FCSP are compressed and densified. From Figure 10.9, it is seen that the central region of the sandwich panels is highly deformed. This is as a result of the spherical blast wave generated during the explosion of TNT, which first interacts with the front skin center point and applies a high intensity of pressure

[24–27]. The value of applied overpressure on the front skin gradually decreases with radial distance as shown in Figure 10.4(b). Hence, the core crushing and skin inelastic deflection gradually decrease from the center of the panel to its fixed edges. Due to increases in the overpressure intensity with a rise in TNT mass, the skin deflection and core crushing increase. Hence, the sandwich panel deformation modes underneath the air blasts include honeycomb core wall buckling, foam core crushing, and skin out-of-plane bending. In Figure 10.9, it is noted that the honeycomb core crushing is greater than the foam core crushing. In the case of FCSP, the front skin spring back effect is high, which causes the front skin to again pull the foam core in an upward direction. However, the front skin deflection of the FCSP is smaller than that of the HCSP. It can also be confirmed from Figures 10.5(a) and 10.6(a).

The equivalent plastic strain (PEEQ) of both honeycomb and foam cores for every blast situation is shown in Figure 10.10. The PEEQ shows how much material was yielded for any region of the structure [12]. The structure failed mostly in the high PEEQ region. From Figure 10.10, it is pointed out that both honeycomb and foam core PEEQ values are high in the central region, which indicates that the core structure first fails in the central region. The PEEQ value for honeycomb core is superior than the

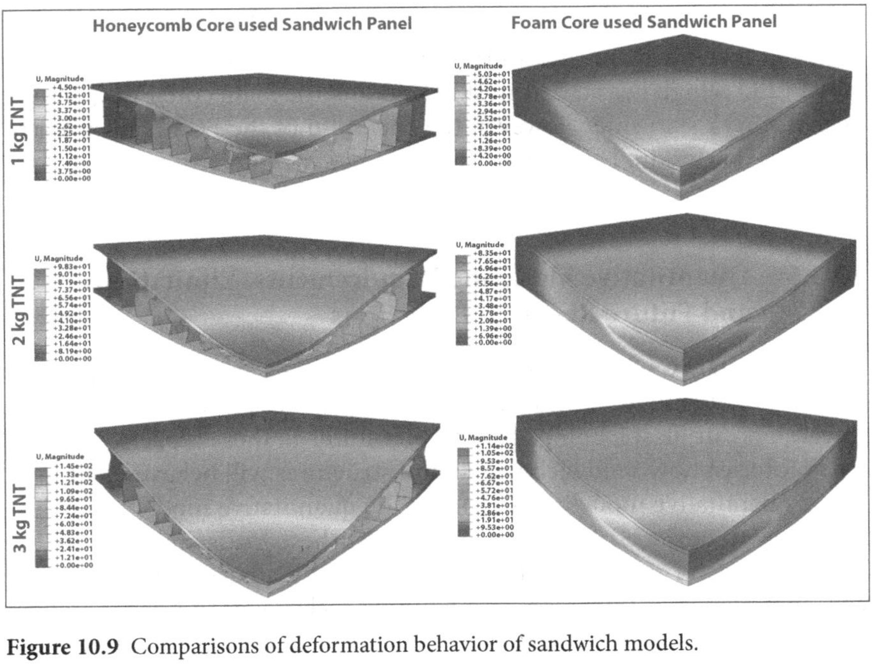

Figure 10.9 Comparisons of deformation behavior of sandwich models.

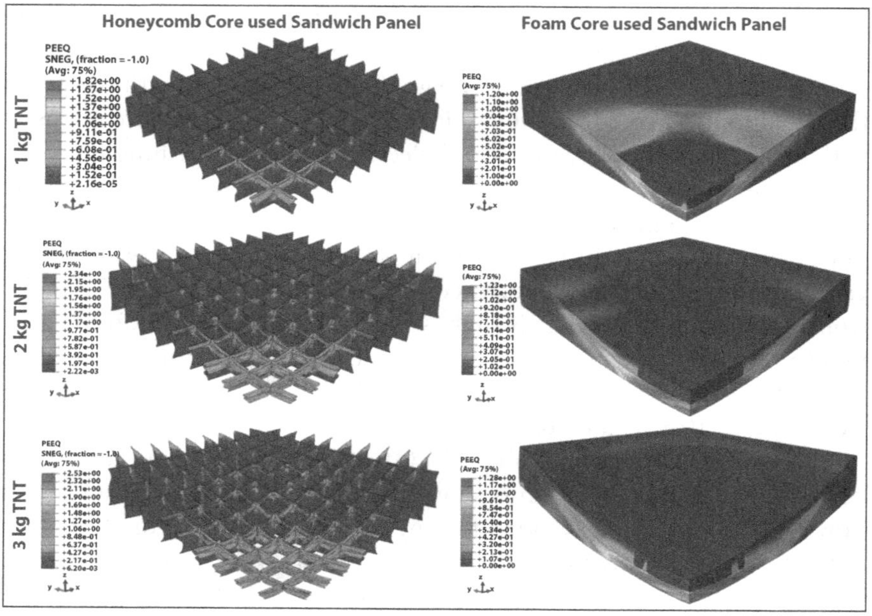

Figure 10.10 Comparisons of the PEEQ of both honeycomb and foam cores under blast loads.

foam core for a fixed blast load. It means the yield of the honeycomb core is greater than the foam core for the same blast conditions. The honeycomb core PEEQ is high near the front skin side, and the foam core PEEQ is high near the back skin side. However, the back skin of the FCSP fails prior to the HCSP back skin. Hence, the HCSP is a better protective structure than the FCSP.

10.3.5 Quantitative Modeling Requirements, Limitations, and Future Research Directions

The performance and safety of panels and materials exposed to explosives depend on quantitative modeling, assessment of sandwich panels, and research on air blast loading. This chapter fulfill all these aspects. To effectively forecast how sandwich panels and structures will behave to air blast loads, quantitative modeling is crucial. The quantitative modeling enables architects to create structures that can resist these loads and evaluate the consequences of various explosion scenarios. By adjusting variables like core material, skin thickness, and bonding techniques, modeling aids in improving the design of sandwich panels to obtain the appropriate blast resistance while reducing weight and expense. By permitting the construction of structures that can lessen the impacts of explosions and lower the danger of harm or damage, quantitative modeling assures the safety of people and property. By using quantitative modeling of the HCSP and FCSP in this study, their blast mitigations capabilities are investigated and found that the use of HCSP is the far better choice than the FCSP due to its good combination of energy absorption and lightweight properties. The quantitative modeling has many limitations. A thorough understanding of blast dynamics, material characteristics, and structural behavior is necessary for the quantitative modeling of air blast loading, which may be a challenging task. It could entail simulations that need a lot of computing. Sandwich panels' consistent blast resistance might provide issues due to variations in material qualities and production techniques. It can be difficult to scale results from small-scale laboratory testing to full-scale applications because of scaling problems and variations in boundary conditions. The cost and time involved in conducting blast testing and creating precise models may prevent them from being used widely. The quantitative modeling has wide area for future research directions. It is expected that research will focus on creating advanced and complex materials with better anti-crack properties to improve the performance of sandwich panels. Research into sustainable practices for sandwich parts can be aligned with global efforts to reduce the company's carbon footprint.

10.4 Conclusions

The current chapter examines the comparisons between the dynamic performance of the HCSP and FCSP quarter finite element models under blast loads. From this quantitative numerical study, the following key findings are obtained:

- For HCSP, when increasing the peak overpressure from 366 MPa to 506 MPa by blast loads, their peak deflections of front skin and back skin, maximum kinetic energy, maximum internal energy, maximum PDE, honeycomb core's SIEA, honeycomb core's SPDE, and maximum PEEQ were enhanced by 3.21, 4.13, 6.47, 7.06, 7.0, 4.8, 4.78, and 1.39 times, respectively.
- For FCSP, when increasing the peak overpressure from 366 MPa to 506 MPa by blast loads, their front skin peak deflection, back skin peak deflection, maximum kinetic

energy, maximum internal energy, maximum PDE, honeycomb core's SIEA, honeycomb core's SPDE, and maximum PEEQ were enhanced by 2.33, 4.3, 5.3, 5.7, 5.76, 3.77, 3.73, and 1.07 times, respectively.

- Core structure densification, region of core crushing and skin stretching of the sandwich panels grew as the blast intensity rose. The core structure's densification and skin stretching are the major failure criteria of the sandwich panels in the blast phenomenon.
- The designed HCSP is 14.6% lighter than the FCSP, and it has 1.9%–5.6% smaller back skin peak deflection, 6.4%–24.1% higher internal energy absorption, 101%–156% higher core's SIEA, 96.2%–151.7% higher core's SPDE, and 2.3%–31.4% higher core's PDE as compared to the FCSP. Hence, the HCSP is the best choice for making protective structures in various manufacturing industries to protect people and expensive objects from any type of blast pressure wave.

The blast proof HCSPs play a crucial role in various industries due to their unique mixture of high strength and lightweight construction as compared to other sandwich panels, which make them highly suitable for various industrial applications, where protection from blast is essential. In defense and military industries, the HCSPs provide lightweight armor military vehicles to withstand blasts. In the oil and gas exploration and production industries, by reducing the damage caused by explosions and the possibility of fire and poisonous gas emissions, the HCSPs improve worker safety. Blast-resistant buildings are frequently used in the mining sector to safeguard personnel and machinery in dangerous subterranean conditions. Mining refuge chambers and blast-resistant shelters can be built using the HCSPs. In the construction sector, secure buildings in high-risk locations, such as embassies, governmental facilities, and financial institutions, can be constructed using blast-resistant HCSPs. Standards for safety and security are helped by these panels. In shipbuilding, the HCSPs can be used as blast-resistant panels to increase the security of offshore platforms, commercial ships, and military ships. They offer defense against underwater impacts and explosions. The blast-resistant HCSPs can be utilized in the building of buses, trains, and subway cars in the transportation industry to increase passenger safety. When explosions, accidents, or terrorist threats occur, they assist in reducing damage and safeguarding passengers. The HCSPs are crucial in all these industries because of their capacity to offer impact and blast defense while retaining a light and structurally sound construction.

This makes it possible for organizations to improve security and safety without compromising usability or performance.

References

1. Patel, M., Patel, S., Ahmad, S., Blast Analysis of Efficient Honeycomb Sandwich Structures with CFRP/Steel FML Skins. *Int. J. Impact Eng.*, *178*, 104609, 2023. https://doi.org/10.1016/j.ijimpeng.2023.104609.
2. Langdon, G.S., Nurick, G.N., Yahya, M.Y., Cantwell, W.J., The response of honeycomb core sandwich panels, with aluminum and composite face sheets, to blast loading. *J. Sandwich Struct. Mater.*, 12, 733–754, 2010. https://doi.org/10.1177/1099636210368470.
3. Patel, M. and Patel, S., Blast mitigation analysis of novel designed sandwich structures using novel approaches. *Mech. Adv. Mater. Struct.*, Advance Online Publ., 2023. https://doi.org/10.1080/15376494.2023.2243614.
4. Chi, Y., Langdon, G.S., Nurick, G.N., The influence of core height and face plate thickness on the response of honeycomb sandwich panels subjected to blast loading. *Mater. Des.*, 31, 1887–1899, 2010. https://doi.org/10.1016/j.matdes.2009.10.058.
5. Patel, M. and Patel, S., Novel design of honeycomb hybrid sandwich structures under air-blast. *J. Sandwich Struct. Mater.*, 24, 2105–2123, 2022. https://doi.org/10.1177/10996362221127967.
6. Dharmasena, K.P., Wadley, H.N.G., Xue, Z., Hutchinson, J.W., Mechanical response of metallic honeycomb sandwich panel structures to high-intensity dynamic loading. *Int. J. Impact Eng.*, 35, 1063–1074, 2008. https://doi.org/10.1016/j.ijimpeng.2007.06.008.
7. Ul Haq, A., Gunashekar, G., Narala, S.K.R., The Dynamic Response of AuxHex and Star-Reentrant Honeycomb Cored Sandwich Panels Subject to Blast Loading. *Arabian J. Sci. Eng.*, 48, 11755–11771, 2023. https://doi.org/10.1007/s13369-022-07564-0.
8. Chai, G.B. and Zhu, S., A review of low-velocity impact on sandwich structures. *Proc. Inst. Mech. Eng., Part L: J. Mater.: Des. Appl.*, 225, 207–230, 2011. https://doi.org/10.1177/1464420711409985.
9. Palomba, G., Epasto, G., Crupi, V., Lightweight sandwich structures for marine applications: a review. *Mech. Adv. Mater. Struct.*, 29, 4839–4864, 2022. https://doi.org/10.1080/15376494.2021.1941448.
10. Zong, Z., Zhao, Y., Li, H., A numerical study of whole ship structural damage resulting from close-in underwater explosion shock. *Mar. Struct.*, 31, 24–43, 2013. https://doi.org/10.1016/j.marstruc.2013.01.004.

11. Cai, S., Liu, J., Zhang, P., Li, C., Cheng, Y., Experimental study on failure mechanisms of sandwich panels with multi-layered aluminum foam/ UHMWPE laminate core under combined blast and fragments loading. *Thin-Walled Struct.*, 159, 107227, 2021. https://doi.org/10.1016/j.tws.2020.107227.
12. Patel, M. and Patel, S., Design and analysis of lightweight stiffened honeycomb metallic sandwich panels under high-intensity air blast. *Proc. Inst. Mech. Eng., Part L: J. Mater.: Des. Appl.*, 238, 22–38, 2024. https://doi.org/10.1177/14644207231180811.
13. Patel, M. and Patel, S., Investigation of air-blast mitigation by single-V and multi-V shape face sheets used honeycomb sandwich panels. *Mater. Today Proc., Advance Online Publ.*, 2023. https://doi.org/10.1016/j.matpr.2023.06.432.
14. Gaur, B., Patel, M., Patel, S., Strain rate effect on CRALL under high-velocity impact by different projectiles. *J. Braz. Soc Mech. Sci. Eng., 45*, 103, 2023. https://doi.org/10.1007/s40430-023-04031-1.
15. Lan, X.K., Huang, Q., Zhou, T., Feng, S.S., Optimal design of a novel cylindrical sandwich panel with double arrow auxetic core under air blast loading. *Def. Technol.*, 16, 617–626, 2020. https://doi.org/10.1016/j.dt.2019.09.010.
16. Patel, M. and Patel, S., Numerical analysis on hexagonal honeycomb sandwich structure under air-blast loading, *Recent Advances in Machines and Mechanisms: Select Proceedings of the iNaCoMM 2021*, Springer Nature, Singapore, pp. 725–734, 2022, https://doi.org/10.1007/978-981-19-3716-3_59.
17. Patel, S. and Patel, M., The efficient design of hybrid and metallic sandwich structures under air blast loading. *J. Sandwich Struct. Mater.*, 24, 1706–1725, 2022. https://doi.org/10.1177/10996362211065748.
18. Grujicic, M., Yavari, R., Snipes, J.S., Ramaswami, S., Use of aluminum foam core sandwich structures to improve the blast-mitigation performance of light tactical vehicle side-vent-channel solution. *Proc. Inst. Mech. Eng., Part L: J. Mater.: Des. Appl.*, 232, 993–1011, 2018. https://doi.org/10.1177/1464420716659094.
19. Novak, N., Vesenjak, M., Duarte, I., Tanaka, S., Hokamoto, K., Krstulovic-Opara, L., Guo, B., Chen, P., Ren, Z., Compressive Behaviour of Closed-Cell Aluminium Foam at Different Strain Rates. *Mater. (Basel)*, *12*, 24, 4108, 2019.
20. Patel, M. and Patel, S., Influence of honeycomb core height on the blast mitigation performance of sandwich panel. *Mater. Today Proc.*, 74, 611–620, 2023. https://doi.org/10.1016/j.matpr.2022.09.530.
21. Patel, M. and Patel, S., A novel design of blast-proof sandwich structure with hybrid skin. *Mater. Today Proc.*, 87, 221-227, 2023. https://doi.org/10.1016/j.matpr.2023.05.191.
22. Gargano, A., Donough, M., Das, R., Mouritz, A.P., Damage to fibre-polymer laminates caused by surface contact explosive charges. *Composites, Part B*, *197*, 108162, 2020. https://doi.org/10.1016/j.compositesb.2020.108162.

23. Jin, X., Wang, Z., Ning, J., Xiao, G., Liu, E., Shu, X., Dynamic response of sandwich structures with graded auxetic honeycomb cores under blast loading. *Composites, Part B*, 106, 206–217, 2016. https://doi.org/10.1016/j.compositesb.2016.09.037.
24. Patel, M., and Patel, S., Effect of honeycomb cell size on the air-blast performance of sandwich panels. *Mater. Today Proc.*, 78, 792–797, 2023. https://doi.org/10.1016/j.matpr.2022.10.283.
25. Patel, M., Patel, S., Ahmad, S., Soares, C.G., Numerical Investigation of Air-Blast Performance of Cross-Filled Honeycomb Sandwich Panel. *Proceedings of the ASME 2023 42nd International Conference on Ocean, Offshore and Arctic Engineering. Volume 2: Structures, Safety, and Reliability, V002T02A005*, 2023. https://doi.org/10.1115/OMAE2023-104526.
26. Patel, M. and Patel, S., Dynamic response optimization of the multistage sandwich structures imperiled to explosive loading. *Proc. Inst. Mech. Eng., Part L: J. Mater.: Des. Appl.*, Advance Online Publ., 2024. https://doi.org/10.1177/14644207241238220.
27. Patel, M., and Patel, S., Impact of Stand-off Distance on the Dynamic Response of Honeycomb Sandwich Panel in Air Blast. Mechanika 2023. *Proceedings of the 27th International Scientific Conference*, 83–87, 2023. https://doi.org/10.5755/e01.2783-5677.2023.

11

Implementation of Industry-Based Perspective in Technical Education: A Qualitative and Quantitative Analysis

Ajit Kumar Singh[1]*, Sachin Akoji Meshram[2], Vinay Khandelwal[3,4], Pinaz Tiwari[3] and Pankaj Singh[3]

[1]*Jaipuria Institute of Management, Jaipur, India*
[2]*G.H. Raisoni University, Amravati, Maharashtra, India*
[3]*Institute of Business Management, GLA University, Mathura, India*
[4]*Stirling Management School, University of Stirling, Stirling, United Kingdom*

Abstract
Technical education is important from the insights of the development of the society and has witnessed a surge in popularity among Indian students after the completion of their intermediate exams. This surge has led a proliferation among the institutions toward offering technical education across the country, considering many insights and perceptions from students, parents, and industries. It is highlighted that, despite having a collective intake capacity of 1,753,436 students across 3,391 institutions, only 846,497 students have enrolled in the year 2023. To address this issue, this chapter aims to bridge the gap between technical education and industry demands by conducting a comprehensive analysis, combining both qualitative and quantitative methods, to be implemented considering an industry-centric perspective under technical education. The chapter presents the industrial requirements from technical institutes and the same are collected and analysed in the present chapter by interviewing HR professionals. Here, a questionnaire survey is conducted to rank these requirements. It is highlighted that the qualitative and quantitative means are utilized that assisted in collecting industry requirements. The results revealed that problem-solving skills are the most important requirement for recruiters, whereas a sports background is the least important. The chapter demonstrated that recruiters place

**Corresponding author*: ajitkumar2101@yahoo.com

Atul Kumar Sahu, Rakesh D. Raut, Rohit Raja, Anoop Kumar Sahu and Nitin Kumar Sahu (eds.) Industrial and Manufacturing Designs: Quantitative and Qualitative Analysis, (321–346) © 2024 Scrivener Publishing LLC

great importance on problem-solving skills when hiring technical graduates. Accordingly, it is highlighted to the technical institutions to focus on developing their students' problem-solving abilities considering industrial perspective. The general key finding of the chapter highlighted the need to technical institutions to offer programs and activities that allow students to apply their technical knowledge in practical and relevant ways to solve real-world problems. Institutions should also focus on providing opportunities to students to participate in internships and industry collaborations, which will give them practical experience and an understanding of industrial needs. The present chapter will be useful for researchers, academics, and practitioners in designing technical education curricula that align with industrial needs. This chapter highlights the crucial theme of implementing an industry-based perspective in technical education and emphasizing the need of both qualitative and quantitative analysis. By incorporating qualitative measures like curriculum relevance and quantitative metrics like job placement rates, the gap between academia and industry can be filled, which ensures the responsive of technical education to real-world needs. Additionally, the chapter presented many qualitative and quantitative parameters that can be used to develop general framework by the technical education apart from industry–institute linkages.

***Keywords*:** Quality management, recruiter's requirement, qualitative analysis, quantitative analysis, industrial perspective, decision-making, technical education

11.1 Introduction

In the dynamic landscape of modern education, the integration of industry-based perspectives into technical education has emerged as a critical pathway toward ensuring the relevance and quality of academic programs. As industries are rapidly evolving, the workforce is needed to be equipped with up-to-date skills and practical knowledge as per the perspective of the industries, which has become more pronounced than ever. The present chapter encapsulates the essence of a comprehensive investigation into the symbiotic relationship between academia and industry, with a focus on enhancing the educational experience and employability of students. Technical education are nowadays encompassing fields such as engineering, technology, and applied sciences, which plays an indispensable role in shaping future professionals who drive innovation, contribute to economic growth, and address complex societal challenges. However, the rapid pace of technological advancements has created a disjunction between traditional curricula and the rapidly changing demands of the job market. This disjunction has led educators, policymakers, and stakeholders to recognize the imperative practices for bridging the gap between academia and industry to better prepare graduates for the multifaceted demands of the real world. The concept of industry–academia collaboration is not novel, but its execution and outcomes have garnered renewed attention due to the evolving dynamics of global industries. This chapter thus delves into both the qualitative and quantitative dimensions to evaluate the concept of industry–academia collaboration and endeavors to shed light on the effectiveness, challenges, and potential improvements in incorporating industry perspectives into technical education.

In today's highly competitive and rapidly changing world, industries require employees who possess a range of skills beyond technical knowledge. Employers expect from technical education institutions to produce graduates who are not only technically competent but also have a range of soft skills, such as communication, problem-solving, teamwork, and leadership. Accordingly, the present chapter utilized qualitative and quantitative means to receive certain facts related with the linkages among industry and technical educational institutes.

The qualitative facet of the chapter will explore the experiences, perspectives, and insights of educators, industry professionals, and students who are engaged in industry-linked educational initiatives. With the utility of through in-depth interviews, surveys, and case studies, qualitative aspects can be defined to unravel the intricacies of how industry involvement enhances curriculum design, faculty development, and student engagement. Furthermore, qualitative analysis can be used to examine the alignment between the skills imparted through technical education and the skill desired by employers, and, thus, qualitative aspects is utilized to understand the extent up to which industry needs are being met by technical bodies.

Complementing the qualitative exploration, the quantitative component is also presented in the present chapter, which involves analyzing data related to student outcomes, employment rates, and performance metrics. The quantification of the impact of industry engagement on measures such as post-graduation placements, job satisfaction, and career trajectories can be done by quantitative assessments. Quantitative dimensions will provide empirical evidence to validate the efficacy of industrial integration in technical education. In essence, the present chapter embarks on a comprehensive journey to uncover the intricacies of integrating industry-based perspectives into technical education considering qualitative and quantitative aspects. By employing a dual approach of qualitative inquiry and quantitative analysis, the chapter aims to contribute valuable insights to educators, institutions, policymakers, and industries striving toward collaborating harmoniously in order to cultivate a skilled workforce that can navigate the complexities of a rapidly evolving professional landscape.

For opting a post in an Indian engineering education system after intermediate examinations, technical education has become a prevalent choice among students. Accordingly, technical education is playing an important role, and this inclination aligns the investment of time and resources that students make in supplementary tutoring, aiming to gain a competitive advantage in entrance exams for getting good technical institute. The surging demand for technical engineering education has prompted the establishment of numerous engineering colleges across the nation. According to data from the All-India Council of Technical Education, there are currently 3,391 sanctioned engineering colleges in India. Among these, 2,943 operate as unaided private institutions, 6 as privately aided, 19 as privately managed by universities, 61 as government-managed by universities, 61 as government-aided, 277 as government-funded, 10 as central universities, 12 as private-deemed universities, and 2 as government-deemed universities. The distribution of institutions across different categories is depicted in Figure 11.1. However, despite the significant expansion in the number of technical colleges and their intake capacity, there has been a decline in student enrolment. This trend raises concerns about the declining quality of engineering (technical) education. In response, this chapter has been conceived with the objective of enhancing the quality of engineering education based on industry perspective. In the context of quality education, the primary focus is to ensure industries' satisfaction with the educational framework adopted by the decision authorities for teaching-leaning. The hypothesis about high-quality education lies in framing right educational framework as per placement demand. Thus, one can assume that, if the students get placed, who are the customers of the educational institutions, then the high-quality education assumption will be right. Therefore, the core purpose of this chapter is to identify the key dimensions for the placement in engineering education. The chapter ascertains the requirements, which can be subsequently used to formulate strategies to enhance the overall quality of engineering education. The quality management approach is disseminated in this chapter, which the senior management

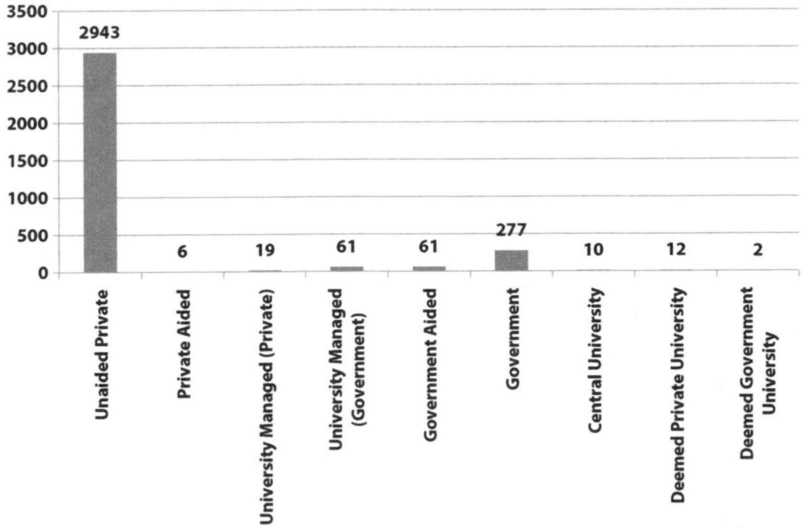

Figure 11.1 Graphical representation of the number of engineering institutions in different categories of institution.

can employ to effectively address the needs of industry. The same thereby will help in attracting a greater number of high-achieving individuals to pursue engineering education.

The present chapter hereby obliges its existence by presenting a directory of few critical qualitative or quantitative parameters that can be used for decision-making. The same can assist top management in preparing eminent policies and strategies for developing quality framework. The same will jointly give support in attracting more number of students toward educational institutes. This chapter will act as a crucial resource by providing an address list of essential qualitative and quantitative factors that can be used for decision-making purposes. These factors are intended to support top management in formulating effective policies and strategies for the development of a quality framework within educational institutions. The same will contribute toward the goal of attracting a greater number of students to these institutes, which thus will enhance their overall appeal and educational quality. The present chapter will address the following questions for engrossing sustainability:

1. How the academic institutions can productively amplify the involvement of diverse age groups onto higher education?

2. What are the momentous quality parameters, which can assist in the judgment toward framing educational policies for a particular alternative of educational institute as per industrial perspective?
3. What industries expect from learning institutes and how the sustainability can be embraced in said field?

The present chapter reciprocated toward above research questions and enlighten many qualitative and quantitative factors that can assist in this regard. The chapter will help academic institutions to foster inclusive practices in their policies by enduring flexible factors to accommodate diverse age groups' schedules. The chapter will help in decision-making, which will create a welcoming environment and demonstrate the unique perspectives that industries expects from learners bring. The dimensions presented here will help to meet the needs of all industrial perceptions that, in turn, can enhance their overall educational experience.

11.2 Literature Review

The integration of industry-based perspectives within technical education has emerged as a pivotal concern in the global academic landscape. As industries evolve rapidly, the need for graduates who possess relevant skills and practical knowledge aligned with real-world demands has garnered substantial attention. This section delves into existing research and erudition surrounding the implementation of industry-based perspectives in technical education, aiming to provide a comprehensive understanding of the subject's nuances, challenges, and potential benefits. The concept of collaboration between academia and industry is not new. Early studies presented by Etzkowitz and Leydesdorff [1] highlighted the importance of knowledge exchange and partnerships between academia and industry sectors and suggested that the symbiotic relationship among them can enhance the curriculum's relevance, which thereby can lead in aligning students with the industrial cultures and assist in ease adapting the rapidly changing demands of the professional world [1]. Researchers like Kift *et al.* [2] emphasized the necessity of blurring the boundaries between academia and industry to foster innovation, knowledge dissemination, and skill development.

It is found significant to conduct qualitative studies for exploring the boundaries of innovation and to receive critical disclosures. Qualitative studies delve to deliver critical understandings that are based on experiences and perceptions of stakeholders involved in industry-linked educational initiatives. Sackney and Mitchell [3] explored the impact of industry involvement on curriculum development toward emphasizing the importance of industrial input in creating programs that meet real-world requirements [3]. Qualitative data are geared toward answering questions like "What value was contributed more?", "Who played a crucial role?", and "When did events transpire?". The answers toward these questions are typically gathered with qualitative data through methods such as direct or participant observation, interviews, focus groups, case studies, and from written documents. Analyzing qualitative data involves processes like scrutinizing, comparing, contrasting, and interpreting patterns within the information. This analysis often involves identifying recurring themes, coding, grouping similar data, and distilling the data into meaningful and significant insights. Qualitative analysis is normally utilized to build grounded theory for building robust decisions in the situation of incomplete information [4]. Qualitative observations plays a crucial role in shedding light on behaviors, as well as the social context and underlying meanings, as they allow the evaluator to witness events directly. Observational techniques for qualitative analysis encompass activities such as closely monitoring participants, programs, videotaping interventions, and recording individuals' perceptions. Here, the participants are encouraged to verbalize their thoughts by engaging them in a task, which is a method commonly referred to as "thinking aloud" [5].

Interviews are a valuable tool in qualitative analysis for delving into intricate matters, which can be conducted individually or in group settings. They can take structured forms within controlled conditions or involve open-ended questions in a more relaxed format. It is often advisable to obtain permissions for recording interviews, as this can aid in the analysis of themes and content to be disclosed publicly. Fruitful information can be extracted from interviews based on qualitative analysis, and the same can help in recognizing a specific focus, such as exploring a detailed critical incident recalled by an individual. Focus groups also assist in qualitative analysis, on the other hand. Focus groups are led by a facilitator who guides discussions among a group of individuals who are chosen based on specific characteristics (e.g., being clients of the program under evaluation). The participants engaged in open-ended conversations in response to questions posed by the facilitator assist in collecting information for evaluation and building a robust theory for execution and management. The strength of focus group method lies in the dynamics of group discussions, which can generate ideas and trigger memories, with topics evolving as the discussion unfolds [6].

11.2.1 Critical Mapping

Qualitative data offer notable advantages, such as supplying contextual information that helps to elucidate intricate matters. They also serve supplementary aid to complement quantitative data by unraveling the underlying "why" and "how" aspects that accompany the "what". However, it is important to acknowledge the constraints associated with using qualitative data for evaluation purposes. Qualitative data collection will help in ease managing the resource-intensive and expensive process of data collection, as well as the intricate and time-consuming nature of data analysis and interpretation [4]. Additionally, studies by Brown and Duguid [7] emphasized on the "communities of practice" concept, which highlighted the importance of incorporating industry professionals as active participants in the educational process to enhance the practicality of learning [7]. A critical dimension of the integration of industry-based perspectives is curriculum design. Research by Smith [8] underscores the importance of aligning curricula with industry needs to ensure that graduates are equipped with skills that enhance employability [8]. Similarly, studies by Billett [9] demonstrated that the development of the technical framework based on practical educational settings in workplaces is needed to effectively organized and leveraged higher education system into new boundaries of innovation to help students to realize the desired outcomes of their courses as per industrial requirements. Today, it is required to prepare students for specific careers upon graduation. The present chapter delves into the question of suggesting a directory of measures that can assist in active engagement of the students. The dimensional factors are presented in this chapter to help in preparing decision-making based on the exploitation of best competencies required for success in student's professions after graduation. The chapter highlights the need of integration of qualitative and quantitative studies in higher education institutions and programs for moulding their alignment under three critical aspects for sustainability, i.e., coping with the demands of professional practices, maintaining the graduates' ability to secure employment in their chosen fields, and the smooth transition of the students into professional practices [9].

The development of industry-based perspectives is required to mould student engagement as per employability, which has also garnered researcher's attention. Research endeavors exemplified by Yorke *et al.* [10] delve deeply into the exploration of the vital factors for seamlessly incorporating employability skills within the framework of technical education. This integration is needed to serve a pivotal bridge for connecting the realms of classroom-based learning with practical, real-world applications [10].

A study by Kuh and Whitt [11] investigated the utilization of the concept of culture within the higher education literature to understand various aspects of faculty life, where culture is defined as the shared, mutually influential patterns of norms, values, practices, beliefs, and assumptions that assist in shape the behavior of individual faculty members and groups effectively and motivatedly. The development of collective cultural parameters is found significant to develop a framework for executing the events and actions related to the campus effectively and sustainably. The study explored the concept of development of subculture to faculty groups for indirect development of the students. Here, the studies on subcultures have been found influenced by two predominant perspectives, i.e., the notion of single, largely uniform professionals with more similarities should be adapted in academics in primary hierarchy and a complex amalgamation of sub-professional of diverse similarities should be there in secondary hierarchy for building robust decision-making. It is important to delve into the enclaves of subcultures parameters including disciplinary subspecialties, women faculty, ethnic practices, racial minority faculty, and part-time faculty to form distinct subcultures in secondary hierarchy to drive primary hierarchy with the purpose to receive concrete judgements and solutions [11]. Research conducted by McKeachie [12] delved into the assessment of the impact of industry-linked projects on students' skill acquisition within an educational context. This research shed light on a notable and constructive correlation that existed between the incorporation of industry-linked projects into the curriculum and the enhancement of students' learning outcomes [12]. Additionally, the studies conducted by Rugarcia *et al.* [13] provided valuable insights into the employment rates and career trajectories of graduates who had experienced industry-integrated education. These findings offered empirical evidence of the advantages of adopting refined educational approach based on industry interactions [13].

It is found significant to conduct qualitative and quantitative studies for reinforcing the abilities of educational sphere, where quantitative studies are needed to quantify the impact of industry engagement in technical education. Quantitative methods involve data that can be quantified and are useful for addressing questions such as "How many people were involved?", "What were the specific outcomes?", "Who participated?", and "What were the costs involved?". These data are in numerical value form and can be gathered through various means, including surveys, questionnaires, pre- and post-tests, and observations, by examining existing documents, databases, and clinical information. Quantitative surveys can be self-administered or conducted through interviews in person, over the phone, via mail, or online. The analysis of quantitative data encompasses a range of statistical techniques ranging from basic descriptive statistics to more complex analyses. Quantitative data serve to gauge the scope and scale of an implementation, such as the number of participants or program completions. Quantitative data can reveal its effects and overall impact. The strength of using quantitative data for evaluation lies in their potential for generating precise results (if the sample represents the larger population). With quantitative data, the relative analysis can be done with consistency, ease, and precisely, when collected reliably. However, limitations of these data may include low response rates in surveys, challenges in obtaining relevant documents, and difficulties in ensuring the validity of measurements. Furthermore, quantitative data may not provide a comprehensive understanding of the program's context and may lack the depth needed to explain complex issues or interactions [14]. Thus, one can conclude to collect qualitative and quantitative data for reinforcing the abilities of system and allied characteristics.

11.2.2 Summary and Concerns

While industry–academia collaboration holds immense promise, many challenges in the form of enrolment rates, better learning tools, technological drivers, adaptations, learning mechanism, students' expectations, locations of institutions, national recognitions, and so on persist. A study conducted by Miles and Loucky [15] have pointed out the potential conflicts and challenges from divergent fields to be integrated for linking academia and industry. The rapid pace of technological change and right utility of technology can integrate linkage between academia and industry for maintaining the relevance of curricula. It is found imperative to exploit future directions related with the exploration of effective models for sustained collaboration, identifying strategies to ensure faculty's industry relevance, and addressing concerns related to intellectual property and confidentiality [15].

In conclusion, the implementation of industry-based perspectives in technical education is multifaceted needed with profound implications for curriculum design, student engagement, and employability. The literature review underscores the evolving nature of industry–academia collaboration and highlights the need for a holistic approach that integrates qualitative insights with quantitative measurements. The syntheses of these approaches are needed to offer a comprehensive understanding of the benefits, challenges, and potential future directions in enhancing technical education through industry integration. The literature showcases numerous authors that addressed the enhancement of technical institutions' quality, but a notable absence of insight toward fulfilling recruiters' needs to facilitate greater employment prospects for graduates is found. Thus, this chapter seeks to bridge this gap by delving into the industrial requirements that can contribute to the quality improvement of technical institutions through various qualitative and quantitative factors.

Today, institutes are in the need of leveraging quality characteristics to ensure that their products and services have the abilities to congregate industrial expectations. Accordingly, researchers thrust to seek many developed criteria focusing on the characteristic like reliability, maintainability, presentation, expectation, and customer satisfaction for attaining success and survival. Researchers also thrust to employ many statistical tools like Six Sigma, factorial assessment, and graph analysis toward identifying areas for improvement and setting benchmarks for higher quality and robust solutions. The right identification of these practices and allied qualitative and quantitative measures is needed to provide valuable reference points for ongoing development and expansion strategies, which thereby can enhance overall business performance. Thus, in the present chapter, the authors have accordingly explored quality measures linked with industrial standpoint to be included in educational institutes to act as explanation for auxiliary developing and expanding the creature of higher education and related sphere. The authors in this chapter have delved into quality measures that are derived from an industrial perspective with the aim to integrate them into academic institutes. The chapter provides a potential solution for fostering the growth and improvement of higher education and its related fields. This alignment with industry standards and practices can enhance the relevance and effectiveness of academic programs.

11.3 Methodology

The primary objective of this chapter is to meticulously identify and delineate the precise requisites sought by recruiters within the realm of technical education. By gaining a profound understanding of these requirements, the overarching aims to empower the abilities of technical institutes to formulate and implement strategic measures that would not only meet but also surpass the industrial demands can be met. The endeavors present in this chapter aims to augment the overall standard of technical education. It is crucial to acknowledge that stakeholder's satisfaction serves as the fundamental yardstick, by which the quality of any product or service is gauged [16–18]. In the context of technical education, this principle holds true with recruiters being a vital category of stakeholders, who profoundly influence the quality of graduates produced by technical institutes. The methodology employed for this study underpinned chains of factors, measures, and metrics that can be used for evaluations to build a robust commitment toward ensuring the contentment of all relevant stakeholders associated with technical institutes, with a particular focus on the gratification of recruiters. A comprehensive compilation of criteria that recruiters anticipate from the students and a structured interview were utilized as a methodology to extract valuable insights directly from the perspectives of recruiters. A structured approach is utilized as a methodology, where the formulation of questionnaires is derived from the compiled criteria. The subsequent sections of this chapter will provide an exhaustive and meticulous exposition about data collection and analysis toward the formulation of actionable strategies and their subsequent implementation. By following the structured and systematic approach presented in this chapter, technical institutes can effectively address the specific needs of recruiters and satisfying their expectations, ultimately enhancing the overall calibre of technical education.

11.3.1 Identification of Customer of Technical Education

Within the realm of higher education, a multitude of research voices have collectively recognized the extensive and varied array of participants that constitute the intricate ecosystem of technical education. The participants that play pivotal roles in shaping the higher educational landscape span a diverse spectrum of individuals ranging from students, parents, recruiters, government bodies, faculty members, support staff, and the broader society at large. Each of these groups of stakeholders wields a distinctive influence and interest in the higher education system and their engagement is crucial in shaping the direction and ensuring its effectiveness. It is acknowledge that the presence of this broad spectrum of participants is essential for comprehending the complex web of relationships, expectations, and interactions that characterize the world of higher education in sustainable boundaries. The interplay among these diverse participants underscores the development of an intricate dynamics and multifaceted nature into higher education, where students seek knowledge and skills, parents invest for their children's future, recruiters demands a talented workforce for recruitment, government bodies regulate and fund these institutions, faculty members impart knowledge and research, support staff maintain operational efficiencies, and the broader society benefits from the outcomes and contributions of higher education.

In recognizing and addressing the multifarious interests and needs, the higher education system should strive for a more inclusive, holistic, and effective approach that ultimately have the capacity to enhance its overall quality and impact [19–21]. It is always crucial to account notable factors that can drive statistics of system [22]. Notable factors that are related with the industrial realm stand out as a pivotal entity in the pursuit of enhancing the quality of education [23, 24]. Thus, it is attributed to address industrial requirements and its fronts into the curriculum of higher education as, notably, employment rates are provided by industrial participants, which indirectly resonate with the satisfaction of the other stakeholder groups. The present chapter's prime focus is directed toward providing chains of factors, measures, and metrics that are related with recruiters in the decision-making framework of the educational streams as the recruiters are the central customers of the higher education system. The chapter recognizes the role of recruiters as critical and found it prominent in determining the employability and subsequent career trajectories of graduates by the institutes. By prioritizing recruiters as the primary customer, the chapter seeks to unravel the direct link between the fulfilment of industrial needs and the enhancement of employment prospects. The chapter advocated toward recognizing recruiters as the key customers to build a pivotal bridge toward shaping the quality of technical education, which thereby echoing the overarching principle that a satisfied customer base translates into the hallmark of quality in any educational institution.

11.3.2 Collecting Recruiter's Requirement

The primary objective of this chapter revolves around the enhancement of educational institute quality. In the pursuit of establishing a robust framework for superior technical education, a crucial prerequisite is a profound understanding of the needs and expectations of recruiters. Therefore, this chapter is dedicated to the meticulous collection and analysis of recruiter's requirements, where a comprehensive compilation of criteria that recruiters anticipate from the students, when hiring them as recent graduates (who have recently completed their studies), are presented. In order to achieve this goal, a structured interview was conducted to extract valuable insights directly from the perspectives of recruiters. The study engaged a total of 29 Human Resource (HR) representatives from various private industries, each contributing unique viewpoints. Through these interviews, a comprehensive compilation of recruiter's expectations was formulated, encapsulating a total of 28 distinct parameters that hold significance streams during the process of hiring final year or pass-out graduates.

The chapter recognizes the variability in nomenclature and phrasing to understand crucial and meticulous ideas to consolidate and harmonize the various parameters. The same will facilitate in the creation of a refined and comprehensive list of recruiter's requirements, which ultimately comprises 19 discernible parameters. To encapsulate the findings of the study, the refined list of recruiter's requirements is presented in Table 11.1. Table 11.1 offers a concise and insightful snapshot of the expectations expressed by recruiters during the recruitment of recent graduates. This aggregation of

Table 11.1 List of recruiter's requirement from a technical institute.

S. no.	Recruiter's requirements
1.	Core technical knowledge
2.	Skills to use technical knowledge to solve real-life problems
3.	Commercial awareness about the industry work cycle
4.	Long retention period
5.	Programming skills
6.	Knowledge of algorithms
7.	Versatility
8.	Desire to continue learning
9.	Sincerity
10.	Problem-solving skills
11.	Familiar background
12.	"Can do" and "Will do" attitude
13.	Truthfulness
14.	Sports background
15.	Confidence
16.	Good communication skills
17.	Divergent thinking
18.	Ethics and morality
19.	Soft behavior

criteria not only illuminates the diverse and multifaceted needs of recruiters but also lays the foundation for aligning educational institutes more effectively with the expectations of the job market.

11.3.3 Ratings Collection Through Questionnaire Survey

The compilation of industrial requirements is presented in Table 11.1. Utilizing these requirements, the subsequent phase of the methodology is utilized on gauging the significance levels to attribute these collected industrial requirements. To accomplish this, a structured approach is utilized, and the formulation of questionnaires is derived from the said industrial parameters. The data related to the same are acquired by conducting the semi-structured interviews. The questionnaire encompasses a comprehensive set of 19 inquiries, each tailored to capture distinct facets of industrial expectations. Upon its meticulous formulation, the questionnaires was transitioned into a quantitative digital format and distributed online. Precisely, it was dispatched to a cohort of 315 HR departments spanning various categories of private enterprises. To ensure a well-rounded perspective, an additional set of 20 HR departments were also approached to participate in the questionnaires offline toward fostering a balanced and comprehensive representation. The utilization of both online and offline modes of questionnaire dissemination aimed to encompass a diverse cross-section of organizations and to contribute toward a more holistic assessment of the industrial needs and the corresponding levels of importance attributed to them. The utilized multi-faceted approach underscores the meticulous methodology and assisted in capturing and quantifying the weightage of each requirement articulated by the industrial stakeholders.

Upon widespread distribution of the questionnaires, a total of 76 online responses from industry professionals and 11 offline responses were successfully garnered. After a rigorous review for any potential gaps in data, six responses were deemed incomplete and consequently excluded from the dataset. Ultimately, a corpus of 81 valid responses emerged as the foundation of the analysis. The effectiveness of the questionnaire dissemination in diverse industries is illuminated through detailed statistics provided in Table 11.2. The insights received from the response rate, quantity of questionnaires dispatched, and the number of verified responses obtained from various sectors within the industry landscape assist in preparing quality facts for planning industrial linked academic policies.

Afterward, the ratings are collected, and the collected importance ratings are evaluated with statistical calculations utilizing SPSS 20 software. This analytical phase aimed to unravel pivotal insights through the examination

Table 11.2 Response rate statistics of industries.

S. no.	Types of industries	Questionnaire distributed	Valid response received	Response rate (%)
1.	Software	140	33	23.57
2.	Core	108	24	22.22
3.	Consultancies	65	16	24.61
4.	Others	22	8	36.36
Total		**335**	**81**	**24.17**

of mean values, standard deviations, and the aggregate count of responses for each of the industrial requirements under scrutiny. The outcomes of this analysis are comprehensively presented in Table 11.3, representing the calculated values of mean, standard deviation, and total response count for each individual industrial demand. These quantitative findings pave the way for a more nuanced understanding of the perceived significance attached to different requirements across diverse industries. The chapter highlighted the refined list of recruiter's requirements in the form of 19 parameters, which are evaluated on the basis of questionnaires.

11.4 Results and Findings

The responses garnered from recruiters assisted in identifying the important ratings that are attributed to industrial expectations and underwent rigorous statistical analysis. The received ratings are represented in a bar chart, as depicted in Figure 11.2. A comprehensive examination of Figure 11.2 highlights the distinct variations in the perceived importance of different parameters, which thereby offering insights into the preferences and priorities of recruiters. The chart effectively communicates "problem-solving skill" as the most highly rated parameter, which boasted an impressive rating index of 4.74. On the other end of the spectrum, "sports background" secured the lowest rating index of 2.63. This range of ratings indicates the degrees of priority that recruiters attach to each parameter in the context of hiring recent graduates.

Upon a thoughtful interpretation of these findings, a critical inference can be drawn for framing institute policies as per these findings. The paramount importance awarded to "problem-solving skill" underscores

Table 11.3 Importance ratings of recruiter's requirements.

S. no.	Recruiter's requirements	Mean	Std. deviation
1.	Core technical knowledge	3.77	1.101
2.	Skills to use technical knowledge to solve real-life problem	4.25	0.662
3.	Commercial awareness about the industry work cycle	3.25	0.662
4.	Long retention period	4.60	0.524
5.	Programming skills	4.11	0.616
6.	Knowledge of algorithms	3.65	1.007
7.	Versatility	4.14	0.609
8.	Desire to continue learning	3.89	0.793
9.	Sincerity	4.62	0.700
10.	Problem-solving skills	4.74	0.443
11.	Familiar background	4.26	0.443
12.	"Can do" and "Will do" attitude	4.51	0.504
13.	Truthfulness	4.62	0.490
14.	Sports background	2.63	0.486
15.	Confidence	4.25	0.434
16.	Good communication skills	4.35	0.513
17.	Divergent thinking	3.62	0.490
18.	Ethics and morality	4.60	0.524
19.	Soft behavior	3.74	1.203

its central role in meeting industrial expectations. This corroborates the notion that students should actively engage in real-life problem-solving scenarios during their project work. Such practical experiences provide a fertile ground for the development and refinement of the essential skills for effectively aligning graduates with the requirements of recruiters. The study underscores the pivotal role of project-based learning and

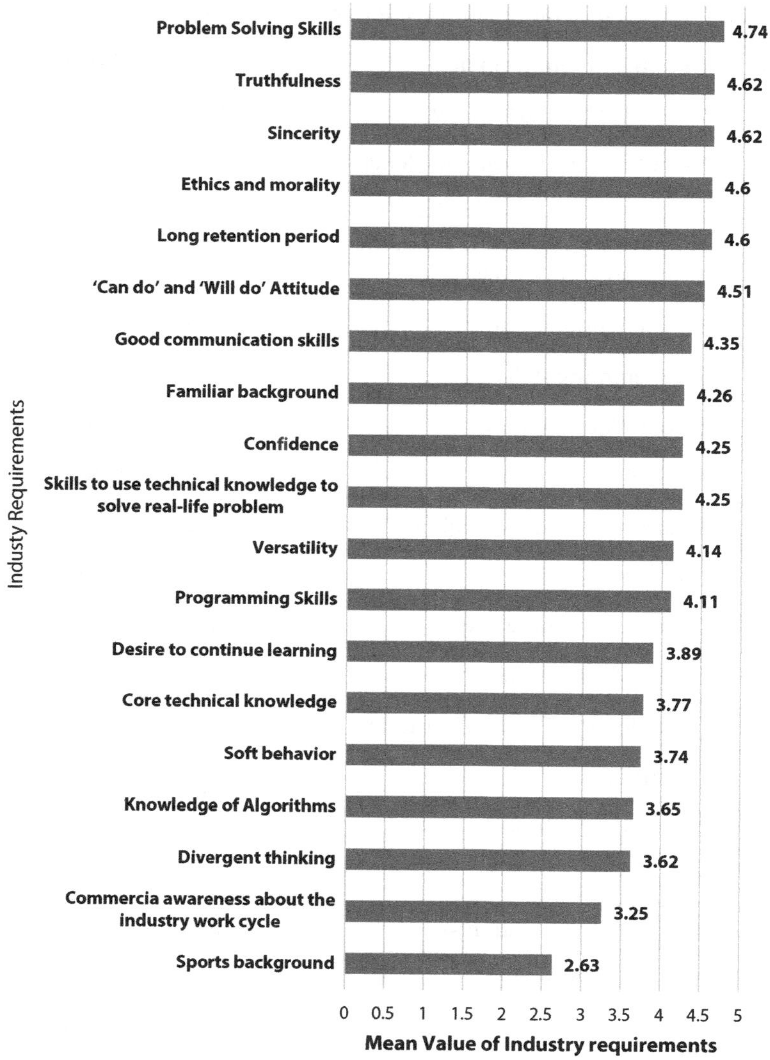

Figure 11.2 Ranking of recruiter's requirement.

advocating its integration within the educational curriculum. By engaging students in real-world challenges, educational institutions can foster the growth of vital skills, thereby enhancing students' suitability in the eyes of recruiters. This strategic alignment not only enriches the academic experience but also bolsters graduates' employability prospects, ensuring their preparedness for the multifaceted demands of the professional landscape.

The finding of the study also indicated sports background as the least important requirement for recruiters. This does not mean that sports activities should be eliminated altogether. Still, they should be viewed as a supplementary and optional component of technical education programs.

11.5 Importance of Qualitative and Quantitative Analysis

It is imperative to evaluate educational practices by the concerned associations based on industrial apprehension for receiving success, revenues, and sustainable dimensions. In today's business landscape, it is crucial to assess and scrutinize number of qualitative or quantitative dimensions related to teaching methods, educational practices, strategies, actions, and ventures for receiving success, revenues, and sustainable dimensions. The evaluation process should be spearheaded by the relevant educational associations or bodies, where it is needed to check various dimensions holding by the educational institutes or not for receiving sustainability. The questions of defining paramount dimensions that can support in assessment and evaluation are rooted in the boundaries of qualitative and quantitative factors selected by the concerns for decision-making. The genuine concerns of the many participants, particularly in multiple terms related with the admission toll for sustainability can be fruitfully managed by the selection and identification of numbers of qualitative and quantitative factors in decision-making. Student perspectives can be enrolled in decision-making framework of the education institutes by the exploitation of qualitative and quantitative measures. The same will lead in embracing sustainability by way of developing robust educational practices to be accepted by the students, parents, industries, and society. One can enhance the relevancy and effectiveness of the educational practices and its adoption by the educational institutes by ensuring the integration and attendance of contemporary global challenges, aspirations of the students, industries, and empowerment of working principal utilizing by the decision authorities, but the same depends on how well the qualitative and quantitative characteristics are collected, identified, and integrated in the decision-making framework. It is advocated that the collection of qualitative and quantitative medium should be done by considering industrial and students' aspirations as imperative to built robust decision. It is always advisable to conduct studies to recognize decisive quality characteristics based on student's desires using qualitative or quantitative or qualitative and quantitative medium for engrossing sustainability [25–27]. The chapter presented many qualitative and quantitative parameters that can be used to develop general framework apart from industry–institute linkage framework. The said parameters are listed in Table 11.4.

Table 11.4 Qualitative and quantitative parameters for modeling.

S. no.	Factors for improvement and decision framework
1	Question paper patterns
2	Assessment medium
3	Placement characteristics
4	Professional traits
5	Criterion of teacher
6	Criterion of clerical staff
7	Closeness of institute to market
8	Personal features of teacher
9	Management policies
10	Technical image of institute under regional domain
11	Fee structure
12	Students exchange program
13	Institutional affiliation infrastructure
14	Classroom management
15	Students and staff control methods
16	Rules and ordinance
17	Teaching skills of teacher
18	Internal facilities
19	Management information system

11.6 Discussions and Implications

Industrial expectations toward learning institutes will be fruitfully managed by the incorporation of qualitative and quantitative measures presented in this chapter, which will help in providing quality education by way of devising polices relevant to real-world needs. It is advocated that sustainability in technical institutes can be embraced through integration of issues, diverse perspectives, student concerns, and fostering a culture of responsible resource to promote sustainable structures for all stakeholders, which rely on various factors, measures, and practices. The same are presented and discussed in this chapter. Here, 19 qualitative and quantitative factors as per industrial perspective and 19 factors to develop a general framework apart from industry–institute linkages are discussed and presented for incorporation in decision-making by the institute authorities. Today, institutes, association, partners, and so on need a number of vital tools for decision-making and for their sustainable progression. For sustainable progression, business institutes seek tools such as data analytics and decisive factors to make informed decisions. Additionally, technologies seek for automation and innovation to build a strong leadership to drive a culture of continuous improvement. Moreover, environmental and social responsibility initiatives are also needed to be essential in every field for a long-term success and to address issues like carbon footprint reduction and ethical business practices. The same can be effectively handled by the exploitation of qualitative and quantitative characteristics and their integration with environmental and social responsibility dimensions for sustainable initiatives. Here, accordingly, the qualitative and quantitative characteristics are presented for reinforcing decision-making capabilities of higher educational domain. The chapter underlines the implementation of industry-based perspective in technical education based on qualitative and quantitative analysis. Additionally, the noteworthy quality factors that can sway a student's choice for selecting an educational institution in their local area are presented, which, in turn, will build an academic excellence and install students' confidence in the institution.

11.7 Significant Aspects and Decision-Making

Currently, amplified competition in the marketplace is required for sinking profit boundaries, and the same is necessary for describing the success of company. Here, factorial evaluation can assist, and, accordingly, it is noteworthy to identify important qualitative or quantitative factors for influencing the operational personality of a business venture. Increased competition in today's scenario is eroding profit margins up to elevated scales, and, thus, implementing decisive strategies is prominent to define a company's success. Thus, it is advocated to exploit deterministic and probabilistic techniques to evaluate the identified qualitative and quantitative factors toward shaping the right decision for handling business' sphere and its operational identity. Reorganization of the key qualitative or quantitative dimensions is crucial for adapting and thriving in a highly competitive business environment. The same will help in deciding technological improvements and developing a thrive environment to be employed by the decision-making authorities for apparent survival and success.

This chapter thus serves this purpose by offering a comprehensive directory of crucial qualitative and quantitative factors to facilitate in data-driven decision-making process. The implementations of these factors are important for top management for crafting effective policies and strategies to enhance the quality framework of educational institutions. Furthermore, the insights derived from these parameters will assist in playing a pivotal role in attracting a greater number of students, fostering growth, and ensuring academic excellence in these institutes under the shadow of industrial perspectives. In this chapter, the authors have meticulously examined quality measures from an industrial perspective and proposing their integration into educational institutions. This integration serves as an elucidation for the advancement and expansion of higher education and its related fields. By aligning with industrial standards and practices, educational institutions can enhance their relevance, efficiency, and competitiveness, which ultimately can contribute toward the growth and evolution of the higher education sector. The authors underscore the importance of this alignment as a transformative step in the pursuit of academic excellence and development in the broader educational sphere. This chapter highlights the crucial theme of implementing an industry-based perspective in technical education and emphasizing the need for both qualitative and quantitative analysis.

11.8 Conclusions

This paper adopts a deliberated approach based on the quantification of the 19 criteria for giving precedence to the requirements outlined by recruiters within the context of a technical institution. The prioritization of the scrutinized criteria is accomplished through a two-fold methodology consisting of interviews and a questionnaire survey. The outcomes of the comprehensive study presented in this chapter distinctly highlight that the attribute of "inculcating problem-solving skill" is the foremost parameter from the insights of industrial professional. The same indicated the highest rating index in terms of importance. This paramount finding profoundly indicates the direction of recommendations for adoption of the technical institutions. The chapter, in conclusion, underscores the critical necessity of integrating problem-solving skills within the educational paradigm of technical institutions. It is highlighted that the reorganization of the aspects related with skill development in the curriculum is needed and plays a critical role in enhancing the overall quality of technical education and ensuring students' employability. The chapter presented an approach that involves students in grappling with real-life problems encompassing various dimensions related with industrial and social challenges. The transformation of educational experience into a dynamic platform can be done based on the consideration of industry-related criteria presented in this chapter for lifelong learning.

This chapter presents a qualitative andquantitative evaluation to cultivate a mindset that will equipped to tackle a diverse array of parameters and to reproduce graduates for the socio-technical transformations for the future. The implications of this study resonate in attaining linkage among industry and academic sphere. It extends its influence to the realms of top-level management, policy formulation, and administrative decision-making within engineering institutions. The insights presented in the chapter will help key stakeholders to strategically develop and implement initiatives that holistically address the expectations and needs of industries, which, consequently, will lead to the gradual but impactful enhancement of the quality of engineering education and aligning it with the ever-evolving demands of the professional landscape. It is advocated that, by fostering a learning environment focused on real-world problem-solving, technical institutions can contribute to the holistic development of individuals and society at large. This chapter stands as a guiding compass for stakeholders to make informed decisions that ultimately uplift the standards of technical education while nurturing individual's capabilities toward contributing positively to complex socio-technical transformations.

Many quantitative and quantitative parameters, i.e., pattern of question papers, evaluation medium, placement characteristics, professional status, qualification of teacher, qualification of clerical staff, closeness of institute to market, personal characteristics of teacher, management policies, technical image of institute under regional domain, fee structure, students exchange program, institutional affiliation infrastructure, classroom management, control methods, rules and ordinance, teaching skills of teacher, internal facilities, institute ethics, and management information system, are presented in this chapter that can be used to develop general framework apart from industry–institute linkages. The chapter additionally presented many qualitative and quantitative factors that can be utilized by the technical institutions for developing their statistics and sustainability.

References

1. Etzkowitz, H. and Leydesdorff, L., The Triple Helix–University-industry-government relations: A laboratory for knowledge-based economic development. *EASST Rev.*, *14*, 1, 14–19, 1995.
2. Kift, S., Nelson, K., Clarke, J., Transition pedagogy: A third-generation approach to FYE-A case study of policy and practice for the higher education sector. *Student Success*, 1, 1, 1–20, 2010.
3. Sackney, L. and Mitchell, C., Postmodern expressions of educational leadership, in: *Second International Handbook of Educational Leadership and Administration*, pp. 881–913, Springer, 2002.
4. Patton, M.Q., Two decades of developments in qualitative inquiry: A personal, experiential perspective. *Qual. Soc. Work.*, 1, 3, 261–283, 2002.
5. Matsuo, M., Wong, C.W.Y., Lai, K., Experience-based learning of Japanese IT professionals: A qualitative research. *J. Strategic Inf. Syst.*, 17, 3, 202–213, 2008.
6. Redmond, R.A. and Curtis, E.A., Focus groups: Principles and process. *Nurse Res.*, 16, 3, 57–69, 2009.
7. Brown, J.S. and Duguid, P., Organizational learning and communities-of-practice: Toward a unified view of working, learning, and innovation. *Organ. Sci.*, 2, 1, 40–57, 1991.
8. Smith, K., *What is the 'knowledge economy'? Knowledge-intensive industries and distributed knowledge bases*, AEGIS, University of Western Sydney, Sydney, Australia, 2000.
9. Billett, S., Learning through work: Workplace affordances and individual engagement. *J. Workplace Learn.*, 13, 5, 209–214, 2001.
10. Yorke, M. *et al.*, *The experiences of part-time students in higher education: A study from the UK*, Higher Education Academy, 2008.
11. Kuh, G.D. and Whitt, E.J., Using the Cultural Lens to Understand Faculty Behavior, Information Analyses, Conference Papers, 1988. https://scholar.google.com/scholar?hl=en&as_sdt=0%2C5&q=Kuh%2C+G.D.+and+Whittt%2C+E.J.%2C+Using+the+Cultural+Lens+to+Understand+Faculty+Behavior%2C+1988.&btnG=
12. McKeachie, W.J., Research on college teaching: The historical background. *J. Educ. Psychol.*, 82, 2, 189, 1990.

13. Rugarcia, A., Felder, R.M., Woods, D.R., Stice, J.E., The future of engineering education: Part 1. A vision for a new century. *Chem. Eng. Educ.*, 34, 1, 16–25, 2000.
14. Garbarino, S. and Holland, J., Quantitative and qualitative methods in impact evaluation and measuring results, GSDRC Emerging Issues Research Service, 2009. https://www.betterevaluation.org/sites/default/files/EIRS4.pdf
15. Miles, J.C. and Loucky, J., Parks, protected areas, and environmental education in North America, in: *Protected Areas and the Regional Planning Imperative in North America: Integrating Nature Conservation and Sustainable Development*, pp. 261–276, 2003.
16. Sahu, A.K., Sharma, M., Raut, R.D., Sahu, A.K., Sahu, N.K., Antony, J., Tortorella, G.L., Decision-making framework for supplier selection using an integrated MCDM approach in a lean-agile-resilient-green environment: evidence from Indian automotive sector. *TQM J.*, 35, 4, 964–1006, 2023.
17. Sahu, A.K., Raut, R.D., Gedam, V.V., Cheikhrouhou, N., Sahu, A.K., Lean–agile–resilience–green practices adoption challenges in sustainable agri-food supply chains. *Bus. Strategy Environ.*, 32, 6, 1–20, 2022. https://doi.org/10.1002/bse.3299.
18. Kang, D., Prabhu, M., Ahmed, R.R., Zhang, Z., Sahu, A.K., Digital-IIoTs spheres approach toward public development: an exploiting fuzzy-grey mathematical modeling of IIoTs spheres. *Grey Syst.*, 12, 2, 389–416, 2022.
19. Owlia, M.S. and Aspinwall, E.M., Application of Quality Function Deployment for the Improvement of Quality in an Engineering Department. *Eur. J. Eng. Educ.*, 23, 1, 105–115, 1998.
20. Singh, A. and Rawani, A., Application of QFD in education sector: A review. *Int. J. Mech. Eng. Technol.*, 9, 3, 592–599, 2018.
21. Singh, A.K. and Rawani, A., Application of quality function deployment for the prioritization of National Board of Accreditation quality parameters. *Qual. Assur. Educ.*, 27, 1, 127–139, 2019.
22. Bag, S., Sahu, A.K., Kilbourn, P., Pisa, N., Dhamija, P., Sahu, A.K., Modeling barriers of digital manufacturing in a circular economy for enhancing sustainability. *Int. J. Product. Perform. Manage.*, 71, 3, 833–869, 2021. https://doi.org/10.1108/IJPPM-12-2020-0637.
23. Hill, F.M., Managing service quality in higher education: The role of the student as the primary consumer. *Qual. Assur. Educ.*, 3, 3, 10–21, 1995.
24. Singh, A.K. and Rawani, A.M., Industry-oriented quality management of engineering education: An integrated QFD-TOPSIS approach. *Int. J. Product. Perform. Manage.*, 13, 2, 904–922, 2022.
25. Bag, S., Viktorovich, D.A., Sahu, A.K., Sahu, A.K., Barriers to adoption of blockchain technology in green supply chain management. *J. Glob. Oper. Strateg. Sourc.*, 14, 1, 104–133, 2021.
26. Sahu, A.K., Sahu, N.K., Sahu, A.K., Laminating STRATH block chain technology-SWOT architectures to endure business strategy between

digital transformation, firms, and supply chains capabilities for sustainability. *J. Cleaner Prod.*, 383, 135531, 2023.

27. Ahmed, K.A., Sahu, A.K., Sahu, A.K., Sahu, N.K., Quantify the Behavior Intention of Individuals to Control SC Performance by Exploring Cloud Storage Services: An Extended UTAUT2 Approach. *Int. J. Technol. Hum. Interact.*, 18, 7, 1–28, 2022. http://doi.org/10.4018/IJTHI.306227.

12

Demonstrating Sustainable Design Under Qualitative and Quantitative Features for Secure Fundraising in Supply Chains Based on Blockchain Technology

Pankaj Chandra*, Santosh Soni, Kunal Raj, Prayas Kumar and Koushal Malviya

Department of Information Technology, School of Studies in Engineering and Technology, Guru Ghasidas Vishwavidyalaya, Bilaspur, Chhattisgarh, India

Abstract

The main purpose of this chapter is to propose a sustainable design to bring a revolutionizing change for secure fundraising in supply chains based on blockchain technology. The purpose of this chapter is to propose a decentralized fundraising platform that utilizes blockchain technology to increase transparency and trust toward organization related with transactional activities, charity, and crowdfunding. The chapter primarily utilizes qualitative and quantitative features to build a secure fundraising platform based on blockchain technology. Here, Web3-based authentication, Next.js, smart contract deployment, bytecode, application binary interface, agile methodology, and code compilation; the role of data-holding places, HD wallet provider, and related working model; as well as its design and implementation are presented for enhancing the performance and user experience with the proposed design. In the present chapter, the importance and theoretical utility of each of the aforesaid qualitative features are discussed precisely to endure the applicability of the developed fundraising platform. The proposed platform presented in this chapter is supported by a Web3-based authentication for managing supply chain system, which is implemented using blockchain technology. Here, smart contracts are used to handle fund transactions and resource allocation without the need for a central authority. The Ethereum blockchain network serves as the source of truth for each transaction's details to prevent any financial

**Corresponding author*: pankaj2684@gmail.com

Atul Kumar Sahu, Rakesh D. Raut, Rohit Raja, Anoop Kumar Sahu and Nitin Kumar Sahu (eds.) Industrial and Manufacturing Designs: Quantitative and Qualitative Analysis, (347–374) © 2024 Scrivener Publishing LLC

discrepancies. The proposed platform is designed by the authors with the aim to increase public trust in charity and crowdfunding organizations by ensuring proper resource allocation and preventing any misuse of funds by entrepreneurs. The present chapter can be useful to many entrepreneurs and practitioners, who are often depositing funds directly into the account of the resource provider. The platform is loaded with the capability to ensure the right allocation of funds properly to the concerns. The proposed platform can be utilized by the supply chain management practitioners for the allocation of the right resources to the entrepreneurs, where smart contracts will prevent hackers from altering transactions. The chapter additionally presented the importance of blockchain technology in supply chain management and importance of agility in supply chain management.

Keywords: Ethereum, Web3 authentication, smart contracts, Goerli test network, Infura, crowd-funding, blockchain, supply chain

12.1 Introduction

Blockchain is a technology that could have a huge impact on the charity sector, finance, banking, healthcare, and philanthropy and can help in the management and distribution of funds securely and transparently. Business enterprises and governments sectors can use blockchain innovations in wide-ranging cases and activities. Blockchain technology can be increasingly applied to various fields, including charity and fundraising systems. Recent studies have shown the potential of blockchain to be utilized in the charity systems and in tracking donation systems [1–3]. Crowdfunding charity platforms can be effectively managed by the application of blockchain [4]. The popularity of charity-based crowdfunding platforms has been increasing in recent years [5], which thereby highlighted the importance of exploring and improving the donation recurrence and donor retention in crowdfunding. Crowdfunding has become an increasingly popular way for entrepreneurs, artists, and philanthropists to raise funds for their projects. With the rise of crowdfunding, there has been a growing interest in understanding the factors that contribute to the success or failure of these campaigns. Several studies have explored various aspects of crowdfunding and the importance of strategic purchasing considerations in reward-based crowdfunding [6], which demands the need of implication of numerous technological aspects, criteria, tools, and techniques for handling crowdfunding platforms [7, 8]. Moreover, with the increasing popularity of charity-based crowdfunding platforms, there is a growing need to develop secure and transparent blockchain-based charity systems [9]. However, trust in charities has been declining, which can hinder the growth of charity-based crowdfunding [10]. In this context, it is essential to develop innovative solutions that can enhance transparency, security, and trust in charity-based crowdfunding platforms.

Crowdfunding has become an increasingly popular method for raising funds for various purposes, ranging from creative projects to charitable causes. Many studies have been conducted to investigate the qualitative and quantitative factors that influence the success of crowdfunding campaigns. Some of these studies have focused on the impact of qualitative factors such as campaign characteristics, social network effects, and platform design. Others have explored the role of trust, honesty, and communication in the success of crowdfunding campaigns. For instance, a systematic analysis of empirical studies is required to be identified to uncover a range of qualitative and quantitative factors that affect crowdfunding success [11, 12]. In addition, there is a need to design guidelines for successful crowdfunding that can help campaign creators to increase their chances of success [13]. Many research studies have focused to examine the driving elements of a successful crowdfunding campaign, which include factors such as campaign duration, funding goal, and rewards [14]. The aforesaid critical insights have been considered by the authors, and the present chapter is presented to propose an industrial design to bring a revolutionizing change for secure fundraising in supply chains based on blockchain technology. Here, the blockchain technology is preferably utilized to create more secure and transparent donation systems for charities [15].

The present chapter stresses on the development of an Ethereum-based decentralized application (dApp), to ensure a secure fundraising platform/website and to ensure zero fraud in the name of crowdfunding, like kickstart.com, using smart contracts. In the present application platform, an entrepreneur has to create a campaign, and, then, it is required to raise a request for meaningful funding. The request gets approved by at least half of the approvers to receive funding for that request only (both the approver and manager of a campaign have to pay in ETHER to create/donate a campaign). The amount does not reside in the entrepreneur's account; it directly goes to the recipient who provides resources. Here, the fund's transaction and resource supply are taken care of by smart contracts without relying on a third central party or any other intermediaries [16], which ensure no fundraising scams. Introduction of blockchain in crowdfunding will make it more reliable, transparent, trusted, decentralized, cost-efficient, and convenient [17]. Currently, there is a website called Kickstarter.com, which possesses the same motive behind fundraising, but it is built upon Web2 technologies, and, unlike our proposed application platform, it puts all the money in an entrepreneur's bank account, after which there is no way that the platform can determine, if the money is used for the purpose for which the campaign was created. Traceability of funds becomes an issue hereafter [2]. Because of this, there are a considerable number of fundraising scams worth millions. We can avoid this shortcoming by using proposed application platform by using the Ethereum smart contract, by directly transferring the money from contributors to the resource provider.

Blockchain is a technology that could have a huge impact on the charity sector and philanthropy and can help in the management of funds securely and transparently, which help in reducing fraud. In addition, blockchain can also increase transparency and accountability in the fundraising process. Introduction of blockchain in crowdfunding will make it more reliable, transparent, trusted, decentralized, cost-efficient, and convenient [17]. The distributed nature of the blockchain ledger ensures that transactions are recorded in a tamper-proof manner, making it easy to track the flow of funds and assures that the funds are being used for their intended purpose. This can help to build trust between donors and fundraisers, leading to increased donations and greater social impact. Another benefit of using blockchain in fundraising is that it can reduce costs by eliminating the need for intermediaries such as banks or payment processors. This can make it easier for smaller charities and nonprofits to raise funds, as they may not have the resources to manage complex payment systems.

Overall, the use of blockchain in fundraising has the potential to revolutionize the way that charities; to assist in raise funds; and to manage valuable supply chain by providing a more efficient, secure, and transparent way to manage and distribute funds. The use of blockchain has evident benefits, but there are some challenges that are also needed to be addressed, such as scalability and regulatory issues. The potential benefits of blockchain can make it an exciting area of innovation for the future. The present chapter will assist the readers in identifying the below-mentioned research questions (RQs):

> RQ1: How can blockchain technology be applied toward fundraising to create a more secure and transparent system?
> RQ2: What are the potential benefits and challenges of using blockchain technology for industrial fundraising, and how the same can these be addressed?
> RQ3: How can the adoption of blockchain-based fundraising solutions impact the efficiency and effectiveness of fundraising processes, and what are the potential implications for the industry?

12.2 Literature Review

Charity and donation systems are becoming increasingly popular on blockchain platforms for creating value chains. The research on blockchain-based charity systems is growing with several approaches, which included a blockchain-based track donation system [2], a crowdfunding charity platform [3], a secured fundraising framework using smart contracts [4], and a blockchain-based charity foundation platform [9]. One reason for the increased popularity of charity-based crowdfunding is the declining trust in traditional charities [10]. Successful crowdfunding campaigns require careful planning, effective communication, and precise technology [14]. Additionally, trusted network and strategic tools are needed to be deployed for success and sustainability [6, 7]. The same are also needed to be documented under empirical boundaries for understanding usability and utility [11]. The use of blockchain technology can provide secure and transparent process and may lead in inducing technical, social, and economic pillars [1]. The technology has the potential to plan financial intermediation by increasing access to financing for charity and supply chain–based organizations [16]. The use of blockchain technology has gained increased attention in recent years by delivering package of benefits related to delivery of right asset to right hands, right allocation, employer satisfaction, enhanced services, and reduced complexity [17]. Several studies have investigated the success factors for blockchain-based crowdfunding as compared to conventional crowdfunding. The use of blockchain technology can increase transparency, reduce transaction costs, and improve trust between donors and recipients [18]. It is found that blockchain-based social crowdfunding can promote social value creation through the creation of trust and the reduction of intermediaries. In addition toward improving transparency and trust, blockchain technology has also been used to address financial irregularities in charitable donations. A blockchain-based charity donation scheme called UpHaaR [19] is proposed to prevent financial irregularities and to ensure that donated funds are used for their intended purpose, where it is suggested to address qualitative and quantitative dimensions to make the blockchain-based system robust and capable. Furthermore, it is found that blockchain technology can be used as a fundraising strategy, such as initial coin offerings [20]. A multiple-case study on success factors indicates that the presence of a strong team and the ability to attract high-profile investors were important factors for successful implementation of blockchain technology [21]. It is found that the use of blockchain technology can be significant for preventing crowdfunding fraud [22].

A method for crowdfunding fraud prevention using blockchain technology is defined in few studies, which involved the creation of a smart contract that would only release funds to the project once certain conditions were met [22]. It is important to note that crowdfunding campaigns need creation of diverse chains and that can be managed by blockchain technology. Donation-based crowdfunding platforms are nowadays required to be developed, where donors do not receive any equity or rewards in return for their contributions [23]. However, those who do use blockchain technology in their crowdfunding campaigns should need to have familiarity with relevant programming languages such as Solidity [24] and tools such as Web3.js [25]. Other factors that have been found to impact crowdfunding success include user behavior [26], the level of financial commitment from entrepreneurs [27], the diversity of information sources and social networks [28], and the level of transparency and security of the crowdfunding platform [29]. In addition, blockchain technology has been proposed as a means of facilitating the crowdsourcing of funds for tertiary education [30].

Overall, the literature suggests that blockchain technology has the potential to improve transparency, reduce transaction costs, and increase trust in crowdfunding campaigns. Accordingly, the authors propose a decentralized fundraising platform in this chapter that utilizes blockchain technology to increase transparency and trust toward organization related with transactional activities, charity, and crowdfunding for creating valuable chain. The chapter presented research directions that are needed to be explored for resolving potential drawbacks and limitations of using blockchain technology.

12.2.1 Web3-Based Authentication

The chapter advocates the utility of Web3-based authentication in a decentralized fundraising platform that utilizes blockchain technology, where users are connected to a special test network through Web3 authentication. Users can communicate with authenticated users after authentication. Therefore, every Web3 dApp needs this authentication (refer to Figure 12.1 for understanding utility of Web3 authentication). The following are some justifications, which explicate "Why Web3 authentication is superior to Web2 or email/password logins".

a) Enhanced security: Here, public key encryption is used, which is a more secure method of proving identity than email or a password. Instead of storing credentials on

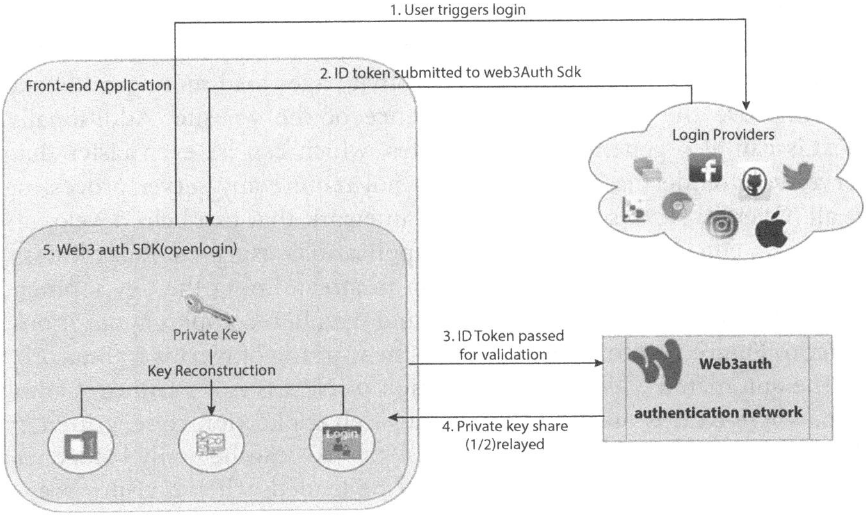

Figure 12.1 Web3 authentication: A blockchain-powered, decentralized solution ensuring secure and user-centric online identity verification.

servers, Web3 wallet or authorizing tools can store them locally on the device. Consequently, the security attack surface is reduced as a result.

b) Enhanced privacy: No email address is needed, passwords cannot be forgotten, and no outside parties are involved.

c) User interface has been made simpler and it takes only one click to complete the login process. They are not required to write down or remember any passwords.

12.2.2 Next.js

The utility of Next.js element is proposed in developed platform. Next.js is a popular framework for building web applications based on the React library. One of the biggest advantages of using Next.js on the frontend is that it can save developers significant amount of time during the coding process. This is because Next.js provides a number of tools and features that simplify the development process. For instance, if a developer wants to add more pages with routing to their web application, then all they need to do is create a new folder called "pages" and add the new pages to it. Next.js will then automatically handle the routing for those pages, making it easy to quickly add new functionality to the application.

Another key feature of Next.js is server-side rendering, which can improve website performance by rendering pages on the server before sending them to the client. This means that pages load more quickly and can improve the overall user experience of the website. Additionally, Next.js can also generate static websites, which can be even faster than server-side rendering because they do not require any server processing at all. Overall, Next.js is a powerful framework that can help developers build fast, efficient, and scalable web applications using the React library. Its built-in tools and features can help in streamlining the development process, thus making it easier to build and maintain complex applications. Refer to Figure 12.2 for understanding the working of Next.js framework.

The automated code splitting function of Next.js is a significant addition, where Next.js mechanically divides web application code written by developers into smaller, independent pieces that can be easily loaded. By lowering the amount of code that must be loaded, when a visitor sees a page, this can enhance website performance. Web developers may guarantee that their apps load swiftly and fluidly even on slower connections by implementing automated code splitting. Additionally, Next.js has a simple routing mechanism and Next.js will automatically create the required routes for those pages. As a result, adding new functionality to a web application is simple and does not need worrying about routing or other low-level issues. Next.js includes a variety of plugins and extensions in addition to its basic functionality, which can aid developers in creating more intricate online applications. Building e-commerce websites or blogs is simple with Next.js' support for integration with well-liked content management

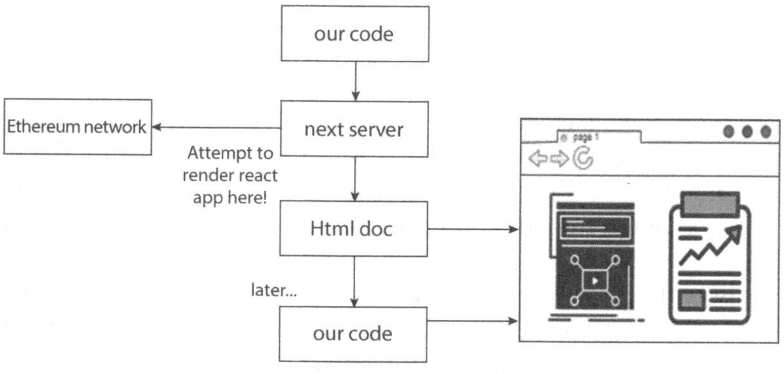

Figure 12.2 The seamless integration and server-side rendering capabilities of Next.js for enhancing the performance and user experience.

systems like WordPress and Shopify. The capability for server-side rendering in Next.js is one of its main advantages. Next.js may render a page on the server and deliver it to the client, when a user wants it and speeding up page loading. For websites with dynamic material, where the information often changes based on user input, this can be extremely helpful. The website's efficiency can be enhanced with Next.js while ensuring that viewers see the most recent material toward server side rendering.

Overall, Next.js is a strong and adaptable framework that may assist programmers in rapidly and simply creating high-performance online apps. It is a well-liked option for creating sophisticated React online apps because of its support for server-side rendering, automated code splitting, and simple routing.

12.3 Methodology and Working Model

12.3.1 System Architecture

12.3.1.1 Smart Contract Deployment

In the proposed framework, smart contract deployment is required, where Solidity is used, which is a programming language designed to write smart contracts on the Ethereum blockchain. However, in order for a Solidity contract to be executed on the Ethereum virtual machine (EVM), it must be compiled first. This is where the solc compiler comes in. The solc compiler takes the high-level Solidity code and translates it into bytecode, which is the machine-readable language used by the EVM. Once the contract has been compiled, it is represented as a JSON object, which contains the bytecode and the application binary interface (ABI). The ABI is a specification for how to interact with the contract, and it defines the functions, arguments, and return types that are available. In order to interact with the contract, the JSON object is converted into low-level ABI calls using the Web3 API references. This allows you to work with JavaScript objects that are smart contracts. The working mechanism of a Solidity compiler to compile smart contract is shown in Figure 12.3.

Once the contract has been compiled and converted into a JSON object with the ABI, it can be deployed onto the Ethereum test network. This is done by sending the bytecode and ABI to the network, which then creates a new instance of the contract. The contract is then stored on the blockchain and can be interacted with by sending transactions to its functions.

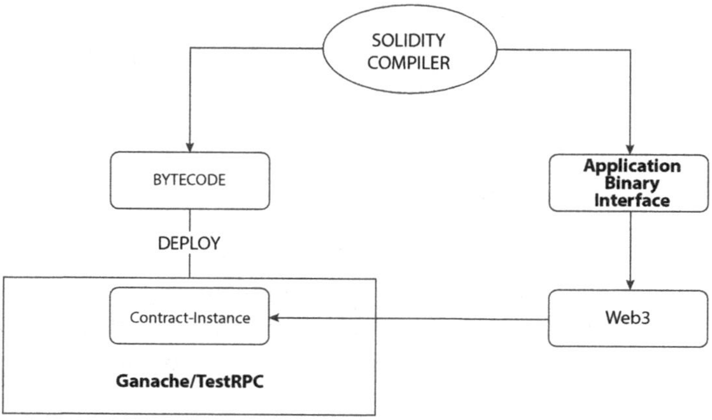

Figure 12.3 Figure showcasing the Solidity compiler's process of transforming smart contracts into executable code for blockchain deployment.

12.3.1.2 Agile Methodology in Code Compilation

In order to carry out our proposed project effectively and efficiently, we created two separate contracts. This decision allowed us to focus on specific functionalities of each contract and to avoid overwhelming complexity that could hinder our progress. Furthermore, as part of our development process, we utilized agile methodology principles to ensure that our project was completed in a timely and iterative manner. This involved implementing a compilation setup that generated two JSON files, each containing bytecode and ABI encoding for their respective contracts. This approach helped us to maintain a clear and organized record of our project's code, making it easier to manage and maintain as we progressed. By adhering to

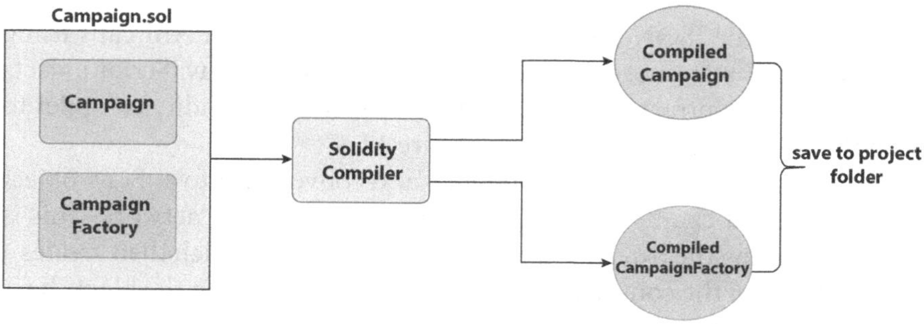

Figure 12.4 Solidity compiler: Streamlining smart contract compilation for efficient development and deployment of agility.

these agile methodology principles, we were able to deliver a high-quality research project that met our goals and objectives. Agile methodologies aim toward deploying critical means for identifying the right allocation for incremental development in functionality based on small cross-functional objects, feedback, and corrections. The systematic compilation of contracts for efficient development and deployment of agility is shown in Figure 12.4.

12.3.1.3 Memory and Storage Data Locations

Memory and storage data locations are prominent for maintaining the effectiveness of the system. Here, in the proposed platform in the EVM, a smart contract's execution, which involves two types of storage named as memory and storage, is utilized. Memory is a temporary storage area that is used during the smart contract execution and is completely cleared off for the subsequent execution when the execution ends. On the other hand, storage is persistent and is used to store data permanently in the smart contract.

Every transaction that is executed on the EVM incurs gas fees, which are paid in ether. The amount of gas consumed during a transaction depends on the complexity of the transaction and the amount of data that it involves. Therefore, it is essential to minimize the gas usage of the smart contract to keep the transaction cost low.

The Solidity code performs better when it consumes less gas. Memory uses less gas per unit of energy than storage does. Therefore, it is always preferable to store the final result in storage and use memory for intermediate calculations or function calls. By using memory instead of storage for temporary storage during the execution, the smart contract can reduce its

Data holding places		
Storage	Holds data between function calls	Pretty much like a computer's hard drive
Memory	Temporary place to store data	pretty much like a computer's ram

Figure 12.5 Showcasing Solidity's storage for persistent data storage and Solidity's memory.

gas consumption, thereby lowering the overall transaction cost. The storage and memory keywords in Solidity are depicted in Figure 12.5.

In summary, efficient use of memory and storage is essential for optimizing the performance of a smart contract and reducing the gas cost of transactions. Using memory for intermediate calculations and finalizing the result in storage is a widely adopted best practice to minimize gas consumption and optimize performance.

12.3.2 Working Model

12.3.2.1 Basic Working of Website

An entrepreneur who wants to raise money for his or her business campaign must first get themselves verified as a businessman by a web-based authentication system. The same is also valid for resource providers and donors who want to contribute to campaigns. The business owner can then launch his own marketing campaign by paying gas prices each time that he makes the same trip. Contributors would see a page full of these campaigns and could donate to any of them based on how successful they thought the campaign might be. The user would need to pay a minimum fee to be

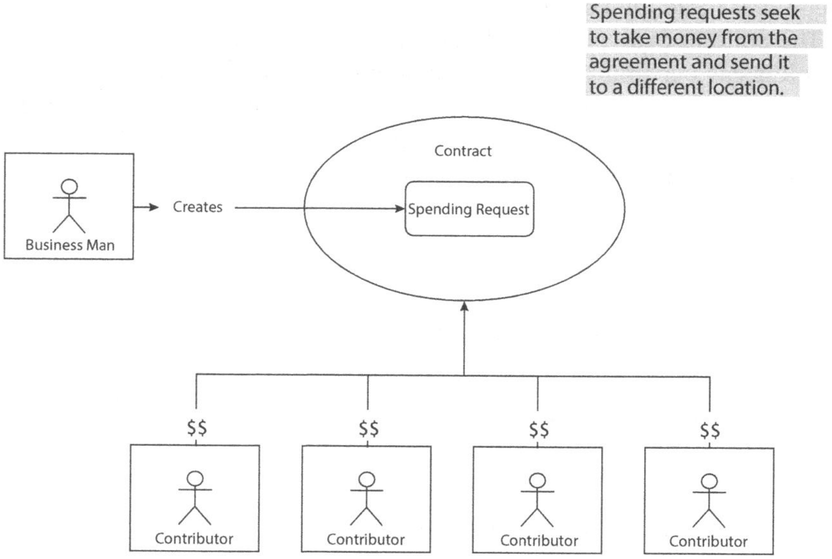

Figure 12.6 Overview of creating of spending request by the entrepreneur and getting approved by contributors.

included in any of the campaigns. Then, a page with numerous funding requests for the same campaign is presented to the users. Requests can be approved by users based on their preferences. The entrepreneur may request to withdraw the entire amount if the number of approvers for a request exceeds half of the campaign donors. The basic working structure of website can be identified in Figure 12.6. The resource provider would receive the withdrawn amount. Who will then distribute the resources needed for that specific request? Until the resources reach the entrepreneur, a supply chain powered by blockchain ensures that they are handled properly.

12.3.2.2 Design and Implementation

A blockchain secure fundraising platform is easy in implementation and is a cutting-edge solution that leverages the power of the Ethereum blockchain to provide a secure, reliable, and transparent platform for fundraising. Ethereum is a popular blockchain platform that enables the creation of dApps using smart contracts. Smart contracts are self-executing pieces of code that run on the Ethereum blockchain and can be programmed to automate various actions, such as transferring funds, verifying transactions, and executing complex business logic. By utilizing smart contracts, the blockchain secure fundraising platform ensures that all transactions are executed in a secure and transparent manner, without the need for intermediaries or third parties.

The platform has been designed with user-friendliness in mind, offering a simple and intuitive interface that enables users to create and manage their fundraising campaigns with ease. This means that even users who are new to blockchain technology can easily create and manage their campaigns without the need for extensive technical knowledge. The blockchain secure fundraising platform also provides investors with a transparent and secure platform to invest their funds. By leveraging the transparency of the blockchain, investors can view the progress of the fundraising campaign in real time, ensuring that they can make informed investment decisions. Overall, the blockchain secure fundraising platform is a powerful solution that offers a secure, transparent, and efficient way to manage fundraising campaigns. By utilizing the Ethereum blockchain and smart contracts, the platform ensures that all transactions are executed in a secure and reliable manner, whereas the user-friendly interface ensures that users of all levels of technical expertise can easily create and manage their campaigns.

12.3.2.3 Utility of Smart Contract

The smart contract is utilized in the proposed platform for keeping track of all the deployed campaigns. The smart contract code is written in Solidity, which consists of two contracts: CampaignFactory and Campaign. These contracts facilitate crowdfunding campaigns on the Ethereum blockchain.

- The "CampaignFactory" contract is responsible for creating instances of the Campaign contract and keeping track of all the deployed campaigns. It maintains an array called "deployedCampaigns" that stores the addresses of the deployed campaigns. The contract provides two functions: "createCampaign" and "getDeployedCampaigns". The "createCampaign" function creates a new instance of the Campaign contract with a specified minimum contribution and the address of the creator as the manager. The address of the new campaign is added to the "deployedCampaigns" array. The "getDeployedCampaigns" function simply returns the array of addresses of all the deployed campaigns.
- The Campaign contract represents a crowdfunding campaign and has various features. It includes a manager, who is the creator of the campaign and can create requests for funding. It also specifies a minimum contribution amount that contributors must meet to become approvers of the campaign. The contract maintains a list of approvers, who are the contributors to the campaign, and a list of requests for funding created by the manager. The requests can be approved or rejected by the approvers.
- The Campaign contract has several fields, including "manager" (the address of the campaign manager), "minContribution" (the minimum contribution required to become an approver), "approversCount" (the number of approvers), "approvers" (a mapping of addresses to Booleans indicating whether an address is an approver), and "request" (an array of Request structs representing funding requests).
- Each Request struct contains fields such as "description" (the request description), "value" (the amount of funding requested), "recipient" (the address of the funding recipient), "complete" (a Boolean indicating whether the request has been fulfilled), "approvalCount" (the number of approvals

for the request), and "approvals" (a mapping of addresses to Booleans indicating approval or rejection of the request).

- The Campaign contract provides several functions. The "contribute" function allows an address to contribute to the campaign and become an approver. The "createRequest" function enables the manager to create a request for funding, specifying the description, value, and recipient address. The "approveRequest" function allows an approver to approve or reject a specific request by its index. The "finalizeRequest" function permits the manager to fulfill a request if it has received more than half of the approvals from the approvers.
- Additionally, the Campaign contract includes functions such as "getSummary" (returns various campaign details like balance, number of requests, minimum contribution, number of approvers, and manager's address), "getAllRequests" (returns all the requests as an array of Request structs), and "getRequestsCount" (returns the number of requests).
- The code also defines two modifiers: "restricted" and "checkMinimumContribution". The "restricted" modifier restricts access to certain functions to only the manager of the campaign. The "checkMinimumContribution" modifier ensures that a contribution is greater than the specified minimum contribution before allowing it.
- In summary, these contracts provide a framework for creating and managing crowdfunding campaigns on the Ethereum blockchain, allowing contributors to become approvers, approving or rejecting funding requests, and enabling the manager to fulfill approved requests.

12.3.2.4 HD Wallet Provider

An HD wallet–enabled Web3 provider is required to sign transactions in the proposed platform. In the proposed platform, your data application can communicate with an Ethereum Node using the Web3 provider (Infura). By sending the request to an HTTP-based server, providers receive JSON-RPC requests and respond with the response. Using a 12- or 24-word mnemonic, an HD wallet–enabled Web3 provider signs transactions for addresses (Figure 12.7). A software interface called Web3 provider enables web applications to communicate with the Ethereum blockchain network. The JSON-RPC protocol through HTTP or WebSocket may be used by the web application to connect to the Ethereum network and to the Web3 provider.

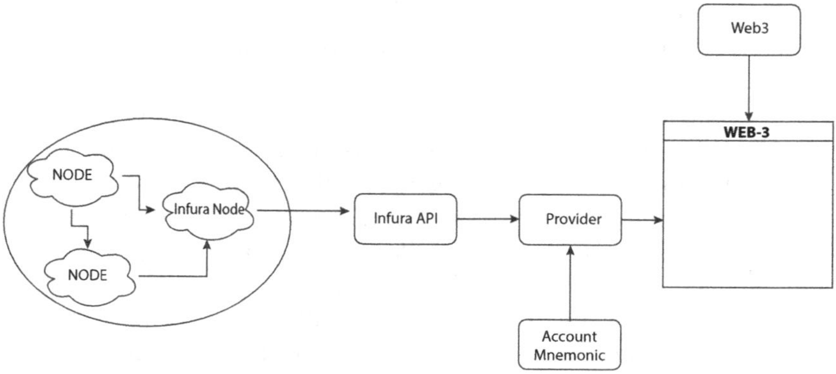

Figure 12.7 The seamless integration of a wallet provider with Infura API, enabling secure, and efficient blockchain transactions.

Popular Web3 service provider (Infura) provides a dependable, scalable, and user-friendly architecture for Ethereum node communication. Authentication is necessary for Web3 services like Infura to sign transactions and interact with the Ethereum network. The master seed for generating private keys is utilized in the authentication process as a mnemonic or secret phrase, consisting of 12 or 24 syllables. The mnemonic is used to generate a hierarchical deterministic (HD) wallet that generates a distinct set of private and public keys for each transaction. HD wallets are superior than traditional wallets in a variety of ways. Because they use cutting-edge encryption techniques to generate a collection of keys from a single seed phrase, they provide a more secure way to handle cryptocurrencies. The user may still get their cash by using their mnemonic even if they forget their password or lose their device. Overall, employing an HD wallet–enabled Web3 provider like Infura provides a flexible and convenient method to manage many transactions across several platforms while also offering a safe and dependable means to engage with the Ethereum network.

A HD wallet provider is a type of cryptocurrency wallet that generates a unique, hierarchical set of private and public keys for each transaction. HD wallets are designed to be more secure than traditional wallets because they use advanced algorithms to create a series of keys that are derived from a single seed phrase. This means that the user only needs to remember one seed phrase to generate an unlimited number of keys, making it easier to manage multiple transactions and reducing the risk of losing funds. HD wallet providers offer a range of services, including secure storage of private keys, automatic backup and recovery of data, and integration with multiple cryptocurrency exchanges. Popular HD wallet providers include Trezor, Ledger, and KeepKey.

12.4 Web Application

12.4.1 Next.js in Frontend

To ensure server-side rendering, Next.js is primarily used in the proposed platform. When using server-side rendering, the server generates the page's HTML for each request. To make the page interactive, the generated HTML along with JSON information and JavaScript functions are sent to the client. Here, JSON data and JavaScript functions are used to make the web page interactive, whereas HTML is used on the client to display a quick, non-interactive page (for example, embedding event handlers to a form). This process is generally referred as hydration.

The fetching of data for frontend pages that have been rendered frequently uses JavaScript functions. Sending the page from the server with the data already filled in makes server-side rendering possible. Search Engine Optimization can benefit from this. The benefits of Next.js in Frontend can be identified from Figure 12.8.

Other benefits of using Next.js in an application include minification, seamless Redux support, high levels of configurable, folder-based routing, and ESLint compatibility.

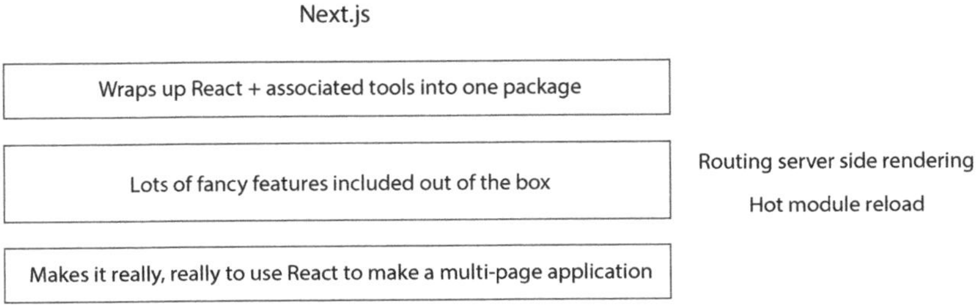

Figure 12.8 Next.js features in server-side rendering, static site generation, API routes, automatic code splitting, etc.

12.5 Data Collection

The dApp platform proposed in this chapter used a variety of Solidity data types for data collection in smart contracts. The various objects used for data collection in smart contract can be understood in Figure 12.9. The various elements related to data collection are as follows:

a) Address: This is an address, where you can send ether because it has a 20-byte value (the size of an Ethereum address).
b) Uint: An unsigned integer (a non-negative value data type). The word "uint" is short for "uint256". By default, this value data type accepts up to 32 bytes. This data type is used in the proposed application platform to store the minimum contribution that donors are required to make and to maintain the approval-Count or the total number of donors who approved a specific request.
c) Struct: This is referred as custom-defined types called structures, which can group various variables. This data type is

Campaign Request

Variables

Manager	**Address**	address of the campaign's supervisor
Min-Contribution	**Uint**	To be considered a contributor or 'approver', a minimum donation is required.
Approvers	**Address[]**	Addresses for everyone who has contributed money
Requests	**Request[]**	List of requests created by the manager

Functions

Campaign	The minimumContribution and the owner are set by the fuction Object() { [native code] } function.
Contribute	When someone wants to donate money to the campaign and become a 'approver', this number is dialled.
CreateRequest	The manager requested that I create a new 'spending request'.
ApproveRequest	Each contributor is summoned to approve a spending request.
FinalizeRequest	When a request has received enough approvals, the manager can use this to have money sent to the vendor.

Figure 12.9 Various user-defined solidity objects used in campaign request of smart contract for data collection.

used in the proposed application platform by the business owner to create the request format.

d) Modifier: Modifiers are employed to automatically check a condition before calling a function. This is used in the proposed application platform to check the donor's minimum contribution.

e) Constructor: Constructor is a function object. A function Object() { [native code] } is a function that is optional and called, when a contract is created. It is used, where the initialization code for the contract is executed. The manager and minimum contribution of the campaign are initialized using this in the proposed application platform each time, when an entrepreneur creates a new campaign.

f) Mapping: Like arrays and struts, mapping is a reference type. (_KeyType => _ValueType) mapping is used to prevent double spending. Our proposed application platform uses mapping to determine whether the donor has already given to that campaign or not.

12.6 Results and Findings

The implementation of the proposed blockchain-secured fundraising system resulted in a decentralized platform, where entrepreneurs can request funding for their business campaigns from sponsors. The proposed platform presented in this chapter is built using a Web3-based authentication for managing supply chain system, which is implemented using blockchain technology. The same will allow secure and transparent handling of funds and resource allocation.

The use of smart contracts in the platform ensured that all transactions were transparent and tamper-proof. This added an extra layer of security to the platform, making it almost impossible for hackers to alter any information. Additionally, the use of a Web3-based authentication system ensured that only authorized users could access the platform. The Ethereum-based blockchain network was utilized in platform with the intention to maintain source of truth for each transaction and to ensure that, in the event of any financial discrepancies, all the details of the transaction could be retrieved from the blockchain. The build platform posses the capabilities to support transparency by allowing all approvers to discuss the business campaign, much like a social media platform.

The Web3-based build platform possesses the advance capabilities to manage supply chain management system by ensuring proper resource allocation between the resource allocator and the entrepreneur. Here, the Web3-based authentication system will assist in adding an extra layer of transparency and security to the platform, ensuring that all funds and resources were properly managed.

Overall, the implementation of proposed blockchain-secured fundraising system can be sustainable future platform for handing multiple activities in multiple supply chains. The platform was secure and transparent and provided a decentralized solution to the traditional fundraising process. The findings suggest that blockchain technology can be used to increase public trust in fundraising organizations and to encourage more giving. Thus, one can say that a crucial development toward a decentralized and secure fundraising platform is the blockchain-secured fundraising system. A platform that is safe, open, and impenetrable can be created using blockchain technology, smart contracts, and Web3-based authentication for managing supply chain management systems.

The transparency that the blockchain-secured fundraising mechanism offers is among its most important benefits. Every transaction on the site is tracked and freely accessible by anybody by the usage of smart contracts and blockchain technology. As a result, it is practically impossible to edit or manipulate any information on the platform because all transactions are accessible and transparent. The attainment of this transparency fosters trust and motivates more individuals to donate to fundraising efforts.

The proposed platform makes processing of money safe on the other side. All transactions are made tamper-proof through the use of smart contracts, and money is only delivered until a set number of conditions have been satisfied. This lowers the possibility of fraud or financial misuse and guarantees that funds are given to the appropriate initiatives.

The blockchain-secured fundraising platform is a major advancement in the field of fundraising as a whole. It offers sponsors access to transparent information about the projects that they are sponsoring while also giving entrepreneurs a decentralized and safe platform to submit funding requests for their initiatives. The platform represents an important step in boosting public confidence in fundraising organizations and enticing more individuals to take part in fundraising initiatives.

12.6.1 Importance of Blockchain Technology in Supply Chain Management

Today, for handling the unstable demands in supply chain management and to attain infinite competencies with limited resources and availabilities, there is an evident need of a designed technology-based platforms [31, 32], where the utility of blockchain technology is needed for managing system efficacy, trust, support, demands, complaints, and supplies under the domain of supply chain management. The principal actions that can influence regular activities in supply chain by blockchain technology are ease accessibility, induced availability, interchangeability of information, and attainment of performance marks [33]. Blockchain technology can present notable paybacks for handling financial assets, critical pace, dignified resources, availabilities, requirements, and so on for making sustainable supply chain designs. But that too needed a wide extent of qualitative and quantitative features for precise outputs [34, 35]. Blockchain technology holds the abilities of handling complex time-related activities with elevated sensitivity and accurateness. The success of supply chain management depends on the efficient administration of the extensive actions connected with the flows of products, processes, information, money, supply, and demands. It is always needed to have a synchronized association between dependents related with customers, suppliers, employers, employees, assets, network, and so on and that also needed effectual harmonization to preserve a value chain for sustainability, which excessively can be received by the implementation of blockchain technology [36].

Blockchain technology is needed in supply chain management for acquiring trust, security, transparency, and the traceability of shared data across a business network. Its implication can assists in capturing the perception of the customers for product development, predicting market behavior, and inter-operational accessibility for reducing risk, uncertainty, and resiliency. The implementation of blockchain technology can deliver package of benefits related to the delivery of right product at right place, right allocation, employer satisfaction, enhanced services, reduced complexity, transparency in sale, scaling, diversification, amplified product range management, identification of, precise preferences, apparent management of availability and supplies, prediction of unsure demand conditions, effortless collaboration between multiple partners, ease handling of many geographic locations, easy survey provisions, and ease coping of variety of intermediaries.

12.6.2 Blockchain Technology for Agility in Supply Chain Management

Blockchain technology can bring agility in SCM. The agility is important for satisfying the swift demands of the customers, trust, and indeed satisfaction [37]. Here, in the presented platform, agility is increased by incorporating varieties of qualitative and quantitative dimensions in Blockchain technology based on Web3 authentication, Next.js, smart contract deployment, bytecode, ABI, code compilation, data-holding places, and HD wallet provider. Blockchain technology ventures a digital decentralized network in comparison to precedent technologies and consequently holds ability to competently support transparency and transactions amid parties [38]. The same will permit business associates, policymaking authorities, government regulatory, and, importantly, customers to check and respond toward uncertainties stirring in the pathway of deliveries in a reactive and documented way, which, in turn, will induce agility. Blockchain technology capably has several success potentials that can bring agility in supply chain, i.e., real-time communications, trust among contacts, certification of transactions, development of secure associations, quicker processing of funds, low manufacturing costs, synchronization of production activities, and reduce lead times. A blockchain technology can provide agile estimation, assist in forecasting demand, reduce bottlenecks functions, and thus will escort in the preservation of funds and realization of recycling actions throughout the supply chain agilely and effectively. Blockchain technology is needed for innovation and development and to attain gears of performance presentations for experiencing elevated outcomes with agility and economically [39], where reduced mismatching of orders, correct delivery, precise handling of orders with minimum logistics, and low energy consumption will lead in deprived environmental loads and will provide agility marks, which can be attained by the said technology with developed blockchain-based integrated framework.

12.7 Limitations

In this chapter, a platform is presented for secure fundraising in supply chains based on blockchain technology. The same will assist to prevent entrepreneurs from misusing donors' funds under decentralized fundraising, but still there is need to develop an assured way to confirm that the resource provider is the actual entrepreneur seeking money from the general public. However, this restriction can be handle by the fact that every time a business owner launches a campaign, they must pay gas fees in cryptocurrency, which may later prove to be expensive depending on market trends and thus prevent the scammer from trying to fraud more than once. Although, the proposed platform aims to increase transparency and trust, it is not completely immune to fraud or misuse of funds. Smart contracts can only handle the conditions that have been pre-programmed into them, and there is still a possibility of errors or bugs in the code. The platform is built on the Ethereum blockchain, which has its own limitations, such as scalability issues and high transaction fees during periods of high network congestion. The proposed platform requires the use of cryptocurrencies, such as Ether, which may not be widely accepted or accessible to all potential users, particularly in developing countries. The platform may also face regulatory challenges, particularly with regard to crowdfunding and charity laws in different jurisdictions.

12.8 Conclusion

Crowdfunding platforms based on Web2 have always faced a major challenge of ensuring authenticity and transparency, which has sparked contentious debates. The application of blockchain technology has proven to be an effective solution to this challenge, as it provides a high level of security for both the campaign creator and approvers. Through the use of blockchain, data manipulation by hackers can be prevented, because transactions are already stored on the Ethereum network. The decentralized nature of blockchain technology ensures that each campaign request undergoes a thorough vetting process, which is further bolstered by requiring a contribution of cryptocurrency from the business owner. In addition, the inclusion of a discussion forum for each campaign request allows users to express their opinions about the company. Decentralized applications (dApps) leveraging smart contracts ensure that transactions between two unknown parties are carried out seamlessly.

The proposed decentralized fundraising platform utilizing blockchain technology has the potential to enhance transparency and foster trust in the charity and crowdfunding sectors. The utilization of smart contracts for fund transactions and resource allocation, along with the Web3-based authentication for managing fundraising supply chain is a remarkable step toward preventing fraud and ensuring the proper allocation of resources. However, further research and development are crucial to overcoming the platform's limitations and facilitating its widespread adoption and success. Overall, the platform holds immense promise for the future of fundraising, and its integration with blockchain technology paves the way for greater security and transparency in the industry. The chapter presented the utility of many features to build a secure fundraising platform by blockchain technology. The importance and utility of many critical features like Web3-based authentication, Next.js, smart contract deployment, bytecode, ABI, agile methodology, and code compilation; the role of data-holding places, HD wallet provider, and related working model; as well as its design and implementation under the bench of blockchain technology are demonstrated in this chapter for attaining performance and user experience. The chapter highlighted the need to implement blockchain technology in supply chain management and also presented how agility in supply chain management can be attained.

References

1. Hu, B. and Li, H., Research on Charity System Based on Blockchain. *IOP Conf. Ser. Mater. Sci. Eng.*, 768, 7, 072020–072029, 2020, doi: 10.1088/1757-899X/768/7/072020.
2. Shaheen, E. *et al.*, A Track Donation System Using Blockchain. *2021 International Conference on Electronic Engineering (ICEEM)*, pp. 1–7, 2021, doi: 10.1109/ICEEM52022.2021.9480649.
3. Saranya, S., Muvvala, S.P., Chauhan, V., Satwik, R., Crowdfunding Charity Platform Using Blockchain. *2022 International Conference on Inventive Computation Technologies (ICICT)*, pp. 1–7, 2022, doi: 10.1109/ICICT54344.2022.9850562.
4. D., M., Raswanth, S.R., Akella, S.V.V.S., Kumar, P., A Secured Distributed Ledger Based Fundraising Framework Using Smart Contracts. *2021 IEEE 4th International Conference on Computing, Power and Communication Technologies (GUCON)*, pp. 1–5, 2021, doi: 10.1109/GUCON50781.2021.9573916.
5. Zhao, H., Recent years have witnessed the rapid development of crowdfunding platforms among which the charity-based ones are becoming increasingly popular", Voice of Charity: Prospecting the Donation Recurrence & Donor Retention in Crowdfunding. *IEEE Trans. Knowl. Data Eng.*, 14, 8, AUGUST 2019.
6. Du, P., Xu, L., Qiao, X., Li, H., Li, D., Post-crowdfunding in reward-based crowdfunding with strategic purchasing consideration. *2017 International Conference on Service Systems and Service Management*, pp. 1–4, 2017, doi: 10.1109/ICSSSM.2017.7996207.
7. Zvilichovsky, D., Inbar, Y., Barzilay, O., Playing Both Sides of the Market: Success and Reciprocity on Crowdfunding Platforms. *SSRN Electron. J.*, 4, Jan. 2013, doi: 10.2139/ssrn.2304101.
8. Forbes, H. and Schaefer, D., Guidelines for Successful Crowdfunding. *Proc. CIRP*, 60, 398–403, 2017, doi: 10.1016/j.procir.2017.02.021.
9. Renat, G., Peresichansky, A., Belenov, A., Barger, A., Karma - blockchain based charity foundation platform. *2021 IEEE International Conference*

on Blockchain and Cryptocurrency (ICBC), pp. 1–2, 2021, doi: 10.1109/ICBC51069.2021.9461072.
10. Garton, J., Why has trust in charities been declining? *Conversat.*, 20, 3, September 2015, 11–13, 2016.
11. Chitsazan, H. and Bagheri, A., Factors Affecting Crowdfunding Success: A Systematic Analysis of the Empirical Studies. *2019 International Conference on Computational Intelligence and Knowledge Economy (ICCIKE)*, pp. 20–24, 2019, doi: 10.1109/ICCIKE47802.2019.9004279.
12. Pollack, J.M. and Bosse, D.A., When do investors forgive entrepreneurs for lying? *J. Bus. Ventur.*, 29, 6, 741–54, 2014 Nov 1.
13. Mollick, E., The dynamics of crowdfunding: An exploratory study. *J. Bus. Ventur.*, 29, 1–16, 2014, https://doi.org/10.1016/j.jbusvent.2013.06.005.
14. Kaartemo, V., The elements of a successful crowdfunding campaign: A systematic literature review of crowdfunding performance Valtteri Kaartemo. *Int. Rev. Entrep.*, 15, 3, 291–318, 2017.
15. An, K.-H. and Seo, H., Donate System Development Using Blockchain Technology. *J. Korea Inst. Inf. Commun. Eng.*, 22, 5, 812–17, May 31, 2018, doi: 10.6109/JKIICE.2018.22.4.812.
16. Cai, C.W., Disruption of financial intermediation by FinTech: a review on crowdfunding and blockchain. *Account. Finance*, 58, 965–992, 2018. https://doi.org/10.1111/acfi.12405.
17. Baber, H., Blockchain-Based Crowdfunding, Blockchain Technology for Industry 4.0, in: *Blockchain Technologies*, R. Rosa Righi, A. Alberti, M. Singh (Eds.), Springer, Singapore, 2020, https://doi.org/10.1007/978-981-15-1137-0_6.
18. Hartmann, F., Grottolo, G., Wang, X., Lunesu, M., II, Alternative Fundraising: Success Factors for Blockchain-Based vs. Conventional Crowdfunding. *2019 IEEE International Workshop on Blockchain Oriented Software Engineering (IWBOSE)*, pp. 38–43, 2019, doi: 10.1109/IWBOSE.2019.8666515.
19. Saraswat, D., Patel, F., Bhattacharya, P., Verma, A., Tanwar, S., Sharma, R., UpHaaR: Blockchain-based charity donation scheme to handle financial irregularities. *J. Inf. Secur. Appl.*, 68, 103245, 2022.
20. Panin, A., Kemell, K.K., Hara, V., Initial coin offering (ICO) as a fundraising strategy: a multiple case study on success factors, in: *Software Business: 10th International Conference, ICSOB 2019Proceedings*, Jyväskylä, Finland, November 18-20, 2019, vol. 10, Springer International Publishing, pp. 237–251, 2019.
21. Nguyen, L.T., Hoang, T.G., Do, L.H., Ngo, X.T., Nguyen, P.H., Nguyen, G.D., Nguyen, G.N., The role of blockchain technology-based social crowdfunding in advancing social value creation. *Technol. Forecast. Soc. Change*, 170, 120898, 2021.
22. Pandey, S., Goel, S., Bansla, S., Pandey, D., Crowdfunding fraud prevention using blockchain, in: *2019 6th International Conference on Computing*

for Sustainable Global Development (INDIACom), 2019, March, IEEE, pp. 1028–1034.
23. Kenton, W., Donation Based Crowdfunding. Investopedia, 2017, December. [online] Available at: https://www.investopedia.com/terms/d/donation-based-crowdfunding.asp [Accessed 10 Oct. 2022].
24. Solidity Documentation, 2018, October. [online] Available at: http://www.solidity.readthedocs.org/en/latest/ [Accessed 10 Oct. 2022].
25. Kurian, A.M., Interacting with Ethereum Smart Contracts through Web3.js, 2018, October. [online] Available at: https://www.medium.com/coinmonks/interacting-with-ethereum-smart-contracts-through-web3-jse0efad17977 [Accessed 10 Oct. 2022].
26. Beier, M. and Wagner, K., User behavior in crowdfunding platforms–exploratory evidence from Switzerland. In System Sciences (HICSS). *2016 49th Hawaii International Conference*, 2016, January, IEEE, pp. 3584–3593.
27. Löher, J., Schneck, S., Werner, A., A research note on entrepreneurs financial commitment and crowdfunding success. *Ventur. Cap.*, 20, 3, 1–14, 2018.
28. Polzin, F., Toxopeus, H., Stam, E., The wisdom of the crowd in funding: information heterogeneity and social networks of crowdfunders. *Small Bus. Econ.*, 50, 2, 251–273, 2018.
29. Mishra, A., Khatter, H., Gupta, G., Jamshed, A., Srivastava, A.K., Enabling secure and transparent crowd funding approach powered by blockchain, in: *Emerging technologies in data mining and information security: Proceedings of IEMIS 2022*, vol. 2, pp. 637–650, Springer Nature, Singapore, 2022.
30. Rashid, M.A., Deo, K., Prasad, D., Singh, K., Chand, S., Assaf, M., TEduChain: A platform for crowdsourcing tertiary education funds using blockchain technology. *arXiv preprint arXiv:1901.06327*, 35, 27–42, 2019.
31. Bag, S., Sahu, A.K., Kilbourn, P., Pisa, N., Dhamija, P., Sahu, A.K., Modeling barriers of digital manufacturing in a circular economy for enhancing sustainability. *Int. J. Product. Perform. Manage.*, 71, 3, 833–869, 2021a. https://doi.org/10.1108/IJPPM-12-2020-0637.
32. Bag, S., Viktorovich, D.A., Sahu, A.K., Sahu, A.K., Barriers to adoption of blockchain technology in green supply chain management. *J. Glob. Oper. Strateg. Sourc.*, 14, 1, 104–133, 2021b.
33. Zelbst, P.J., Green, K.W., Sower, V.E., Bond, P.L., The impact of RFID, IIoT, and Blockchain technologies on supply chain transparency. *J. Manuf. Technol. Manage.*, 31, 3, 441–457, 2019.
34. Wang, W., Huang, L., Zhu, Y., Jiang, L., Sahu, A.K., Sahu, A.K., Sahu, N.K., Decision support system toward evaluation of resilient supplier: A novel fuzzy gain-loss computational approach. *Kybernetes*, 49, 6, 1741–1765, 2019a. https://doi.org/10.1108/K-05-2019-0345.
35. Wong, L.W., Tan, G.W.H., Lee, V.H., Ooi, K.B., Sohal, A., Unearthing the determinants of blockchain adoption in supply chain management. *Int. J. Prod. Res.*, 58, 7, 2100–2123, 2020.

36. Sahu, A.K., Katyayan, A., Khandey, U., Jangde, P., Sahu, A.K., Sahu, N.K., Adaptation of block chain technology in SCM for steering managerial strategies: investigative study under Indian context. *J. Int. Logist. Trade*, 21, 1, 18–40, 2023a. https://doi.org/10.1108/JILT-06-2022-0020.
37. Sahu, A.K., Sahu, N.K., Sahu, A.K., Laminating STRATH block chain technology-SWOT architectures to endure business strategy between digital transformation, firms and supply chains capabilities for sustainability. *J. Clean. Prod.*, 383, 135531–135540, 2023b. https://doi.org/10.1016/j.jclepro.2022.135531.
38. Vaio, D.A. and Varriale, L., Blockchain technology in supply chain management for sustainable performance: Evidence from the airport industry. *Int. J. Inf. Manage.*, 52, 102014–102029, 2020. https://doi.org/10.1016/j.ijinfomgt.2019.09.010.
39. Leduc, G., Kubler, S., Georges, J.P., Innovative blockchain-based farming marketplace and smart contract performance evaluation. *J. Clean. Prod.*, 306, 127055–127074, 2021. https://doi.org/10.1016/j.jclepro.2021.127055.

13

Manufacturing of Polymer Matrix Composite Under Qualitative and Quantitative Designs: Investigations of Different Methods and Properties of Composites for Societal Applications

Alok Agrawal[1]*, Vivek Mishra[2], Gaurav Gupta[3] and Pravat Ranjan Pati[4]

[1]Department of Mechanical Engineering, Sagar Institute of Research and Technology, Bhopal, India
[2]Department of Aeronautical Engineering, Tulsiramji Gaikwad-Patil College of Engineering and Technology, Nagpur, India
[3]School of Mechanical Engineering, Vellore Institute of Technology, Vellore, India
[4]Department of Mechanical Engineering, Graphic Era (Deemed to be University), Dehradun, India

Abstract
The manufacturing industry of composites constantly understood the significance of early forms of material and processing. Over the past few decades, material chemistry improved significantly including reductions in resin viscosity and delayed curing of resin that is effortlessly accordant. The industry has also introduced new processes and resorted to more science-based practices. For the development of any category of composites, the manufacturing processes as well as qualitative and quantitative factors are significant to govern the quality and cost of composites. In this chapter, the basic manufacturing processes and qualitative and quantitative factors for developing polymeric composites, which are in use by the scientific community, are discussed in detail. The various available methods for polymer composite manufacturing are discussed. Among them, the important fabrication methods, which are mainly used, i.e., injection moulding method, compression moulding method, resin transfer moulding method, vacuum-assisted resin

**Corresponding author*: alokag03@gmail.com

Atul Kumar Sahu, Rakesh D. Raut, Rohit Raja, Anoop Kumar Sahu and Nitin Kumar Sahu (eds.) Industrial and Manufacturing Designs: Quantitative and Qualitative Analysis, (375–398) © 2024 Scrivener Publishing LLC

transfer moulding method, and hand layup method, are widely discussed in this chapter. The abovementioned fabrication techniques are explained in this chapter concerning the method of manufacturing in regard to the qualitative and quantitative aspects. Along with it, the effect of different manufacturing techniques on the various key mechanical properties of polymeric composites is presented on the basis of the past work performed by the scientific community.

Today, polymer matrix composites are extensively developing by the researchers and are extending their utility in numerous applied fields like automobiles, packaging, furniture, and aerospace industries because of the reason of having good mechanical properties, lower density, lightweight, and so on. The desired characteristics and properties offering by the composite materials have nowadays declared them as sustainable materials with immense caliber and characteristics to be used in different segments of the work activities in industrial as well as societal applications. The chapter advocated toward evaluation of qualitative and quantitative factors for the development of polymer matrix composites with reinforced capacities for different applications. The chapter rationalizes its importance in defining several factors and manufacturing techniques for composites. It is observed that many qualitative and quantitative factors in the form of fracture toughness, raw components, volume ratio, axial tensile strength, flexural tensile strength, ultimate deformations under axial compression, ultimate deformations under axial tension, and elastic modulus are significant for modeling and evaluating polymer matrix composites. The study presented in this chapter tried to provide a complete possible aspect to be used to evaluate the quantitative and qualitative aspects of polymer matrix composites.

***Keywords*:** Polymer matrix composites, manufacturing method, hand layup, resin transfer moulding, injection moulding, compression moulding

13.1 Introduction

Composites are composed of two or more components that have been combined. The combining material must have chemically different ingredients on a macroscale and have a dissimilar interface extrication them. These composite materials are often stronger than their parts, making them ideal for many applications. Among the combining parts, one phase is called the matrix that is also called the continuous phase and the other phase is known as reinforcement, which is known as the discontinuous phase. Based on the matrix, these are classified as metal, ceramic, and polymer matrix composites. Among all, polymer matrix composites are in high demand because of their multiple advantages, which include qualitative and quantitative characteristics in the form of high strength-to-weight ratio, ease of fabrication, resistance to corrosion, flexibility in design, dimensional stability, and durability. Apart from that, the polymers are easily available at a relatively low cost as compared to other matrix materials. Polymers are basically of two types, i.e., thermoset and thermoplastic. Thermoset polymers do not change their shape by applying heat or pressure, whereas thermoplastics are re-moulded by applying heat and pressure. Polymer composites are also classified as fiber-reinforced or particulate-filled polymer composites based on the types of reinforcement. Both categories of composites have their respective advantages and disadvantages; according to that, they are used in different applications. Further, the fibers can be continuous fibers and discontinuous fibers, and particles are categorized as microparticles or nanoparticles. Continuous fiber and nanoparticles are rooted in the polymer to enhance the mechanical properties like strength and stiffness of the polymers, whereas microparticles are more into the improvement of thermal properties and toughness of the polymers.

The fabrication route adopted has an impactful effect on the properties of the material. The main aim of manufacturing is to derive the best properties from the combining phases and to minimize their weaknesses. The manufacturing industries of composites are continuously engaged in the development of manufacturing techniques to fabricate the composite with superior quality and lower cost. The main aim of the chapter is to provide detailed information on the various commonly used manufacturing techniques for the development of different types of polymeric composites. The chapter rationalizes its importance in defining several factors and manufacturing processes for polymer matrix composites. The chapter advocated toward evaluation of qualitative and quantitative factors for modeling polymer matrix composites for different applications. The various fabrication techniques mostly implemented for the fabrication of polymer composites are the injection moulding method, compression moulding method, resin transfer moulding (RTM) method, vacuum-assisted RTM method, and hand layup (HLU) method. In the following section, the abovementioned fabrication techniques are discussed in detail concerning the procedure followed and the effect of fabrication on the properties of the composites. The study presented in this chapter tried to provide complete possible aspects to be used to evaluate the quantitative and qualitative manufacturing techniques of polymer matrix composites.

13.2 Injection Moulding Method

Injection moulding is one of the most popular methods of fabricating polymer composite material. Among the different types of injection moulding, most industries are using screw-type injection machines. This injection machine can be electrically operated or hydraulically operated. It is a powerful moulding method through which complicated shapes with precision can be moulded with the help of a well-designed dye. This technique of moulding applies toward fabricating both thermoplastic polymer composites and thermoset polymer composites. In this section of the chapter, the complete information on the injection moulding method is discussed, which includes the equipment, mould, and process used for the preparation of the polymer composites. Further, a few recent works on the development of composite along with their useful properties will also be discussed.

A schematic of a typical injection moulding machine is provided in Figure 13.1. The machine is mainly divided into two units, i.e., injection unit and clamping unit. The injection unit consists of a hooper, a barrel, a heating part, a reciprocating screw, and a nozzle. In the clamping unit, a stationary platen and a moving platen are present to clamp the mould. Apart from that, a hydraulic system, electrical control system, and other auxiliary parts like a cooling system and feeding system are integral parts of the machine. Among the different elements of an injection moulding machine, the injection unit is the most important part. The first function of the unit is to melt the raw material with the help of a heating device and by shearing action provided by the screw. It is called the plasticizing system of the unit. After that, the injection system of the unit forces the melted material inside the mould using the screw at a fixed pressure and speed. This injection system has a pressurizing device and a driving device. The pressurizing device provides energy to the screw. The driving member delivers the energy to rotate the screw that is generally provided by the hydraulic motors. It also has a hopper to feed the raw material [1].

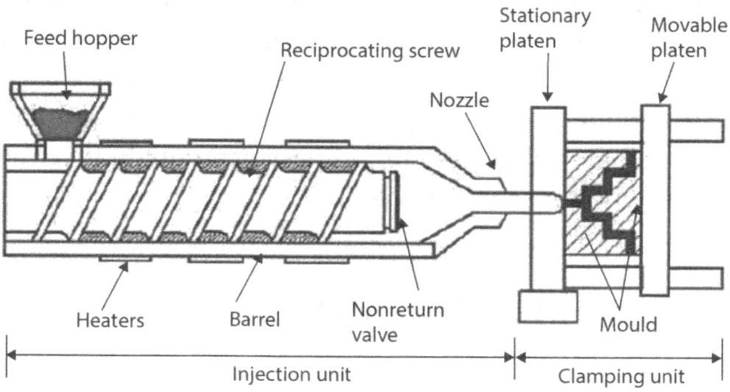

Figure 13.1 Layout of injection moulding machine.

The second unit of the injection moulding machine is the clamping unit, and its function is to clamp the mould and ensure the proper development of the product. It has to support the mould along with the smooth operation during the opening and closing of the mould to avoid vibration. Also, it needs to provide the necessary force, so that the mould will not open because of the pressure of the molten material. The quality of the product majorly depends upon the performance of the clamping unit. It is also provided with the arrangement for removal of the finished product from the mould cavity. The product obtained from the injection moulding method requires very little post-production work as the part obtained looks very finished upon ejection. The product developed from the injection moulding method can be of complex shape with high quality and precision [2]. This technique discussed above for the fabrication of composite materials is mainly used for the fabrication of thermoplastic polymer composites. A wide range of thermoplastic polymer composites that include particulate fillers and short fibers can be manufactured using this technique. Apart from that, some work has also been reported, where the scientific community manufactured polymer composites reinforced with woven mats successfully. A few examples of each are discussed below.

Joseph and Panneerselvam [3] developed a series of high-density polyethene (HDPE) composites filled with varied content of tungsten powder using the injection moulding method. Before using the injection moulding machine, the HDPE and tungsten powder were melted and compounded in the twin screw extruder. The uniformly distributed filler in the HDPE matrix thus obtained is then converted into the composite material using the injection moulding method. The fabricated composites are tested for their mechanical and abrasive wear behavior. Mark *et al.* [4] also explored the preparation of particulate composite using the injection moulding method. They prepared composite polypropylene as the base matrix and micro-sized coconut shell particulates as filler material. In their analysis, they prepared the composites by varying the size of the filler from 63 microns to 425 microns and found that the performance of the injection moulding machine does not depend upon the size of the filler material until the diameter of the particles is less than that of the diameter of the nozzle. They prepared the composite sheets by melt-blending the polypropylene and filler material in the injection moulding machine. During the fabrication of the composite sheets, the injection pressure is maintained at around 100 MPa, and the peak nozzle temperature reaches a value of 260°C, whereas the hooper temperature is maintained at 195°C, which is just above the melting temperature of the polypropylene. They observed that the melting and mixing time is limited to only 35 s in their work, which shows a very low cycle time. From the analysis, they found that the composite fabricated using injection moulding is free from defects, and, therefore, it is found that the quantitative properties under investigation improve with the addition of the filler material. The quantitative properties improved are tensile strength, flexural strength, and hardness of the material. In general, it was observed that the inclusion of filler generally results in a decrement in such properties. They observed enhancement in the properties, which signifies that the filler material is uniformly distributed inside the matrix when the composites are fabricated using the injection moulding method. They also observed that the composite prepared with smaller size particles shows better mechanical properties for a given filler loading.

Koffi *et al.* [5] used the injection moulding method for the fabrication of fiber-reinforced composites. They prepared the composites with HDPE as matrix material and birch fiber as a filler material. For the preparation of the composites, the components were first blended mixer. Here, first, the HDPE was melted on the rollers by maintaining a roller temperature of around 170°C. Later, the wood fibers were added and then blended for another 7 min. Once the blending is complete, the material is peeled off from the roller, which was again blended multiple times to get the composite sheets. These sheets are converted in the form of stripes with the help of scissors. The sheets are then converted into the form of granules with the help of an industrial grinder to feed them into the injection moulding machine for getting the composite specimen. The fabricated samples were investigated for the quantitative properties like tensile properties, impact strength, and hardness of the material. From the experimental work, they found that the composite fabricated via the injection moulding route will increase the tensile strength of the matrix material by 19.7%. Similarly, the tensile modulus of the material increased by 27.2% and a remarkable improvement of 48.26% took place in the value of the hardness.

Jeong *et al.* [6] used the injection moulding method for the first time for fabricating woven fabric-reinforced thermoplastic composites. They implemented an innovative injection moulding method for fabricating this category of composite body as such composites are difficult to fabricate using a two-step conventional method of fabrication. This method of injection moulding will help to reduce the cost of manufacturing and the cycle time as compared to the fabrication of a similar composite via the conventional injection moulding method. Also, the conventional method took more time and energy. In their work, the composites are prepared through a one-step process by keeping the temperature of the mould very high. The layout of the injection moulding method used in their work is presented in Figure 13.2. It can be visualized from the figure that they used a new structure and incorporated an extra mesh, so that the woven fiber will be impregnated with the thermoplastic matrix. The schematic view of the woven fabric and the mesh is present in Figure 13.3 through which the combination of the two is very clear. The fabricated samples were tested for their tensile properties, which include tensile strength and tensile modulus. From the analysis, they found that the new method of fabrication of composite provides better results as compared to the conventional method of injection moulding used for the fabrication of such composites. It was reported that the properties of the composites are increased by three to four times as compared to the properties of virgin polymers used in their work.

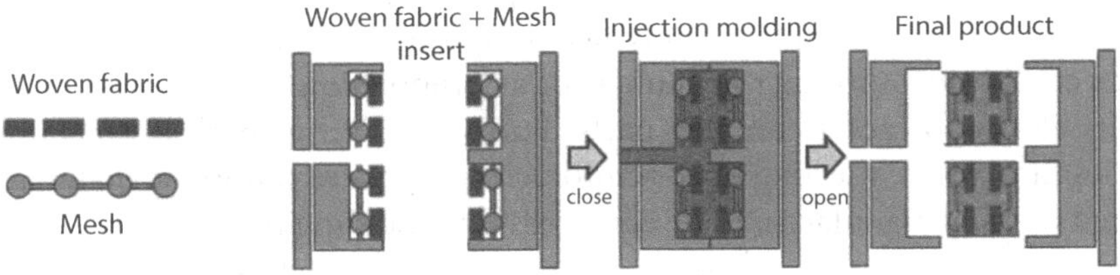

Figure 13.2 Mesh-inserted injection moulding [6].

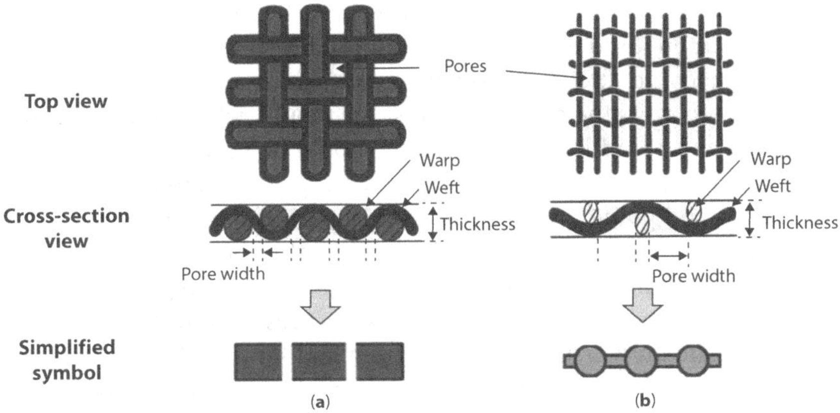

Figure 13.3 Different view of (a) woven fiber and (b) Mesh [6].

13.3 Compression Moulding Method

Compression moulding is the most common technique for fabricating the polymer composite material. The technique of compression moulding is very old and was used to mould thermoset powder at the beginning of the 20th century. It was actually in the middle of the 20th century when it gained acceleration and the technique started being used by the automobile industries for the manufacturing of different components and industries working on the manufacturing of appliances. At present, this method is the most used method for the fabrication of polymeric composites. The reason for the growing interest toward this method of fabricating composites by the industries or the scientific community because it is a relatively simple process that involves pressing a material charge, which can be deformed by the application of heat and pressure between two halves of the mould so that the transformation of material into the finished product took place after cooling. The compression moulding process is equally useful for the preparation of thermoplastic polymer composites and thermoplastic polymer composites. Another advantage of a compression moulding machine is its ability to create complex parts. Further, one drawback of injection moulding, where the long fiber-reinforced composites cannot be manufactured is easily manufactured by compression moulding method. This helps in developing stronger and stiffer materials. In this section of the chapter, the complete information on the compression moulding method is discussed, which includes the equipment and process used for the preparation of the polymer composites. Further, a few recent works on the development of composite along with their useful properties will also be discussed.

The equipment used for compression moulding is called a compression moulding machine. It should contain a press with a capacity as per the requirement to press the material against the mould, where mould has two halves, i.e., upper and lower. The mould should be precisely produced to get the desired shape and size product and has provision to accommodate the material and guide them to reach every part of the mould. There is a provision for heating the material. It should either be a separate heating chamber or integrated within the mould. Its main aim is to provide molten material. The general layout of the machine is presented in Figure 13.4. It has a base plate on which the heated platen and lower half portion of the mould is mounted. The function of the heated platen is to provide heat in the lower half portion of the mould to raise the temperature of the raw material. Another upper half of the mould is attached to the upper moving plate of the machine with a heated plate in between the two. The moving

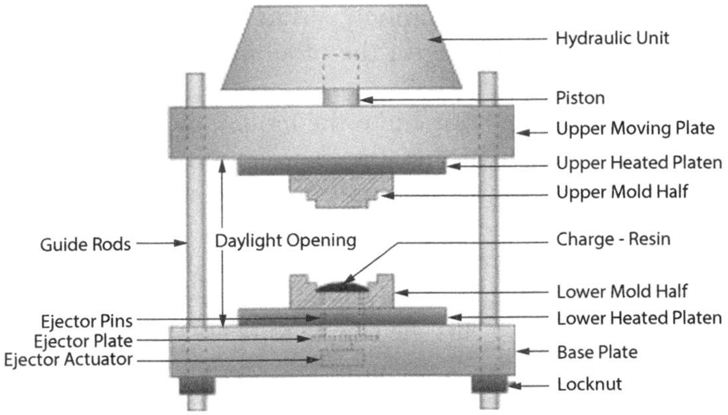

Figure 13.4 General layout of compression moulding machine.

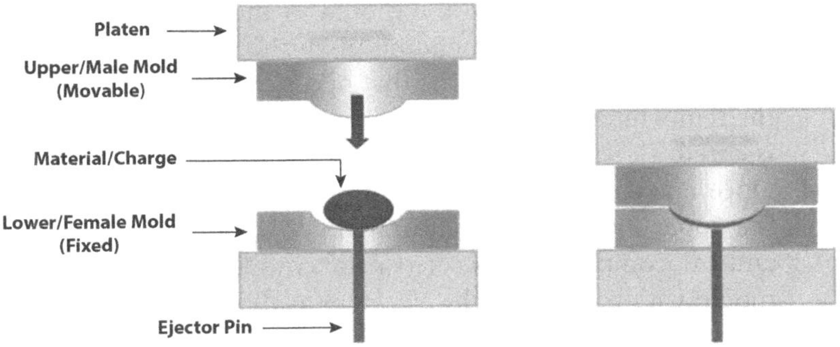

Figure 13.5 The compression moulding process.

plate is guided with the help of guide rods that are powered by a hydraulic unit. The charge is placed on the lower half mould. Apart from that, the ejector system of the machine is also placed at the bottom for the easy removal of the moulded material from the mould. Here, the amount of material, which is required to be moulded, is kept between the two parts of the mould, which is pressed to obtain the desired material as shown in Figure 13.5.

The compression moulding process consists of four basic steps [2]:

1. The step consists of the preparation of the charge. It is the moulding material, which is to be placed in the mould and from which the composite material is formed. The placement of the charge is an important part as it governs the properties

of the composites. Commonly, a stack of moulding material is prepared for placing in the mould and is called a sheet moulding compound.

2. After placing the charge into the mould, the next step is to close the mould. In this, the lower mould is in a stationary position, whereas the upper mould moves down rapidly until it touches the charge present in the lower mould. After that, the movement of the upper mould is kept slow to compress the charge. Slowly, the pressure inside the mould increases, which helps the charge to flow into every part of the mould. In the process of mould closure, care should be taken that the air should escape from the mould to provide a defect-free product. The mould's temperature and the upper mould's speed are the key quantitative factors that help in determining the properties of the finished product.
3. The next step is to keep the mould in the closed position for a specific period depending upon the type of material, the thickness of the material that needs to be moulded, and the temperature of the mould. During this step, the matrix is cured.
4. Once, the curing is completed, and the part gets solidified, the upper mould will again move back to open the mould, and the composite so formed is taken out from the lower mould with the help of the ejector pin after the mould temperature decreases.

The complete cycle of composite fabrication using a compression moulding machine takes around 3–6 min depending on qualitative factors like the thickness of the material and curing time. A wide range of thermoplastic as well as thermoset polymer composites with varieties of reinforcing material (particulates, short fibers, long fibers, woven mats, etc.) can be fabricated using a compression moulding machine. When the compression moulding is used for the preparation of thermoplastic polymer composites, the matrix and the reinforcement are first mixed with the help of a melt-blend extruder, whereas, for the preparation of the thermoset polymer composites, the matrix and the reinforcement are mixed with hand stirring in case of particulate or short fiber composites, and it is poured over the reinforcement in case of long fiber or mats. A few examples of such composites fabricated using compression moulding are discussed below along with the properties obtained.

Siraj *et al.* [7] used a combination of HDPE in combination with silica particles for the preparation of the thermoplastic composite body. Before the compression moulding, the combination of HDPE and silica particles is mixed in the twin-screw melt-blend extruder at a temperature of 170°C for 15 min. During this process, the uniform mixing of the filler and the matrix took place. The temperature and mixing time will vary depending on the melting temperature of the matrix used. The uniform mixed material is taken out from the extruder and cut into the form of pallets of small size, which are then put into the compression moulding between the hot press to obtain the composite sheet of uniform thickness. The sheet is later cut for obtaining specimens of different sizes as per the standard for conducting various characterization tests. The specimen is then put into the machine to obtain the properties of the material. The complete process involved in their study is presented in Figure 13.6.

The quantitative properties evaluated by them are, i.e., tensile strength, toughness, elastic modulus, and ductility of the material with respect to filler content and filler size. In their analysis, they found that the inclusion of silica particulates in the limited content gainfully progresses the various

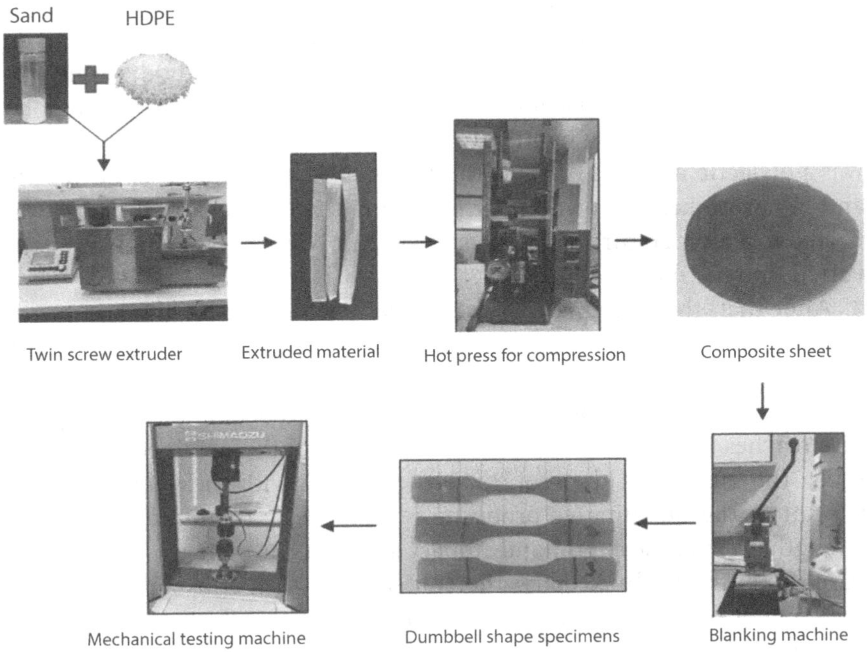

Figure 13.6 Experimental process [7].

quantitative properties. Further, they found that the composite prepared with smaller particulates delivers better properties.

Ju *et al.* [8] used recycled plastic obtained as a matrix material in combination with coal ash particles as a filler material for the preparation of the composite body. The recycled plastic consists of municipal plastic waste (MPW) and recycled HDPE. The MPW comprises polypropylene, low-density polyethene, and HDPE. First, the waste plastic is converted into pallets. Later, the pallets and the filler material are blended in a screw extruder for the proper mixing of the different phases, which was later converted into the form of pallets. After that, the pallets were fed into the compression moulding machine to obtain the required sheet of the composite material. The properties evaluated in the study were quantitative in nature, i.e., tensile, flexural strength, and compressive strength, which acts as a function of filler loading. In their analysis, they found that the various quantitative properties evaluated by them improved with the addition of the filler material. In their analysis, it was observed that the composite prepared using the compression moulding method produces less defective material, which results in the improvement of the properties of the material. Further, it is observed that, via adopting these routes, the recycled polymer composites can be prepared with relatively less effort.

Thangavel *et al.* [9] present that the compression moulding technique is also useful for the preparation of the hybrid composite, where more than one filler material can be incorporated within the matrix body. They also show that the uniform distribution of the filler material in the molten polymer can be obtained by hand stirring. The matrix material used by them was polylactic acid (PLA), and the fillers used were micro-sized fly ash and tamarind kernel powder. For preparing the composites, the PLA was heated to a temperature of 235°C, so that the polymer was completely converted into the molten state.

After that, the filler material was added into it one by one and hand-stirred. The uniformly mixed combination is poured into the mould, which was compressed in the machine to obtain the specimen sheet. The combination is kept under the pressure of about 6 bars until the temperature reaches back to the room temperature. From the obtained composite sheet, the samples of the required size were cut with the help of a water jet cutter. The quantitative properties evaluated by them are, i.e., tensile strength, flexural strength, and impact strength. In their work, they prepared the samples by keeping the content of fly ash constant and varied the content of tamarind kernel powder to study its effect on the properties under investigation. The results obtained by them are presented in Figure 13.7. The control name is designated for pure polymer, whereas, in

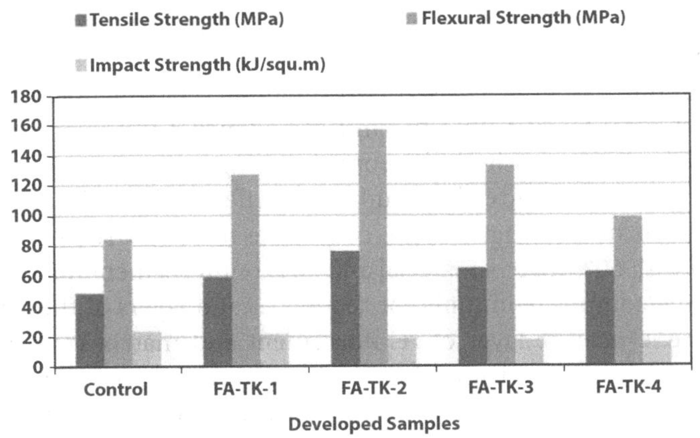

Figure 13.7 Properties of composites as a function of tamarind kernel powder [9].

other compositions, the tamarind kernel powder increased from 1 wt. % to 4 wt. % as designated. From the analysis, they found that the maximum tensile and flexural strength is obtained for the composite with 2 wt. % tamarind kernel powder, whereas impact strength decreases as the content of filler increases.

Rajakumar *et al.* [10] used the compression moulding method for the fabrication of fiber-reinforced composite material. For fabrication, they used abaca fiber with length of 100 mm and reinforced it in polyethene polymer. The composites are fabricated at varied fiber loading and with unidirectional fiber orientation. The matrix material is in the form of powder. For fabricating the composites with long fiber, the combination of matrix and fiber was directly put in the mould rather than first mixing it in the extruder as it is not possible in the case with long fiber. For fabricating the long fiber composites, the first step is to place the fiber and the matrix powder in the mould after measuring the weight of the two. For fabricating unidirectional fiber composites, proper care should be taken while placing the fiber in the mould in a particular direction. They created a composite sheet of 3 mm thickness. The sheets are prepared by heating the mould at 190°C temperature and maintaining the temperature for 5 min. Later, the mould is kept at atmospheric condition for around 30 min, so that the temperature of the mould will come back to room temperature. The samples are then cut from the prepared sheet of the composite specimen. They tested the fabricated samples for quantitative tensile strength and hardness. In their analysis, they found that the properties under investigation increased with the increase in the content of fiber loading confirming that the long fiber-reinforced composites can be successfully fabricated using the compression moulding technique.

Davallo and Pasdar [11] used two different types of fiber (synthetic fiber and natural fiber) with a thermoset polymer, i.e., epoxy matrix and developed a new class of hybrid composite using the compression moulding method. In their study, they studied the effect of hybridization on the tensile and flexural strength of the composite material. The composite is fabricated with 60% matrix material and 40% reinforcement material. They prepared the composite in different sequences and layers of the reinforcement phase. Once the layer of reinforcement and matrix were arranged in the mould, the mould was closed with a specified pressure and for a pre-determined temperature and time. In their analysis, they found an improvement in the various properties of the composite under investigation and confirmed that it is possible to fabricate laminated composites by compression moulding technique.

13.4 Resin Transfer Moulding

Resin transfer moulding (RTM) is implemented for producing plastic fiber products. Here, a reactive resin is decanted into dry, semi-finished fiber pieces that are dipped by applying pressure in a closed vessel. A typical layout of the method is represented in Figures 13.8 (a) and (b). The fibers/fabrics are kept in between to match mould surfaces, i.e., one female and another male. After that, the mould is closed and properly clamped. The thermoset resin is then inserted under reasonable pressure (ranging between 60 psi and 110 psi) in the cavity of the mould through one or more ports in the mould. The injected resin fills all voids in the mould set. The reinforcing material may be taken in different forms like continuous fibers, woven type, or mat type [11]. RTM is an excellent technique for producing well-controlled parts with complex geometries that are difficult to fabricate through the HLU technique consistently. The main advantages of the RTM process include reduced cycle time, high volume repeatability, and placement of performs to suit complexity. However, the main disadvantage of this process is the cost of tooling.

Jamir *et al.* [12] performed a relative study on the tensile behavior of sisal fiber-reinforced polyester composites fabricated through RTM and compression moulding. Samples with five different fiber loadings as well as five different fiber lengths were prepared. They observed that the tensile properties increase up to fiber loading of around 40 vol. %. However, properties decrease when the fiber loading increases further. Low fiber

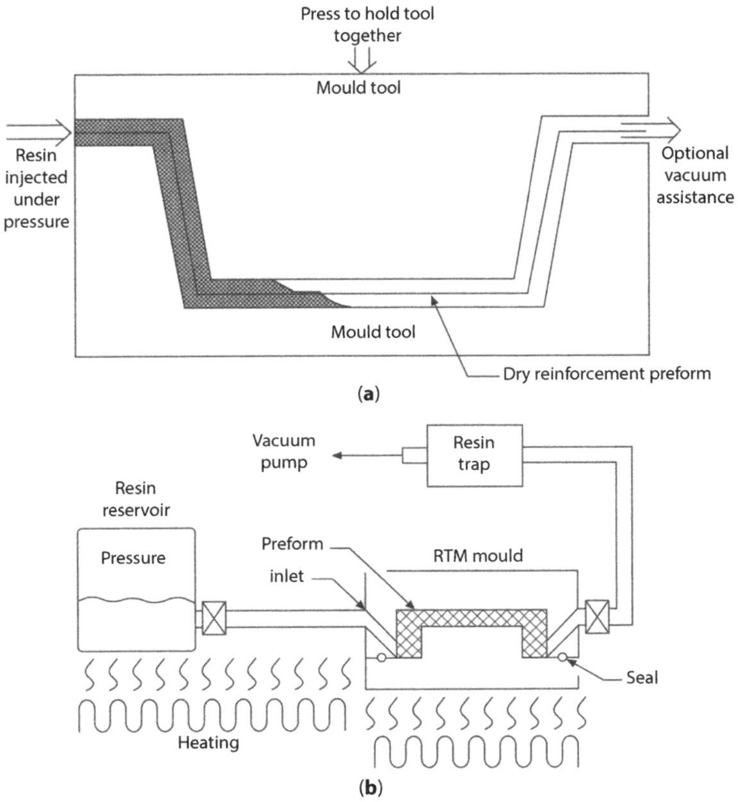

Figure 13.8 (a) Resin transfer moulding and (b) vacuum-assisted resin transfer moulding processes [11].

loading results in very poor fiber dispersion, which prevents proper stress transfer. There is a significant tendency for fiber-fiber contact at increased fiber loading. This results in inadequate fiber wetting and fiber dispersion. At larger loadings, crack initiation and propagation in such systems will be easier. When compared to composites made via compression moulding (CM), RTM-prepared composites in this instance also demonstrated higher mechanical properties.

Tensile properties are higher for the RTM method as compared to compression moulding. This is because the composite made with RTM has a lower void content and higher wettability. In this procedure, the mould is vacuumed before the resin and is put into it at the right pressure. If any air content was left in the mould, then the resin advances by pushing it. It is evident from the study that compression moulded samples have a higher void content than composites made via RTM. Due to hydrostatic pressure,

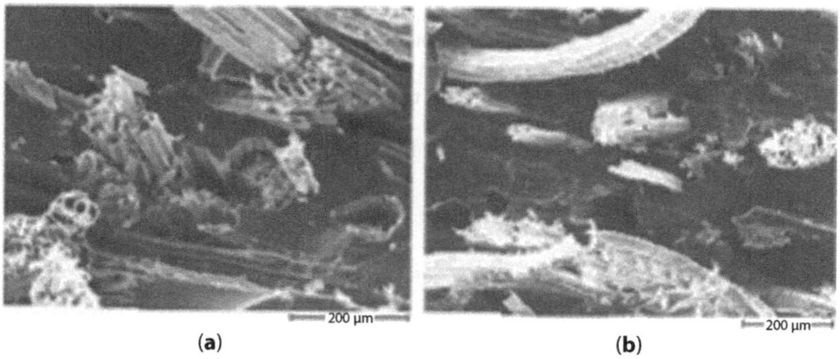

Figure 13.9 (a, b) Scanning electron microscopic (SEM) images of the tensile fracture surface of sisal/polyester composites made by compression moulding and resin transfer, respectively, with fiber loadings of 43 vol. % and 42 vol. % [12].

entrapped air in RTM shrinks, when the resin passes around the fibers. To understand the fracture mechanism, scanning electron micrographs of the fracture surface were taken. The micrographs of samples with RTM and CM techniques are shown in Figures 13.9 (a) and (b). The presence of fiber pullout, fiber breakage, and fibrillation is utilized as qualitative for evaluation and is clearly observed from the morphological images of the fractured tensile test specimens.

Kaynak and Kas [13] investigated the flexural properties of glass-polyester composites prepared by RTM and HLU technique. The matrix material in fiber-reinforced composites has a much lower strength than the fiber. As was seen in the cases of both HLU and RTM laminate composites, failure usually appears as matrix cracking before delamination or fiber breakage. However, as the amount of loading increased, crack formation quickly led to the creation of a significant damage zone through the thickness of the composite by translaminar fiber fracture rather than delamination, resulting in a significant reduction in load at comparatively modest stresses. The load drop related to a specimen's critical failure is affected by the moulding method and quantitative factors, i.e., specimen thickness and fiber volume fraction.

Mouritz [14] conducted a study to investigate the effect of varying injection pressure in RTM on the mechanical properties of epoxy composites reinforced with woven carbon fibers. They used a small laboratory-sized RTM machine to fill flat moulds containing layers of woven carbon fibers with an amine-curing epoxy resin. The composites were produced with five different levels of resin injection pressure, i.e., 1 atm, 2 atm, 3 atm, 4 atm, and 5 atm. The specimens that had been moulded at 2 atm had the maximum flexural strength. This represents that, by way of variation in quantitative values, the effects in output can be changed. Due to partially cured portions on their plates, the specimens that were moulded at 1 atm displayed a 26% reduced flexural strength. Flexural strength decreased as injection pressure was raised over 2 atm. Specimens moulded at 3 atm had an 11% lower flexural strength than specimens moulded at 2 atm. The morphologies of the specimens moulded at 3 atm and 2 atm could not be distinguished by microscopic analysis. No voids were seen within the fiber bundles or between bundles. The drop in flexural strength in samples moulded at higher pressures was most likely due to thicker plates containing more resin, which resulted in a smaller percentage of fibers. In comparison to specimens moulded at 2 atm, specimens generated at 4 atm exhibited a 17% lower flexural strength. After microscopic inspection, several pores were found between the fiber bundles, indicating void formation brought on by the rapid resin flow and the thicker plate layers. This indicated the utility of balancing resin flow and the thicker plate layers based on qualitative or quantitative analysis for avoiding voids. The specimens moulded at 5 atm had a 26% lower flexural strength than specimens moulded at 2 atm. This large drop in strength could be attributed to void formation both between and within fiber bundles, as well as the specimens' significantly thicker structure.

Panneerdhass *et al.* [15] worked to forecast the temperature profile in an RTM mould during curing using a one-dimensional heat conduction equation. The models and experiments are in strong agreement. The processing time can be optimized by using this model along with mould-filling factors. The RTM-produced hemp fiber composites were discovered to have a remarkably homogenous structure with no obvious flaws. To conduct the optimization study, the whole RTM process was divided into four steps, which are pre-pressing, injection, curing, and post-curing. The pre-pressing step takes about 5 min, and, for best results, the duration should not be decreased. The injection takes 31 min with 3.8×10^5 Pa pressure having 35% fiber content. According to the results of this permeability investigation, the injection time might be significantly decreased by enlarging the injection holes or by including more intake ports. All of the samples' tensile, three-point flexural, and Izod impact properties were assessed. With over 11% fiber content, the tensile strength improves linearly. Although the mean value with an 18.5% fiber content initially seems high, more samples were examined at this fiber percentage, and a wider range of findings were obtained. Tensile strength is likely to continue rising until a maximum and then start to decline. The maximum fiber content that could be achieved in this investigation was 35%. Through the study, it became abundantly evident that higher fiber content can result in greater tensile strength. As the fiber content rises above 11%, the flexural properties rise linearly. It is also expected that these characteristics will peak and then begin to decline.

13.5 Hand Layup Technique

The HLU technique is one of the oldest techniques used for polymer composite manufacturing. Initially, dried fibers are manually inserted in the mould, and the continuous phase is applied over them using a brush.

After that, it is rolled with hand rollers to maintain a uniform distribution of the matrix phase. It also ensures improved contact between the two phases and attains the right thickness. At last, the laminates are kept to cure under atmospheric conditions. The technique of curing involves hardening the fiber-reinforced resin composite without the need for outside heat and to obtain a high-quality product [16]. The schematic diagram of the HLU process is represented in Figure 13.10.

Roopa *et al.* [18] worked to understand the effect of reinforcements and fabrication methods on the mechanical behavior of basalt and glass-reinforced epoxy matrix composites. Woven fabric mats of basalt and glass were reinforced with epoxy matrix to prepare laminate composites using vacuum-assisted resin infusion method (VARIM), vacuum bagging, and HLU with compression (HLC). When comparing with the different manufacturing processes, HLU and compression moulding yield the products with the maximum tensile strength. According to the field emission scanning electron microscopic image, the removal of trapped air is caused by the combined action of the rollers and the compaction pressure. The HLC technique exhibited the highest impact strength when compared with

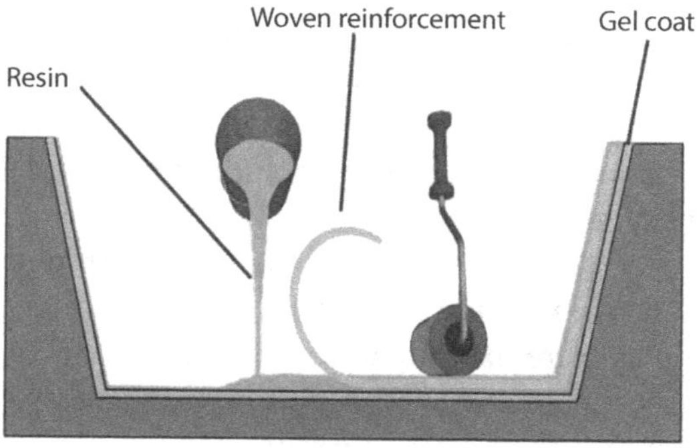

Figure 13.10 Schematic diagram of hand layup process [17].

other techniques used for manufacturing in this study. The elimination of trapped air bubbles in the spaces between the fabric layers is responsible for the improvement in impact strength. The field emission scanning electron microscopy analysis of basalt epoxy composites made from VARIM revealed the presence of voids and bubbles in the samples. In contrast, the microscopic images of composites prepared using the HLC method showed negligible voids in the samples. The entrapped air is effectively eliminated through a combined action of rollers during the HLU process and the application of compression pressure. The results showed that basalt fabric-reinforced epoxy polymer composites had better mechanical characteristics than glass fabric-reinforced epoxy polymer composites made using any of the three techniques.

The performance of vinyl ester/glass and polyester/glass laminates manufactured by RTM and HLU is compared by Rouison *et al.* [19]. To prepare the laminates, a resin transfer mould was created. In comparison to HLU specimens, void content was significantly lower in its counterparts. The use of RTM resulted in a reduction in void content. The resin was injected at a specific pressure and temperature in RTM, ensuring that the fibers were sufficiently wetted and that overflow was given to eliminate the excess entrapped air. Improved fiber wetting was seen as a result of the resin's increased fluidity due to a greater heating rate, which also decreased the resin's viscosity. RTM specimens outperformed HLU in terms of different properties investigated by them. As a result of improved wetting of the fibers and reduced void content than HLU, the experimental findings demonstrated that the RTM method was superior to HLU.

Sharba *et al.* [20] investigated the effect of qualitative and quantitative measures, i.e., moisture content, processing method, and type of polymer on the physical and mechanical behavior of natural fiber-based composites. A remarkable reduction of 8% moisture content was observed when the kenaf fiber temperature was increased over the glass transition temperature of the lignin. When compared to vacuum infusion, the results of HLU with cold press revealed a significant improvement in tensile and flexural strength and moduli for all types of polymers employed, in ascending sequence from unsaturated polyester and epoxy to vinyl ester. The scanning electron microscopic image of a tensile specimen of kenaf fiber-reinforced unsaturated polyester composites is shown in Figure 13.11. Morphological findings showed that the HLU with a cold press technique provided better fiber-matrix bonding with minimal void contents, which improved mechanical properties. Fiber pullout and matrix debonding are the modes of failure in the tensile test as observed from micrographic images.

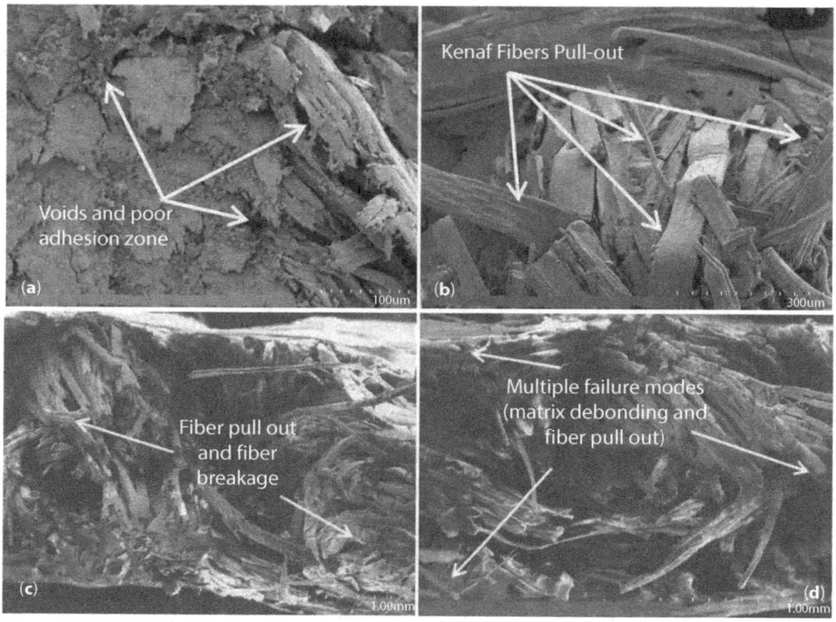

Figure 13.11 SEM images of tensile section failure of kenaf reinforced UP composites (a, b) using the VI method and (c, d) using the HCP method [20].

Sreekumar *et al.* [21] investigated the mechanical properties of chopped randomly oriented luffa fiber-reinforced epoxy polymer composite. Both treated and untreated fiber-reinforced composites samples with different volume fractions, i.e., 30%, 40%, and 50%, were prepared by the HLU technique. The different mechanical properties of the prepared composites were determined. The interfacial bonding between the matrix and fiber was improved due to the alkali treatment of the fibers, thus exhibiting superior properties to the untreated composite samples. It was noticed that the treated fiber composites with 40% fiber content showed the best mechanical properties. The lack of adequate wetting of the fiber with the matrix causes mechanical characteristics to decline after 40% fiber content. It has been noted that treated natural fiber composites have significantly lessened water absorption. The SEM analysis of the fractured surface revealed low-order fiber pullout, which may be the cause of enhanced mechanical characteristics at 40% fiber content.

Accordingly, after understanding the literatures related to the aforementioned techniques in different sections, it is understood that it is always needed to manufacture composites considering various qualitative and quantitative criteria and factors for attending good construction potential. The composite materials can offer many desired characteristics and properties, and, thus, they are sustainable materials with immense caliber and characteristics, which can be used in different segments of the work activities in industrial as well as societal applications.

13.6 Qualitative and Quantitative Evaluation in Manufacturing of Polymer Matrix Composite

The various methods of manufacturing polymer composites discussed in this chapter are the injection moulding method, compression moulding method, RTM, and HLU method. The chapter advocated toward evaluation of qualitative and quantitative factors for modeling polymer matrix composites with reinforced capacities for duty applications. The different fabrication methods discussed are mainly on the basis of qualitative and quantitative means. Considering qualitative and quantitative analysis, it has been observed that, while most of the methods outlined are applicable to both thermoset and thermoplastic composites, the HLU approach is only applicable to thermosetting polymers. Based on qualitative and quantitative analysis, common flaws in composite production are identified, which include voids, insufficient fiber wettability, and fiber debonding, emphasizing the importance of stringent quality control procedures. The chapter found injection moulding as the most promising of the variety of techniques. It has been observed from qualitative and quantitative analysis that the injection moulding method outperforms the other techniques in terms of structural integrity and consistency. This method of manufacturing the composite body results in high-quality products along with a low manufacturing time. Hence, the products obtained by this method have low defects, and a large quality of the product can be manufactured.

13.7 Conclusions

This chapter concludes with a thorough examination of the manufacturing processes that is used to create polymer composites. It is found important to recognize critical qualitative and quantitative drivers and factors to attain a unique set of benefits and drawbacks with the manufacturing process. Accordingly, one can understand that the scientific community intentionally uses a variety of manufacturing processes based on unique project requirements. The chapter provided a unique knowledge of various manufacturing processes. This thorough knowledge of various manufacturing processes gives researchers and professionals the information they need to make wise decisions, promoting innovation and effectiveness in the field of polymer composite fabrication. Future efforts in composite material science are needed to position to achieve new levels of accuracy and performance by utilizing these discoveries.

It is highlighted that composite materials are nowadays needed to be applied in numerous fields like automobiles, packaging, furniture, and aerospace industries because of the reason of having good mechanical properties, lower density, lightweight solutions, and so on. Thus, it is always imperative for the manufacturer of composites to consider various criteria and factors for attaining good construction potential. The chapter rationalizes its importance in defining several factors and manufacturing processes for polymer matrix composites. It is advocated that many qualitative and quantitative factors can be implemented for modeling and evaluating polymer matrix composites. The study presented in this chapter tried to provide complete possible aspects to be used to evaluate the quantitative and qualitative aspects of polymer matrix composites. The chapter highlighted that there is a need to arouse interest in using polymer matrix composites in many applications to achieve sustainability, which is demanding to lead significant research on modeling the scientific problem of quantitative and qualitative aspects of polymer matrix composites. The chapter aimed to ascertain and characterize the manufacturing as well as quantitative and qualitative features of polymer matrix composites.

References

1. Fu, H., Xu, H., Liu, Y., Yang, Z., Kormakov, S., Wu, D., Sun, J., Overview of injection moulding technology for processing polymers and their composites. *ES Mater. Manuf.*, *8*, 4, 3–23, 2020.
2. Advani, S.G. and Hsiao, K.T. (Eds.), Introduction to composites and manufacturing processes, in: *Manufacturing techniques for polymer matrix composites (PMCs)*, pp. 1–12, Woodhead Publishing Series in Composites Science and Engineering, Cambridge, United Kingdom, 2012.
3. Joseph, E.J. and Panneerselvam, K., Effect of particulate fillers on mechanical, metallurgical and abrasive behaviour of tungsten reinforced HDPE composites: a Taguchi approach. *Mater. Today Proc.*, *39*, 1228–1234, 2021.
4. Mark, U.C., Madufor, I.C., Obasi, H.C., Mark, U., Influence of filler loading on the mechanical and morphological properties of carbonized coconut

shell particles reinforced polypropylene composites. *J. Compos. Mater.*, *54*, 3, 397–407, 2020.

5. Koffi, A., Koffi, D., Toubal, L., Mechanical properties and drop-weight impact performance of injection-moulded HDPE/birch fibre composites. *Polym. Test.*, *93*, 106956, 1–11, 2021. https://doi.org/10.1016/j.polymertesting.2020.106956.
6. Jeong, E., Kim, Y., Hong, S., Yoon, K., Lee, S., Innovative injection molding process for the fabrication of woven fabric-reinforced thermoplastic composites. *Polymers*, *14*, 8, 1577, 2022.
7. Siraj, S., Al-Marzouqi, A.H., Iqbal, M.Z., Ahmed, W., Impact of micro silica filler particle size on mechanical properties of polymeric based composite material. *Polymers*, *14*, 22, 4830, 1–18, 2022. https://doi.org/10.3390/polym14224830.
8. Ju, S., Yoon, J., Sung, D., Pyo, S., Mechanical properties of coal ash particle-reinforced recycled plastic-based composites for sustainable railway sleepers. *Polymers*, *12*, 10, 2287, 1–15, 2020. https://doi.org/10.3390/polym12102287.
9. Thangavel, P., Vignesh, S., Sureshkumar, R., Gowrishankar, A., Girimurugan, R., Study on mechanical properties of polylactic acid matrix added with fly ash and tamarind kernel powder micro fillers. *Nanoworld J.*, *9*, S1, S626–S630, 2023.
10. Rajakumar, I., Raguraman, D., Isaac, J.S., Suthan, R., Bhattacharya, S., Seikh, A.H., Raghavan, I.K., Mechanical properties of polymer composites reinforced with alkaline-treated natural fibre. *Adv. Polym. Technol.*, 2022, 1458547, 1–7, 2022. https://doi.org/10.1155/2022/1458547.
11. Davallo, M. and Pasdar, H., Comparison of Mechanical Properties of Glass-PolyesterComposites Formed by Resin Transfer Moulding and HandLay-Up Technique. *Int. J. Chemtech Res.*, *1*, 3, 470–475, 2009.
12. Jamir, M.R.M., Majid, M.S.A., Khasri, A., Natural lightweight hybrid composites for aircraft structural applications, in: *Sustainable Composites for Aerospace Applications*, pp. 155–170, Elsevier, Duxford, United Kingdom, 2018, https://doi.org/10.1016/B978-0-08-102131-6.00008-6.
13. Kaynak, C. and Kas, Y.O., Effects of Injection Pressure in Resin Transfer Moulding (RTM) of Woven Carbon Fibre/Epoxy Composites. *Polym. Polym. Compos.*, 14, 1, 55–64, 2006. https://doi.org/10.1177/096739110601400105.
14. Manufacturing of fibre–polymer composite materials, in: *Introduction to Aerospace Materials*, pp. 303–337, Elsevier, Cambridge, United Kingdom, 2012, https://doi.org/10.1533/9780857095152.303.
15. Panneerdhass, R., Baskaran, R., Rajkumar, K., Gnanavelbabu, A., Mechanical Properties of Chopped Randomly Oriented Epoxy - Luffa Fibre Reinforced Polymer Composite, in: *Applied Mechanics and Materials*, vol. 591, pp. 103–107, 2014, https://doi.org/10.4028/www.scientific.net/AMM.591.103.
16. Raajeshkrishna, C.R. and Chandramohan, P., Effect of reinforcements and processing method on mechanical properties of glass and basalt epoxy composites. *SN Appl. Sci.*, *2*, 5 2020, https://doi.org/10.1007/s42452–020-2774-4.

17. Raji, M., Abdellaoui, H., Essabir, H., Kakou, C.-A., Bouhfid, R., Qaiss, El Kacem, A., Prediction of the cyclic durability of woven-hybrid composites, in: *Durability and Life Prediction in Biocomposites, Fibre-Reinforced Composites and Hybrid Composites*, pp. 27–62, Elsevier, Duxford, United Kingdom, 2019, https://doi.org/10.1016/B978-0-08-102290-0.00003-9.
18. Roopa, T.S., Murthy, H.N., Sudarshan, K., Nandagopan, O.R., Kumar, A., Krishna, M., Angadi, G., Mechanical properties of vinylester/glass and polyester/glass composites fabricated by resin transfer molding and hand lay-up. *J. Vinyl Add. Tech.*, *21*, 3, 166–173, 2015. https://doi.org/10.1002/vnl.21393.
19. Rouison, D., Sain, M., Couturier, M., Resin transfer molding of hemp fibre composites: optimization of the process and mechanical properties of the materials. *Compos. Sci. Technol.*, *66*, 7–8, 895–906, 2006. https://doi.org/10.1016/j.compscitech.2005.07.040.
20. Sharba, M.J., Salman, S.D., Leman, Z., Sultan, M.T.H., Ishak, M.R., Azmah Hanim, M.A., Effects of processing method, moisture content, and resin system on physical and mechanical properties of woven kenaf plant fibre composites. *Bioresources*, *11*, 1, 1466–1476, 2016.
21. Sreekumar, P.A., Joseph, K., Unnikrishnan, G., Thomas, S., A comparative study on mechanical properties of sisal-leaf fibre-reinforced polyester composites prepared by resin transfer and compression moulding techniques. *Compos. Sci. Technol.*, *67*, 3–4, 453–461, 2007. https://doi.org/10.1016/j.compscitech.2006.08.025.

Index

Printed in the USA/Agawam, MA
September 16, 2026

910661.131